High Resolution Numerical Modelling of the Atmosphere and Ocean

Kevin Hamilton · Wataru Ohfuchi
Editors

High Resolution Numerical Modelling of the Atmosphere and Ocean

Kevin Hamilton
International Pacific Research Center
University of Hawaii at Manoa
Honolulu, Hawaii 96822
USA

Wataru Ohfuchi
Earth Simulator Center
Japan Agency for Marine-Earth
Science and Technology
Yokohama, Japan

Library of Congress Control Number: 2007930354

ISBN-13: 978-0-387-36671-5   e-ISBN-13: 978-0-387-49791-4

© 2008 Springer Science+Business Media, LLC

All rights reserved. This work may not be translated or copied in whole or in part without the written permission of the publisher (Springer Science+Business Media, LLC., 233 Spring Street, New York, NY 10013, USA), except for brief excerpts in connection with reviews or scholarly analysis. Use in connection with any form of information storage and retrieval, electronic adaptation, computer software, or by similar or dissimilar methodology now known or hereafter developed is forbidden.

The use in this publication of trade names, trademarks, service marks, and similar terms, even if they are not identified as such, is not to be taken as an expression of opinion as to whether or not they are subject to proprietary rights.

Printed on acid-free paper.

9 8 7 6 5 4 3 2 1

springer.com

Contents

1. **Numerical Resolution and Modeling of the Global Atmospheric Circulation: A Review of Our Current Understanding and Outstanding Issues** 7
   Kevin Hamilton
   1.1 Introduction – Global Atmospheric Simulations 7
   1.2 Effect of Horizontal Resolution on Simulations of Tropospheric Circulation 9
   1.3 Effects of Vertical Resolution on Simulations of Tropospheric Circulation 12
   1.4 Explicit Simulation of Mesoscale Phenomena 13
   1.5 Changing Subgrid-Scale Parameterizations with Model Resolution 16
   1.6 Middle Atmosphere 18
   1.7 Coupled Global Ocean–Atmosphere Model Simulations and Climate Sensitivity 20
   1.8 Summary 22

2. **The Rationale for Why Climate Models Should Adequately Resolve the Mesoscale** 29
   Isidoro Orlanski
   2.1 Introduction 30
   2.2 The Role of the High Frequency Wave Activity in Climate Variability 31
   2.3 The Performance of the Eddy Activity in Three Climate Models . . . 34
   2.4 The Cyclone-Frontal System 39
   2.5 Summary and Conclusions 42

3. **Project TERRA: A Glimpse into the Future of Weather and Climate Modeling** 45
   Isidoro Orlanski and Christopher Kerr
   3.1 Introduction 45
   3.2 High Resolution Results 46
   3.3 The Versatility Offered by Nonhydrostatic GCM's 46
   3.4 Computational Requirements 47
   3.5 Summary and Conclusion 49

4 An Updated Description of the Conformal-Cubic Atmospheric Model 51
John L. McGregor and Martin R. Dix

- 4.1 Introduction 51
- 4.2 Dynamical Formulation of CCAM 52
  - 4.2.1 Primitive Equations 52
  - 4.2.2 Semi-Lagrangian Discretization 54
- 4.3 Physical Parameterizations 60
- 4.4 Parallel Aspects 61
  - 4.4.1 Grid Decomposition 61
  - 4.4.2 Treatment of the Semi-Lagrangian Interpolations 63
  - 4.4.3 Helmholtz Solver 64
  - 4.4.4 Performance 64
- 4.5 Examples of CCAM Simulations 65
  - 4.5.1 Held-Suarez Test 65
  - 4.5.2 Aquaplanet Simulation 65
  - 4.5.3 AMIP Simulation 65
  - 4.5.4 Simulation of Antarctic Snow Accumulation 72
- 4.6 Concluding Comments 72

5 Description of AFES 2: Improvements for High-Resolution and Coupled Simulations 77
Takeshi Enomoto, Akira Kuwano-Yoshida, Nobumasa Komori, and Wataru Ohfuchi

- 5.1 Introduction 77
- 5.2 Dynamical Processes 79
  - 5.2.1 Formulation 79
  - 5.2.2 Modifications to the Legendre Transform 81
- 5.3 Physical Processes 85
  - 5.3.1 Radiation Scheme mstrnX 85
  - 5.3.2 Dry Convective Adjustment 86
  - 5.3.3 Emanuel Convective Parametrization 87
  - 5.3.4 Other Modifications 94
- 5.4 Concluding Remarks 95

6 Precipitation Statistics Comparison Between Global Cloud Resolving Simulation with NICAM and TRMM PR Data 99
M. Satoh, T. Nasuno, H. Miura, H. Tomita, S. Iga, and Y. Takayabu

- 6.1 Introduction 99
- 6.2 Model and the Experimental Setup 101
- 6.3 Precipitation Distribution 102
- 6.4 Precipitation Frequency 103
- 6.5 Spectral Representations of Rain-Top Height 105
- 6.6 Summary 109

7 Global Warming Projection by an Atmospheric General Circulation Model with a 20-km Grid 113
Akira Noda, Shoji Kusunoki, Jun Yoshimura, Hiromasa Yoshimura, Kazuyoshi Oouchi, and Ryo Mizuta
- 7.1 Introduction 113
- 7.2 Methods 115
  - 7.2.1 Models 115
  - 7.2.2 Experimental Design 116
- 7.3 Results 116
  - 7.3.1 Change in Tropical Cyclones 116
  - 7.3.2 Change in Baiu Rain Band 120
- 7.4 Discussion and Concluding Remarks 125

8 Simulations of Forecast and Climate Modes Using Non-Hydrostatic Regional Models 129
Masanori Yoshizaki, Chiashi Muroi, Hisaki Eito, Sachie Kanada, Yasutaka Wakazuki and Akihiro Hashimoto
- 8.1 Introduction 129
  - 8.1.1 Forecast Mode: JPCZ and the Formation Mechanism of T-Modes 130
  - 8.1.2 Climate Mode: Changes in the Baiu Frontal Activity in the Future Warming Climate from the Present Climate 134
- 8.2 Summary 138

9 High-Resolution Simulations of High-Impact Weather Systems Using the Cloud-Resolving Model on the Earth Simulator 141
Kazuhisa Tsuboki
- 9.1 Introduction 141
- 9.2 Description of CReSS 142
- 9.3 Optimization for the Earth Simulator 144
- 9.4 Localized Heavy Rainfall 145
- 9.5 Typhoons and the Associated Heavy Rainfall 146
- 9.6 Snowstorms 150
  - 9.6.1 Idealized Experiment of Snow Cloud Bands 150
  - 9.6.2 Snowstorms Over the Sea of Japan 154
- 9.7 Summary 155

10 An Eddy-Resolving Hindcast Simulation of the Quasiglobal Ocean from 1950 to 2003 on the Earth Simulator 157
Hideharu Sasaki, Masami Nonaka, Yukio Masumoto, Yoshikazu Sasai, Hitoshi Uehara, and Hirofumi Sakuma
- 10.1 Introduction 157
- 10.2 Model Description 159

10.3 Overview of the Simulated Fields 160
    10.3.1 Variations of Global Mean Values 160
    10.3.2 Global Upper-Layer Ocean Circulations 162
    10.3.3 Improvements over the Spin-Up Integration 163
    10.3.4 Remaining Problems 165
10.4 Variability at Various Timescales 166
    10.4.1 Intraseasonal Variability 166
    10.4.2 Interannual Variations 168
    10.4.3 Decadal Variability 175
10.5 Concluding Remarks 180

11 Jets and Waves in the Pacific Ocean 187
Kelvin Richards, Hideharu Sasaki, and Frank Bryan
11.1 Introduction 187
11.2 Jets 188
11.3 Waves 190
11.4 Impact on the Transport of Tracers 193
11.5 Conclusions 194

12 The Distribution of the Thickness Diffusivity Inferred from a High-Resolution Ocean Model 197
Yukio Tanaka, Hiroyasu Hasumi, and Masahiro Endoh
12.1 Introduction 197
12.2 GM Parameterization 199
12.3 Model Description 200
12.4 Results 201
    12.4.1 The Distribution of the EBFC 201
    12.4.2 The Distribution of the Thickness Diffusivity 203
12.5 Summary and Discussion 204

13 High Resolution Kuroshio Forecast System: Description and its Applications 209
Takashi Kagimoto, Yasumasa Miyazawa, Xinyu Guo, and Hideyuki Kuwajiri
13.1 Introduction 209
13.2 Description of Forecast System 211
    13.2.1 Numerical Model 211
    13.2.2 Nesting 214
    13.2.3 Surface Forcings 214
    13.2.4 Observational Data 216
    13.2.5 Data Assimilation 218
    13.2.6 Quality Control 223
    13.2.7 Incremental Analysis Update 224
13.3 Predictions of the Kuroshio Large Meander 226
    13.3.1 Case for April to June 2003 226

13.3.2 Case for May to July 2004 228
  13.3.3 Sensitivity of the Forecast to Parameters 229
 13.4 Toward the Kuroshio Forecast Downscaling for Coastal Oceans
  and Bays 231
 13.5 Summary 234

**14 High-Resolution Simulation of the Global Coupled
Atmosphere–Ocean System: Description and Preliminary
Outcomes of CFES (CGCM for the Earth Simulator)** 241
*Nobumasa Komori, Akira Kuwano-Yoshida, Takeshi Enomoto,
Hideharu Sasaki, and Wataru Ohfuchi*
 14.1 Introduction 241
 14.2 Coupled Atmosphere–Ocean GCM: CFES 242
  14.2.1 Atmospheric Component: AFES 2 242
  14.2.2 Oceanic Component: OIFES 243
  14.2.3 Coupling Method 243
 14.3 Preliminary Results 245
  14.3.1 Simulation Setting 245
  14.3.2 Global View of Snapshots 246
  14.3.3 Local View Around Japan 247
  14.3.4 Annual-Mean Surface Climatologies 248
  14.3.5 Seasonal Cycle of Tropical SST and Polar Sea-Ice Extent . . 254
 14.4 Concluding Remarks 256

**15 Impact of Coupled Nonhydrostatic Atmosphere–Ocean–Land
Model with High Resolution** 261
*Keiko Takahashi, Xindong Peng, Ryo Onishi, Mitsuru Ohdaira, Koji Goto,
Hiromitsu Fuchigami, and Takeshi Sugimura*
 15.1 Introduction 261
 15.2 Model Configuration 262
  15.2.1 MSSG-A Non-hydrostatic Atmosphere Global
   Circulation Model 262
  15.2.2 MSSG-O; Non-hydrostatic/Hydrostatic Ocean Global
   Circulation Model 263
  15.2.3 Grid System Configuration 264
  15.2.4 Differencing Schemes 265
  15.2.5 Regional Coupled Model, Coupling and Nesting Schemes . . 266
 15.3 Results of High Resolution Simulations 267
  15.3.1 Validation of the MSSG-A 267
  15.3.2 Preliminary Validation Results of the MSSG-O 270
  15.3.3 Physical Performance of the MSSG 271
 15.4 Conclusion and Future Work 272

Color Plates 275

Contributors

Frank Bryan
National Center for Atmospheric Research, P.O. Box 3000, Boulder CO 80307 USA

Martin R. Dix
CSIRO Marine and Atmospheric Research, PB1 Aspendale, Vic. 3195 Australia

Hisaki Eito
Meteorological Research Institute
1-1, Nagamine, Tsukuba 305-0052 Japan

Masahiro Endoh
Center for Climate System Research
University of Tokyo
Chiba, Japan

Takeshi Enomoto
Earth Simulator Center
Japan Agency for Marine-Earth Science and Technology

Hiromitsu Fuchigami
NEC Informatec Systems LTD
3-2-1 Sakato, Takatsu-ku Kawasaki-shi Kanagawa, 213-0012, Japan

Koji Goto
NEC Corporation, 1-10 Nisshin-cho Fuchu-shi Tokyo
183-5801, Japan

Xinyu Guo
Center for Marine Environmental Studies, Ehime University
Matuyama, Ehime, Japan

Kevin Hamilton
International Pacific Research Center
University of Hawaii at Manoa
Honolulu, Hawaii 96822, USA

Akihiro Hashimoto
Advanced Earth Science
and Technology Organization
MRI, 1-1 Nagamine
Tsukuba 305-0052, Japan

Hiroyasu Hasumi
Center for Climate System Research
University of Tokyo, Chiba, Japan

S. Iga
Frontier Research Center for Global Change/Japan Agency for Marine-Earth Science and Technology

Takashi Kagimoto
Frontier Research Center for Global Change/JAMSTEC
Yokohama, Kanagawa, Japan

Sachie Kanada
Advanced Earth Science
and Technology Organization MRI
1-1, Nagamine, Tsukuba 305-0052
Japan

Hideyuki Kawajiri
Frontier Research Center for Global Change/JAMSTEC
Yokohama, Kanagawa
Japan

Christopher Kerr
Geophysical Fluid Dynamics
Laboratory/NOAA
Forrestal Campus Princeton University
Princeton NJ 08540

Nobumasa Komori
Earth Simulator Center
Japan Agency for Marine-Earth
Science and Technology

Shoji Kusunoki
Climate Research Department
Meteorological Research Institute
Japan Meteorological Agency
1-1, Nagamine, Tsukuba
Ibaraki 305-0052, Japan

Akira Kuwano-Yoshida
Earth Simulator Center
Japan Agency for Marine-Earth
Science and Technology

Yukio Masumoto
Frontier Research Center
for Global Change
Japan Agency for Marine-Earth
Science and Technology
3173-25, Showa-machi
Kanazawa-ku, Yokohama
Kanagawa, 236-0001, Japan

and

Department of Earth
and Planetary Science
Graduate School of Science
University of Tokyo, 7-3-1, Hongo
Bunkyo-ku, Tokyo, 113-0033, Japan

John L. McGregor
CSIRO Marine and Atmospheric
Research, PB1 Aspendale, Vic. 3195
Australia

H. Miura
Frontier Research Center for Global
Change/Japan Agency for Marine-Earth
Science and Technology

Yasumasa Miyazawa
Frontier Research Center for Global
Change/JAMSTEC, Yokohama
Kanagawa, Japan

Ryo Mizuta
Advanced Earth Science and Technology
Organization, 1-1, Nagamine, Tsukuba
Ibaraki 305-0052, Japan

Chiashi Muroi
Meteorological Research Institute
1-1 Nagamine, Tsukuba 305-0052, Japan

T. Nasuno
Frontier Research Center for Global
Change/Japan Agency for Marine-Earth
Science and Technology

Akira Noda
Climate Research Department
Meteorological Research Institute
Japan Meteorological Agency
1-1 Nagamine, Tsukuba
Ibaraki 305-0052, Japan

Masami Nonaka
Frontier Research Center for Global
Change, Japan Agency for
Marine-Earth Science and Technology
3173-25, Showa-machi, Kanazawa-ku
Yokohama, Kanagawa, 236-0001, Japan

Mitsuru Ohdaira
Earth Simulator Center, JAMSTEC
3173-25 Showa-machi, Kanazawa-ku
Yokohama 236-0001, Japan

Contributors

Wataru Ohfuchi
Earth Simulator Center
Japan Agency for Marine-Earth
Science and Technology

Ryo Onishi
Earth Simulator Center, JAMSTEC
3173-25 Showa-machi, Kanazawa-ku
Yokohama 236-0001, Japan

Kazuyoshi Oouchi
Advanced Earth Science and Technology
Organization, 1-1, Nagamine, Tsukuba
Ibaraki 305-0052, Japan

Isidoro Orlanski
Geophysical Fluid Dynamics
Laboratory/NOAA
Forrestal Campus Princeton University
Princeton NJ 08540

Xindong Peng
Earth Simulator Center, JAMSTEC
3173-25 Showa-machi, Kanazawa-ku
Yokohama 236-0001, Japan

Kelvin Richards
International Pacific Research
Center, University of Hawaii
1680 East West Road, Honolulu
HI 96822, USA

Hirofumi Sakuma
Frontier Research Center for
Global Change, Japan Agency for
Marine-Earth Science and Technology
3173-25, Showa-machi, Kanazawa-ku
Yokohama, Kanagawa, 236-0001, Japan

Yoshikazu Sasai
Frontier Research Center for Global
Change, Japan Agency for
Marine-Earth Science and Technology
3173-25, Showa-machi, Kanazawa-ku
Yokohama, Kanagawa
236-0001, Japan

Hideharu Sasaki
Earth Simulator Center
Japan Agency for Marine-Earth
Science and Technology, 3173-25
Showa-machi, Kanazawa-ku
Yokohama Kanagawa
236-0001, Japan

M. Satoh
Frontier Research Center for Global
Change/Japan Agency for Marine-Earth
Science and Technology and Center
for Climate System Research
University of Tokyo

Takeshi Sugimura
Graduate School of Science
of Nagoya University
Furou-cho Chikusa-ku Nagoya
464-8602
Japan

Keiko Takahashi
Earth Simulator Center, JAMSTEC
3173-25 Showa-machi
Kanazawa-ku Yokohama
236-0001
Japan

Y. Takayabu
Center for Climate System Research
University of Tokyo

Yukio Tanaka
Frontier Research Center for Global
Change, Japan Agency for
Marine-Earth Science and Technology
3173-25 Showa-machi
Kanazawa-ku Yokohama
236-0001, Japan

H. Tomita
Frontier Research Center for Global
Change/Japan Agency for Marine-Earth
Science and Technology

Kazuhisa Tsuboki
Hydrospheric Atmospheric Research
Center (HyARC), Nagoya
University/ Frontier Research
Center for Global Change, Furo-cho
Chikusa-ku, Nagoya, 464-8601, Japan

Hitoshi Uehara
Japan Agency for Marine-Earth
Science and Technology, 3173-25
Showa-machi, Kanazawa-ku, Yokohama
Kanagawa, 236-0001, Japan

Yasutaka Wakazuki
Advanced Earth Science
and Technology Organization MRI
1-1 Nagamine, Tsukuba 305-0052, Japan

Hiromasa Yoshimura
Climate Research Department
Meteorological Research Institute
Japan Meteorological Agency
1-1, Nagamine, Tsukuba
Ibaraki 305-0052, Japan

Jun Yoshimura
Climate Research Department
Meteorological Research Institute
Japan Meteorological Agency
1-1, Nagamine, Tsukuba
Ibaraki 305-0052, Japan

Masanori Yoshizaki
Meteorological Research Institute
1-1, Nagamine
Tsukuba 305-0052, Japan

Introduction

Numerical simulation models of atmospheric and oceanic circulation are simply finite numerical approximations to the continuous governing differential equations. In many branches of applied mathematics, researchers will routinely show that their finite numerical solutions to a particular problem have converged with increasing resolution. In the study of atmospheric and oceanic circulation one rarely has the luxury of such straightforward demonstrations of convergence. The standard practice for attacking difficult problems has been to truncate the model employed at some finite horizontal, vertical, and time resolution, and perform time integrations for a specified period. In operational forecast applications the final state is the prediction, which will eventually be verified against simultaneous observations. In climate applications typically the statistical properties of the flow over the period simulated are analyzed. In either case, it is understood that the model integrations will have deficiencies simply associated with the fact that significant aspects of the real circulation will be unresolved in the finite numerical approximation employed.

Much effort has been expended on parameterizing the effects of subgrid-scale motions on the larger-scale flow, but it has also been a goal of researchers to explicitly resolve as many scales of the actual circulation as possible. For example, in studies of the global ocean circulation, a long-standing concern has been the issue of adequately resolving eddies that have horizontal scales of the order of the Rossby radius (the oceanic analogues of synoptic scale waves in the midlatitude atmosphere). In global and regional atmospheric models, a key issue has been resolving the mesoscale circulations that organize moist convection. With the recent advent of a new generation of high-performance computing systems such as the Earth Simulator, some notable thresholds in terms of model resolution have been approached or, in some cases, surpassed. For example, very recently the first long integrations with genuinely eddy-resolving, or at least eddy-permitting, global ocean models have been reported. On the atmospheric side, decadal integrations using global models with effective horizontal resolution of $\sim$20 km have now become possible, and very short integrations of models that explicitly resolve scales approaching those of individual convective elements have just been reported. These developments in global models have been paralleled by rising research activity with increasingly fine resolution regional atmospheric models for both climate and short-range forecasting applications. It is thus an opportune time to review progress in the field and consider outstanding issues and the prospects for further advances.

An international workshop to address issues in high-resolution modeling was held on September 21–22, 2005 at the home of the Earth Simulator supercomputer, the JAMSTEC (Japan Agency for Marine–Earth Science and Technology) Yokohama Institute for Earth Sciences. Twenty-two speakers were invited from Australia, Canada, Japan, the UK and the USA, and more than 60 scientists attended the workshop. Following the workshop the speakers and some other selected colleagues were invited to submit papers related to the topic of high-resolution modeling of the atmosphere and ocean (but not limited to the material discussed in the workshop) to the present volume. The papers submitted spanned a variety of topics and described ocean, atmosphere, and coupled models with both global and regional domains.

The first group of papers in this volume relate to various aspects of high-resolution global atmospheric modeling. First, Hamilton reviews previous results concerning the dependence of atmospheric simulation on the horizontal and vertical resolution employed. He notes that even the very large-scale aspects of the simulated flow have not necessarily completely converged even at horizontal resolutions of approximately tens of kilometers, both in the troposphere and, more particularly, in the middle atmosphere. Encouragingly, some recent high-resolution simulations have successfully represented many observed features in the mesoscale circulation and, importantly, a realistic overall energy level of mesoscale motions. Hamilton notes some critical open questions relating to the appropriate scaling of physical subgrid scale parameterizations with the model numerical truncation.

Orlanski discusses the relevance of fine numerical resolution to simulation of large-scale atmospheric dynamics. He particularly concentrates on simulation of the extratropical circulation and points out the critical importance of adequately representing the interaction of the mesoscale and synoptic scales. In a companion paper, Orlanski and Kerr discuss results of a brief (24 h) experimental integration of a global version of the Geophysical Fluid Dynamics Laboratory nonhydrostatic zeta-coordinate model.

McGregor and Dix report recent technical improvements to, and simulations with, the finite difference Conformal-Cubic Atmospheric Model (CCAM) that has been developed at the Australian Commonwealth Scientific and Industrial Research Organization (CSIRO). Developing such a model may provide one approach for ultra-high-resolution models to avoid the high computational cost of Legendre transforms used in conventional spectral atmospheric general circulation models (AGCMs), and also to avoid the singularity at the poles in conventional latitude–longitude finite difference AGCMs. McGregor and Dix describe in detail technical aspects, such as the implementation of the semi-Lagrangian scheme and grid decomposition for parallel computing, and they show results from idealized and realistic simulations with CCAM.

Enomoto et al. describe the second generation of the AFES (AGCM for the Earth Simulator) global atmospheric model. The first version of the AFES has been applied in a number of published studies in configurations with resolution up to T1279L96 (i.e., triangular truncation at total spherical wavenumber 1,279, and 96 levels in the vertical). The new version differs from the original AFES notably in incorporating upgrades to the radiation scheme, the convective scheme and the cloud scheme. Enomoto et al. show the effects these changes make on the long-term climate simulated by T79L48 versions of the model. They also include a discussion of the technicalities involved in the calculation of accurate Legendre transforms at very high order. They note that previous methods of this calculation become problematical at about T1800 and higher truncation limits, and they advance their own solution to this problem which may allow the standard pseudospectral models to be efficiently extended to very high truncation limits.

Satoh et al. discuss results of brief integration of a global atmospheric model with 3.5 km horizontal resolution. This study employed the Nonhydrostatic ICosahedral Atmospheric Model (NICAM) that was developed mainly at the Frontier Research Center for Global Change (FRCGC) of JAMSTEC specifically for such ultra-fine-resolution global simulations. The model is run without a convective parameterization and the main focus in the initial analysis of the results is in seeing how the model flow interacts with, and acts to organize, tropical convection. Results are compared with high spatial resolution rainfall estimates from the Tropical Rainfall Measuring Mission (TRMM) satellite.

One of main goals for the Earth Simulator was to provide sufficient computing resources to allow projections of future global warming that might represent a qualitative improvement over other current forecasts in the explicit representation of key processes. Noda et al. give a summary of their recent studies of twenty-first century global warming using time-slice experiments with a 20-km mesh AGCM. With this ultrahigh resolution, they can reasonably simulate some mesoscale phenomena, such as tropical cyclones (TCs) and the Baiu front, though their model employs the hydrostatic approximation and cumulus convection parameterization. They obtained the difference in sea surface temperatures (SST) between 1979–1998 and 2080–2099 from a warming projection performed using with the Japan Meteorological Agency Meteorological Research Institute coupled GCM with a much coarser horizontal resolution. The ultrahigh resolution present-day time slice was run for a decade with observed SST averaged from 1982 to 1993. Then the warmed time slice was run for a decade with the SST used for the present-day run superimposed with the SST difference described above. Noda et al. report that in the warmed climate the number of TCs decreases but their intensity increases. Also the Baiu front lasts longer and the strength of the mesoscale disturbances associated with this front intensify in the twenty-first century integration.

The next group of papers discusses limited-area high-resolution atmospheric models. Yoshizaki et al. apply regional nonhydrostatic models (NHMs) on the Earth Simulator for both weather forecast-type and climate change-type simulations. For the former, they use an NHM with horizontal resolution of 1 km (2,000 × 2,000 grid points) and 38 vertical levels, covering the Sea of Japan. This short-term simulation reproduces the detailed structures of the Japan Sea polar air mass convergence zone. For the climate change application, Yoshizaki et al. use an NHM with horizontal resolution of 5 km (800 × 600 grid points) and with 48 vertical levels, covering the Japan area. They utilize the result from time-slice global warming projection reported in the above Noda et al. paper by employing a spectral boundary coupling method. June and July cases are simulated for both present-day and globally warmed conditions. Yoshizaki et al. conclude that in the warmed climate the activity of Baiu front will increase over southern Japan and the frequency of heavy rainfall will generally increase over all of Japan.

Tsuboki gives a brief description of a nonhydrostatic regional atmospheric model, CReSS (Cloud Resolving Storm Simulator), developed by his group at Nagoya University, and discusses results of some simulations. CReSS has been optimized for the Earth Simulator and now is capable of large-scale simulations. Tsuboki concentrates on short-term (typically a day or so) simulations of high-impact weather events, such as localized heavy rainfalls, tropical cyclones, and snowstorms, using resolutions of 300–1,000 m (typically of order of 1,500 × 1,500 grid points) and 50–400 m (~40–70 vertical levels) in horizontal and vertical directions, respectively. These high-impact weather systems have hierarchical structures from about 100–1,000 km for the whole system down to about 1–10 km. Tsuboki demonstrates the capability of CReSS to simulate the multiscale structures.

While it is easy to see small features of the real atmospheric circulation in satellite and radar imagery, it is very difficult to observe detailed features of the ocean interior because electromagnetic waves do not propagate efficiently in sea water. Thus there may be a particular need for models that can credibly simulate the small-scale features of the ocean circulation. The next four papers discuss fine structures of the ocean currents simulated by ocean models. Sasaki et al. discuss results obtained with the OFES (Ocean GCM for the Earth Simulator), an ocean model specifically developed to run efficiently on the Earth Simulator. They show results from a hindcast experiment forced by daily atmospheric reanalysis data covering the second half of the twentieth century. The domain is quasi-global, from 75° S to 75° N, and this version is run with horizontal resolution of 0.1° and with 54 vertical levels. This experiment was designed to see how well the observed oceanic circulation variations from intraseasonal to decadal timescales are reproduced in a model with realistic time-varying forcing. The analysis considers the simulation of El Niño and the Indian Ocean Dipole events, the Pacific and the Pan-Atlantic Decadal Oscillations, and the intraseasonal variations in the equatorial Pacific and Indian Oceans, among other phenomena. Comparisons are made with an earlier experiment in which the same model was forced with climatological (but still seasonally varying) atmospheric forcing.

Richards et al. report on the spatial and temporal structures of jets and waves simulated in the ocean interior of fine resolution global ocean models. These jets have elongated structures in the longitudinal direction, but a rather small latitudinal scales of ~300–500 km. These features seem rather robust in high-resolution but low-dissipative ocean models and have been increasingly recognized in satellite altimeter data and also by in situ measurements. Richards et al. discuss possible wave-mean interactions that may define the spatial and temporal structures of the jets and waves, and implications for tracer transport.

Tanaka et al. discuss results from very fine resolution simulations in an ocean model with domain confined to the Southern Ocean (from 20° S to 75° S). They consider results from several different horizontal resolution versions, with the finest being 1/12 × 1/8 degrees latitude–longitude. It is fair to say that this very fine resolution permits, if not completely resolves, the midlatitude eddies that are ocean counterparts of atmospheric synoptic-scale cyclones and anticyclones. In more moderate resolution ocean models, the effects of these eddies on resolved-scale momentum and energy have been treated via a parameterization, such as the widely used Gent-McWilliams scheme. Tanaka et al. use their finest resolution simulation to estimate the diffusion coefficient of isopycnal thickness appropriate for low-resolution ocean models. They note that the areas with large inferred effective diffusivity coefficient correspond well with those with strong baroclinic eddies.

Kagimoto et al. introduce the Kuroshio forecast system with about 10 km horizontal resolution, named the Japan Coastal Ocean Predictability Experiment (JCOPE). The forecast is conducted with a limited-area ocean model covering about 60° longitude by 50° latitude. They show, in particular, one very successful forecast of the development of a large Kuroshio meander with about a month lead time. Kagimoto et al. describe not only specific forecast cases but also the detailed technical aspects of the forecast system. They also report on a preliminary study aimed at further downscaling their results for application to coastal ocean areas.

Finally two papers discuss high-resolution coupled global ocean–atmosphere models. Komori et al. describe the CFES (Coupled GCM for the Earth Simulator) model which couples versions of the AFES model with an extended version of the OFES model which includes a sea ice component. CFES uses a novel approach to coupling the ocean, sea ice, and atmospheric simulations and this minimizes communication overhead in the parallel computing environment of the Earth Simulator. Komori et al. discuss preliminary results from a multiyear CFES integration conducted at much finer resolution than has been typical in coupled global GCMs, specifically T239L48 in the atmosphere and 1/4 degree in the ocean. Results are encouraging in that the climate spins up to a realistic state including a reasonable representation of the seasonal cycle of the ocean circulation.

The paper by Takahashi et al. describes the formulation of nonhydrostatic atmospheric and oceanic models which can be run in coupled mode. Their formulation is sufficiently flexible that higher-resolution regional models can be nested within the global model. It is particularly exciting that the code allows either regional atmosphere, ocean, or coupled regional models to be nested. So far the nesting is one-way, but the authors note that they are working on implementing a two-way nesting. The paper goes on to discuss results from some preliminary integrations – short forecasts with the atmosphere-only model run at 5.5 km horizontal resolution, a 15-year run of the ocean-only model (for the North Pacific basin) with observed atmospheric forcing, and then another short forecast run of the coupled ocean–atmosphere model (with an enhanced grid around Japan).

All the papers in this volume were peer reviewed. We would like to thank the anonymous reviewers for their efforts. The workshop on which this book is loosely based was sponsored by the Japan Society for the Promotion of Science. We would like to thank the convener of the workshop, Dr. Tetsuya Sato, Director General of Earth Simulator Center (ESC), JAMSTEC. We thank also ESC staff, especially Dr. Shinya Kakuta and Ms. Reiko Itakura, for their dedicated help.

Wataru Ohfuchi, Co-Editor
Earth Simulator Center
Japan Agency for Marine-Earth Science and Technology

Kevin Hamilton, Co-Editor
International Pacific Research Center
University of Hawaii at Manoa
November 2006

Chapter 1
Numerical Resolution and Modeling of the Global Atmospheric Circulation: A Review of Our Current Understanding and Outstanding Issues

Kevin Hamilton

Summary This chapter presents a survey of published literature related to the issue of how the simulation of climate and atmospheric circulation by global models depend on numerical spatial resolution. To begin the basic question of how the zonal-mean tropospheric circulation in atmospheric general circulation models (AGCMs) vary with changing horizontal and vertical grid spacing is considered. The appropriate modification of subgrid-scale parameterizations with model resolution is discussed. Advances in available computational power have recently spurred work with quite fine resolution global AGCMs, and the issue of how well such models simulate mesoscale aspects of the atmospheric circulation is considered. Experience has shown that the AGCM simulated circulation is particularly sensitive to resolution in the stratosphere and mesosphere, and so studies related to the middle atmospheric circulation are considered in some detail. Finally, the significance of atmospheric model resolution for coupled global ocean–atmosphere models and the simulated climate sensitivity to large-scale perturbations is discussed.

1.1 Introduction – Global Atmospheric Simulations

The first attempt at integrating a multilevel comprehensive atmospheric general circulation model (AGCM) including treatment of the hydrological cycle was that of Manabe et al. (1965). This employed coarse resolution in both the horizontal (∼500 km grid spacing) and the vertical (nine levels from the ground to the model top near 10 hPa). For many research applications (and also for very long timescale climate forecasts), a majority of projects in subsequent decades have employed AGCMs with typically only about twice the horizontal and vertical resolution of this original Manabe et al. model. Until recently, research groups typically devoted their ever increasing computer power principally to making longer integrations and incorporating more sophisticated parameterizations into their AGCMs.

Most of the early efforts to run global models with particularly fine resolution were undertaken by major operational forecasting centers, which generally have state-of-the-art computing facilities and also a strong practical incentive to fully use their resources in producing the best possible deterministic short-range predictions. The horizontal and vertical resolutions used in the global deterministic forecast runs at two leading operational centers, those of the USA (currently the National Centers for Environmental Prediction, NCEP) and Europe (European Center for Medium Range Weather Forecasts, ECMWF) are regularly increased as computational resources permit. Horizontal resolution has improved by roughly a factor of 10 in these runs over the last two decades while vertical resolution has improved by about a factor of 5. Given that time steps for integration are usually scaled with the horizontal resolution, this represents about a 5,000-fold increase in the computational burden over about 20 years. Currently the operational model at the ECMWF uses a triangularly truncated spectral representation (T799) with smallest wavelength resolved of about 40 km, corresponding to an effective horizontal grid spacing of about 20 km.

On the climate side, a pioneering effort to apply substantial supercomputer resources to integration of a very fine resolution AGCM was begun in the 1980s at the Geophysical Fluid Dynamics Laboratory (GFDL). Mahlman and Umscheid (1987) describe simulations with a version of the GFDL "SKYHI" grid-point model with ~100 km horizontal resolution and 40 levels in the vertical. This effort continued over the next decade with simulations performed using grid spacing as fine as 35 km and with versions with up to 160 levels (Hamilton and Hemler, 1997; Hamilton et al., 1999, 2001; Koshyk and Hamilton, 2001).

Recently the efforts to run global AGCMs at fine resolution have attracted interest and participation from a wider range of research groups and have been assisted by substantial investments in development of major supercomputers. Conaty et al. (2001) discuss aspects of the synoptic and mesoscale circulation features appearing in a seasonal integration with a version of a global AGCM with ~100 km grid spacing. In a recent chapter Shen et al. (2006) discuss several 5-day integrations with a $1/8°$ version of the NASA finite-volume GCM. The Atmospheric Model for the Earth Simulator (AFES) is a spectral AGCM which has been adapted to run very efficiently on the Earth Simulator. Ohfuchi et al. (2004, 2005) report on brief (~2 weeks) simulations performed using AFES versions with triangular-1,259 truncation (corresponding roughly to 10 km grid spacing) and 96 levels in the vertical. A global version of the GFDL:ZAETAC nonhydrostatic grid-point model with ~10 km horizontal resolution was integrated for 24 h at GFDL (Orlanski and Kerr, 2007). Finally, in a very ambitious project, Tomita et al. (2005) report on brief integrations using the Earth Simulator of a nonhydrostatic AGCM with roughly 3.5 km horizontal grid spacing.

There have been increasing efforts at using high-resolution models even for long-period climate change predictions. In the current round of very extensive integrations from coupled global models submitted for consideration in the IPCC Fourth Assessment Report the horizontal resolution of the atmospheric components of the models ranges up to T106. Mizuta et al. (2005) discuss an even more ambitious experiment conducted on the Earth Simulator. They performed both a 10-year present-day control simulation and a 10-year late-twenty-first century time-slice global warming simulation using a T959 atmospheric global model in which the sea surface temperatures (SSTs) were taken from comparable periods of a low-resolution coupled GCM global warming scenario experiment.

This chapter is an informal review of research on the question of how horizontal and vertical resolution affects the simulation of the global atmospheric circulation. The focus is on the ability of models to simulate realistic circulations when run from essentially arbitrary initial conditions (i.e., in "climate mode"). Not covered here are related questions concerning the effects of model resolution on the performance of short-range weather forecasts. Mullen and Buizza (2002) and Roebber et al. (2004), among others, provide discussions of the role of model resolution in practical short-range weather forecasts.

The outline of this chapter is as follows. Section 1.2 reviews the results obtained by various groups concerning the basic dependence of the simulated large-scale circulation on horizontal model resolution. In this section only results relevant to the troposphere will be considered. Section 1.3 reviews the rather less extensive published work on the dependence of the basic tropospheric simulation on vertical model resolution. Historically, most of the studies of resolution-dependence of simulated circulation have been performed with model suites that extend up to only relatively modest horizontal resolution (typically grid spacings larger than 100 km). However, the recent efforts at running significantly finer-resolution models raise other issues of convergence, namely whether these mesoscale-resolving (or mesoscale "permitting" to adapt a term from ocean modeling, e.g., Griffies and Hallberg, 2000) global models are producing realistic mesoscale motions. The explicit simulation of the tropospheric mesoscale within global models is reviewed in Sect. 1.4. Section 1.5 reviews the somewhat limited literature related to appropriate scaling of parameterizations with changing model spatial resolution. Section 1.6 considers the dependence of the simulated circulation on resolution for the stratosphere and mesosphere. Section 1.7 reviews the literature on the effect of model resolution on coupled atmosphere–ocean global simulations and on modeled climate sensitivity to large-scale radiative perturbations. Conclusions are summarized in Sect. 1.8.

Throughout this chapter, for simplicity, the effective horizontal grid resolution of a spectral model with triangular truncation at total spherical wavenumber n is taken to roughly the circumference of the earth divided by $2n$. Lander and Hoskins (1997) offer a more detailed discussion of the effective equivalent resolution in grid and spectral representations.

1.2 Effect of Horizontal Resolution on Simulations of Tropospheric Circulation

A basic issue in global modeling is how the overall large-scale and regional features of the simulated climate depend on numerical resolution. This issue has been investigated systematically for a range of horizontal resolution in a number of studies using a variety of models. Held and Suarez (1994), Boer and Denis (1997), and Pope and Stratton (2002) discussed the convergence of the results from idealized dry-dynamical core (DDC) models. The DDC models have no topography, no moisture, and employ radiative heating specified as a function of the latitude, height, and local temperature. Such studies have also been performed with several full AGCMs employing either spectral dynamics (Boer and Lazare, 1988; Boville, 1991; Boyle, 1993) or grid-point dynamics (Hamilton et al., 1995, 2001; Pope and Stratton, 2002). A common result in all the spectral model studies is that even the largest scales of the mean circulation change very significantly as spectral resolution is increased from ~T21 to ~T42, notably with increased poleward eddy fluxes of eastward momentum along with increased midlatitude surface westerlies and corresponding meridional surface pressure gradients. As horizontal resolution is increased still further, these changes in the zonal-mean circulation continue, but at a much slower rate. Similar trends are observable in grid-point model simulations. The changes seem not to have completely converged even at the highest resolution considered in these studies (e.g., T63 for Boville, 1991; T106 for Boyle, 1993; ~35 km grid spacing for Hamilton et al., 2001; ~90 km grid spacing for Pope and Stratton, 2002). Continuing modest changes in the zonal-mean winds and temperatures and also in eddy statistics such as the zonal-mean of the eddy kinetic energy are apparent in these studies, even as the model resolution reaches these relatively fine values.

Williamson (1999) compared some aspects of simulations in a conventional spectral AGCM run at T63 and T106 horizontal resolution in the full model, and in versions in which the subgrid-scale physics parameterizations were performed on a reduced T42 grid. That is, the full resolution spectral fields produced by the dynamical model were truncated to T42, the tendencies due to physics parameterizations were then computed on the appropriate T42 transform grid and expanded into the spectral space. The resulting tendencies were then applied in the full model dynamics. Williamson notes that the strength of the tropical Hadley circulation does not converge in the standard model (even at T170 resolution), but there is convergence when the resolution of the subgrid-scale physics is held at T42. By contrast, the statistical properties of the extratropical storm tracks do change significantly between T63 and T106, even in the version with fixed resolution for the subgrid-scale physics.

A more complex issue is the dependence of simulated regional climatology in realistic models as resolution is improved. One complication is that higher horizontal resolution models typically employ finer-scale topography, and by itself this may be expected to change the simulations. The overall impression obtained by reviewing the studies cited above is that increasing horizontal resolution generally leads to improved regional climatology for such quantities as seasonal-mean sea level pressure or seasonal-mean precipitation. An interesting example is provided by Hamilton et al. (1995) who evaluated the boreal summer and boreal winter precipitation simulations obtained with the GFDL SKYHI grid-point AGCM when run with ~300, ~200, and ~100 km grid spacing. The seasonal-mean results in each case were averaged on $5° \times 5°$ latitude–longitude areas and correlated with observed climatology over the globe. Although the precipitation simulations had some fairly obvious deficiencies (e.g., in the summer South Asian monsoon) at all three resolutions, the objective measure of pattern correlation with observations was reasonably high ($\sim$0.7–0.8) and it increased with improved model resolution. A similar conclusion concerning the global correlation of simulated and observed rainfall patterns was reached by Kobayashi and Sugi (2004) in simulations with different resolution versions of the Japan Meteorological Agency (JMA) spectral AGCM. Pope and Stratton (2002) calculate the rms differences from observations in December–February mean sea level pressure simulated by their grid-point model when run at $\sim$275 km resolution versus $\sim$90 km resolution. They find that the rms error drops very substantially from 3.5 to 2 hPa at the higher resolution. They also ran a version of their high-resolution model with the low-resolution topography, and find that much (but not all) of the improvement in the simulation of sea level pressure at high resolution can be attributed to the finer topography rather than simply the improved resolution of the atmospheric dynamics.

There have been some systematic studies of the horizontal resolution dependence of the AGCM simulation of Asian monsoon circulations and associated rainfall. Sperber et al. (1994) find significant deficiencies in the T42 simulation of the monsoon by the ECMWF model, some of which are alleviated at T106 resolution. Stevenson et al. (1998) compare summer monsoon simulations in T21, T31, T42, and T63 versions of an AGCM. Stevenson et al. found that the large-scale features such as the lower tropospheric westerly jet, the upper tropospheric tropical easterlies, the Tibetan High were simulated by the model at all resolutions. As the resolution was increased the core of the low-level westerly jet moved toward Somalia and became more realistic. However, the model simulated excessive rainfall over the equatorial Indian Ocean and over the southern slopes of the Tibetan plateau, and these errors actually became accentuated at finer resolution. Kobayashi and Sugi (2004) examine the Asian monsoon simulation in prescribed SST simulations with the JMA Global Spectral Model model with horizontal resolution varied between T42 and T213, all L40. Even a large-scale feature such as the seasonal-mean Tibetan High is stronger (and more realistic) at T213. Many smaller scale climatological features are better represented at high resolution as well, notably the location and strength of associated precipitation of the Baiu front.

We can conclude that, while there have been a number of studies addressing the issue of how simulated tropospheric circulation changes with model horizontal resolution, there is nothing definitive that allows a determination of the resolution needed for a particular degree of convergence in the simulated climate. There has been little work along these lines performed at finer model resolution (say effective horizontal grid spacings significantly less than 100 km). The possibility that employing still finer horizontal resolution may significantly improve global model simulations of the mean tropospheric climate cannot be discounted.

1.3 Effects of Vertical Resolution on Simulations of Tropospheric Circulation

The issue of appropriate scaling of the vertical and horizontal resolution of numerical models of the atmospheric circulation has been a concern for some decades, but a clear and general determination of how simulations are affected by the vertical resolution has not been achieved. Lindzen and Fox-Rabinovitz (1989) argued that in order to simulate quasigeostrophic motions in the troposphere, a model should employ a ratio of horizontal grid spacing (Δx) to the vertical grid spacing (Δz) of the order of 300 in the extratropics and at least an order of magnitude larger near the equator. In practice, various atmospheric simulation models have been designed with an enormous range of ratios of the horizontal to vertical grid spacing, almost all significantly smaller than those advocated by Lindzen and Fox-Rabinovitz. For typical global climate GCMs we may have $\Delta x \sim 300$ km and $\Delta z \sim 1$–2 km in the midtroposphere (enhanced vertical resolution near the ground is common of course) for a ratio of $\sim$150–300. The operational global forecast models referred to in Sect. 1.1 have finer horizontal and vertical resolutions, but all have $\Delta x / \Delta z$ ratios of this order, as well. In limited-area mesoscale models the $\Delta x / \Delta z$ ratio is typically much smaller; for example Janjic et al. (2001) describe simulations with a nonhydrostatic mesoscale model with $\Delta x \sim 8$ km and $\Delta z \sim 0.5$ km, or a ratio of $\sim$15. In cloud-resolving calculations it is sometimes the case that Δx will be taken to be almost as small as Δz and so the ratio can be $\sim$1. In general these choices seem to be motivated by a widespread belief that once Δz is down to $\sim$0.5–1 km there is more to be gained by increasing the horizontal resolution than in reducing Δz further. The empirical and theoretical basis for this belief appears not to be as developed as one may like, but there have been a few published relevant studies of how the vertical resolution affects AGCM simulations which seem to support this view.

Boville (1991) discussed a set of simulations with an AGCM run at T21 horizontal resolution and vertical level spacings varied from $\sim$2.8 km down to $\sim$0.7 km. He found little difference in these simulations except in the behavior of vertically propagating equatorial waves (more of an issue for the stratosphere than the lower atmosphere). Some more recent results studying models with different vertical resolution suggest that the largest sensitivity may be in the tropics and may be most significant for the simulation of upper tropospheric water vapor. Tompkins and Emanuel (2000) studied results of a single-column atmospheric model formulated with equal pressure difference between model levels; this model was run to a radiative–convective equilibrium for tropical conditions. They found that the vertical structure of temperature and water vapor was sensitive to improving vertical resolution at least until the level spacing was reduced to $\sim$25 hPa (corresponding to $\Delta z \sim 500$ m in the midtroposphere). Inness et al. (2001) analyzed control climate simulations performed with 19 and 30 level versions of an AGCM. They find modest, but significant, differences between the simulations in terms of the mean temperature and humidity structure and also in the behavior of tropical intraseasonal oscillations.

Roeckner et al. (2006) have performed a systematic investigation of the global rms errors in seasonal-mean fields in an array of simulations with the ECHAM5 AGCM as the resolution varies from T21L19 to T159L31. Consistent with earlier studies, the authors find that at L19 vertical resolution there is an improvement in the simulation with increasing horizontal resolution up to T42, but little improvement beyond that. With the L39 vertical resolution, however, the improvement of the simulation with horizontal resolution in most respects continues through T159 truncation.

These AGCM studies have dealt with modest horizontal resolution models only, and the question of optimum vertical resolution for very fine horizontal resolution global models has not been systematically addressed. This issue also obviously is connected with the performance of subgrid-scale parameterizations, notably those for cloud processes and turbulence.

1.4 Explicit Simulation of Mesoscale Phenomena

While increasing resolution past a certain point may lead to only modest changes in the large-scale circulation, higher resolution models have at least the possibility to explicitly simulate mesoscale circulations. Such features may be very significant for both weather forecasting and climate applications. As climate model simulations are run at ever finer resolution it will become more important to evaluate the mesoscale aspects of these simulations.

Perhaps the most basic question is whether the mesoscales in the simulated flow are realistically energized. It has been known that moderate resolution AGCMs can simulate a realistic spectrum of horizontal variance of the horizontal wind and temperature. These are often referred to as the kinetic energy (KE) and available potential energy (APE) spectra, respectively (Boer and Shepherd, 1983; Boville, 1991; Koshyk et al., 1999). Model results can be projected onto spherical harmonics and horizontal spectra then expressed as a function of total wavenumber, n, of the spherical harmonic (a rough equivalent wavelength is $40,000 \,\text{km}/n$). Observations of tropospheric circulation show a kinetic energy spectrum with a broad peak around $n \sim 5$ and then a roughly n^{-3} regime out to $n \sim 80$. Most AGCMs are truncated within this n^{-3} regime, but observations show that past $n \sim 80$ (or horizontal wavelengths shorter than about 500 km) the kinetic energy spectrum becomes much shallower (e.g., Nastrom and Gage, 1985; Lindborg, 1999).

The simulated horizontal KE spectrum has been examined in a number of earlier studies using relatively modest horizontal resolution AGCMs (Boville, 1991; Koshyk et al., 1999). These studies showed that GCMs can reproduce a realistic n^{-3} regime in the troposphere but, due to the limited horizontal resolution, these models did not allow simulation of a significant range of the shallower mesoscale regime.

It appears that various current very high-resolution AGCMs perform rather differently in terms of their ability to simulate a realistically shallow mesoscale kinetic energy spectrum. Palmer (2001) notes that the ECMWF GCM, when run at fine resolution, actually simulates flow with a KE spectrum that steepens rather than shallows in the mesoscale. However, Koshyk and Hamilton (2001) found that the SKYHI AGCM can simulate a realistically energized mesoscale. In particular, they analyzed results from a control simulation with a ~35 km horizontal resolution, 40-level version of the SKYHI model and found that their fields did reproduce the shallow horizontal KE spectra observed by Nastrom and Gage (1985) in the upper troposphere, down to the smallest model-resolved wavelength (~70 km). Recently Takahashi et al. (2006) analyzed results from T639 AFES model control simulations. With an appropriate choice of subgrid-scale mixing parameter the model can reproduce quite well the observed upper troposphere KE and APE spectra. The experiment was also repeated in a DDC version of the model. This version also simulated a shallow mesoscale range, supporting the view that the mesoscale regime in the atmosphere is energized, at least in part, by a predominantly downscale nonlinear spectral cascade.

Hayashi et al. (1997) examined the space–time structure of low-latitude precipitation in versions of a grid-point AGCM run with horizontal grid spacings of ~50, ~100, and ~300 km. At the finer horizontal resolutions, grid-scale precipitation, which is thought to roughly represent the precipitation associated with cloud clusters, is organized into larger-scale superclusters. The westward propagation of cloud clusters and eastward propagation of superclusters is much more apparent in the high-resolution experiments. These basic conclusions are also found from the results of Yamada et al. (2005), who examined the space–time spectra of equatorial precipitation in versions of a global spectral AGCM with horizontal resolutions varying from T39 to T159 and L48 in the vertical. Yamada et al. considered a simplified "aquaplanet" case with all ocean surface and prescribed SSTs a function only of latitude. They found that as resolution is increased the eastward-propagating precipitation clusters and westward-propagating organizing structures become more clearly defined.

In addition to the analysis of overall energy content in mesoscale motions, there have been efforts aimed at characterizing the simulation of particular features in the circulation. One important challenge has been the simulation of the quasipermanent Baiu frontal zone that appears over East Asia and the far western Pacific region in the May–July period. This is a case where a reasonable simulation of local weather variability requires a good representation of the fairly narrow frontal zone and the mesoscale weather systems that disturb it. A number of studies have demonstrated the difficulty in simulating this feature realistically with moderate resolution global models (Yu et al., 2000; Zhou and Li, 2002; Kang et al., 2002). Kawatani and Takahashi (2003) had some success in Baiu front simulation with a T106 AGCM, but many more details of the front and typical disturbances were successfully captured by Ohfuchi et al. (2004) with their T1279L96 AGCM.

One aspect of mesoscale meteorology in global models that has attracted considerable attention is their ability to simulate tropical cyclones. The great practical interest in forecasting how global change may affect the climatology of tropical cyclone numbers, tracks and intensities is one of the main motivations for pursuing very fine

1.4 Explicit Simulation of Mesoscale Phenomena

resolution AGCM modeling. It has been known for some time that global AGCMs run in climate mode will spontaneously generate tropical depressions and tropical cyclones. Of course, mature intense tropical cyclones (hurricanes and typhoons) in the real world have rather small sizes (peak winds typically ~50 km from the center) and cannot be adequately resolved except by a very fine scale model. However, the ability of AGCMs with various horizontal resolutions to simulate a somewhat realistic climatology of tropical cyclone occurrence and motion has been documented (e.g., Bengtsson et al., 1995; Tsutsui, 2002). While moderate resolution models may be able to reproduce some aspects of the observed tropical cyclone climatology, they are unable to simulate the most intense storms observed in the real atmosphere. For example, in multiyear control simulations using global models with ~300 km grid spacing described by Broccoli and Manabe (1990) and Tsutsui (2002), the deepest central surface pressures in the tropical cyclones that develop are about 980 hPa. In a control simulation using a global model with ~100 km effective grid spacing reported by Bengtsson et al. (1995) the most intense tropical cyclone appearing had a minimum central pressure of 953 hPa and peak surface winds of ~45 m s^{-1}. Peak surface winds of somewhat less than ~50 m s^{-1} are also apparent in the 10-year control run performed using a model with ~100 km effective grid spacing described by Sugi et al. (2002). Hamilton and Hemler (1997) described results from a single season of control integration with a global grid-point atmospheric model with spacing about 35 km. They reported one Pacific typhoon with minimum pressure of 906 hPa and peak winds in the lowest model level ~70 m s^{-1}, comparable to the strongest typhoon that might typically be observed in a given year, but still weaker than the strongest typhoon ever observed (Typhoon Tip in 1979 which had an estimated central pressure as low as 870 hPa according to Dunnavan and Dierks, 1980).

Ohfuchi et al. (2004) and Yoshioka et al. (2005) discuss some aspects of tropical cyclones seen in brief integrations of a T1279L96 AGCM. Ohfuchi et al. (2004) discuss the properties of four west Pacific typhoons in their simulation. Yoshioka et al. (2005) use the fine resolution simulation of intense tropical cyclones to examine the interaction between tropical cyclones and the diurnal cycle. A full global model is needed for first-principles simulation of the atmospheric tidal response to diurnal heating (e.g., Zwiers and Hamilton, 1986; Tokioka and Yagai, 1987) and very fine horizontal resolution is needed to provide a first-principles simulation of intense tropical cyclones, so only recently have models appropriate for study of this interaction been available.

Oouchi et al. (2006) examined the tropical cyclones simulated in 10 years of integration with a T959 AGCM using SSTs taken from the control run of a much lower version of the AGCM coupled to a fully interactive ocean. The model is able to generate a few tropical storms with maximum winds of nearly 50 m s^{-1}. Overall the number and distribution of tropical cyclone occurrences in the model simulation is reasonably realistic, although there is a significant underprediction (factor of ~2) of the number of tropical cyclones in western North Pacific and an overprediction of the occurrence of South Indian Ocean tropical cyclones.

1.5 Changing Subgrid-Scale Parameterizations with Model Resolution

It is generally appreciated that the subgrid-scale parameterizations need to be adjusted as the explicit resolution of a model in changed. Overall, however, this is not an area that has been very deeply explored. One issue that has forced itself on the modeling community is the scaling of subgrid-scale horizontal mixing parameterizations with horizontal resolution. Smagorinsky (1963) proposed a second-order mixing parameterization in which the eddy diffusivity (and eddy viscosity) varied as the inverse square of the model horizontal grid spacing. In more recent times higher order hyperdiffusion (and hyperviscosity) formulations have generally been favored. For idealized one-layer quasigeostrophic models Yuan and Hamilton (1994) found that a simple scaling of the fourth-order (biharmonic) viscosity and diffusivity parameters with the fourth power of the grid spacing worked well (i.e., keeping the diffusion timescale of the smallest resolved scale constant), and led to simulations in which the horizontal variance spectra of the winds appeared consistent as model resolution is changed. However, when the same scaling was used in a one-layer primitive equation (shallow water) model the results were not satisfactory, in the sense that the horizontal variance spectra of the winds was not consistent as the model resolution was changed.

In many studies with full AGCMs the investigators seem to have chosen horizontal diffusivities in a somewhat arbitrary manner. Two studies that tried systematically to examine the dependence of the appropriate diffusivity as a function of resolution are those of Boville (1991) and Takahashi et al. (2006). Both studies used rather standard spectral AGCMs with fourth-order hyperdiffusivity and hyperviscosity parameterizations. Boville examined results with simulations performed at T21, T42, and T63 resolution, while Takahashi et al. considered simulations at T39, T79, T159, T319, and T639. In each a case the diffusivity parameter was adjusted by trial-and-error to produce results in which the end of the horizontal velocity variance spectra follows a power law and in which the spectra were consistent as the model resolution was changed. Both Boville and Takahashi et al. found that the diffusivity coefficient needs to be scaled at about the inverse third power of the spectral truncation (i.e., the diffusion timescale of the smallest resolved scale must drop with finer resolution).

While the need to change the horizontal subgrid-scale mixing parameterizations with model resolution is well appreciated and has attracted some systematic investigation, the comparable issue with vertical subgrid mixing has been less studied. Typically the vertical mixing in AGCMs depends on some measure of the vertical stability based on resolved vertical gradients of temperature (or virtual temperature) and horizontal wind, and most modelers have not seen any necessity to scale this with the vertical grid spacing. One exception is the work of Levy et al. (1982) who developed a scheme in which the mixing across numerical levels mixing depends on the resolved Richardson number in the expected manner, namely that the mixing becomes very strong rapidly as the Richardson number falls below some threshold. They note that in the real world subgrid-scale variability would introduce smaller scale variations in the Richardson number. Thus some mixing would be expected to occur even before the resolved-scale Richardson number appears to be unstable. Levy et al. noted that the modification to the Richardson number dependence to account for this effect should itself depend on the explicit vertical resolution, and they derive a proposed vertical resolution scaling of the Richardson number criterion, based on observations of typical vertical variability at small scales.

A particularly problematic issue in subgrid-scale parameterization is the treatment of moist convective processes. It is has been shown that the space–time variability of simulated precipitation in moderate-resolution AGCMs depends strongly on which parameterization scheme is used for moist convection (Ricciardulli and Garcia, 2000). One might naively expect that, as model resolution is made finer, the results obtained with different subgrid-scale schemes will converge and converge toward a realistic result. Unfortunately what evidence exists suggests that the differences in the behavior among convective parameterization schemes, and some unrealistic aspects of convective simulation, may actually be exacerbated at fine model resolution. Ricciardulli and Sardeshmukh (2002) find that the moist convective adjustment scheme employed in the $\sim$35 km grid version of the SKYHI model produced an unrealistically noisy tropical precipitation field. Enomoto et al. (2007) discuss results obtained with different schemes in the AFES with fine resolution. They note that for resolutions finer than about T639 the Arakawa–Schubert (Arakawa and Schubert, 1974) scheme behaves very unrealistically in that it produces very little convective rain, and the model nonconvective parameterization takes over the production of tropical rain. Enomoto et al. (2007) find a better behavior at fine resolution when employing a version of the Emanuel scheme (Emanuel and Zivkovic-Rothman, 1999). Of course, as grid spacing becomes smaller there are potential conceptual problems with at least some convective parameterizations as currently formulated. The parameterizations are generally regarded as describing the statistical effects of a collection individual convective updrafts and downdrafts assumed to occupy the grid box. As the box becomes smaller such a statistical treatment may not make sense. For example Enomoto et al. (2007) note that the Arakawa–Schubert scheme assumes that (at most) a small fraction of a grid-box is occupied by strong convective updrafts. This is a reasonable assumption for large grid-boxes, but Enomoto et al. question whether it is still appropriate for grid boxes of the order of 10 km horizontal dimension.

Yamada et al. (2005) examined the vertical resolution dependence of the tropical rainfall in their aquaplanet simulations. Interestingly, they find significant differences in the rainfall behavior even between two versions with modest horizontal (T39) and reasonably fine vertical resolution, L48 and L96. In particular, they find that the rainfall rates are typically weaker but more widespread in the L96 version, possibly because the fine resolution opens up additional possibilities for very thin convectively unstable regions to form.

As resolution is increased to a sufficiently fine degree it may be reasonable to expect models to explicitly resolve individual convective updrafts and downdrafts. Certainly there have been some impressive successes with such "cloud-resolving" or "cloud system resolving" limited area models (e.g., Randall et al., 2003a). Typically such models employ roughly 1 km grid spacing in the horizontal, or even finer resolution (Randall et al., 2003a). The very recent work by Tomita et al. (2005), mentioned earlier, showed that reasonable results for organized convection (in many respects, at least) can be obtained with a sufficiently fine resolution global nonhydrostatic model without any convective parameterization, just bulk microphysics parameterizations. Such a model would presumably have no need to change the cloud-related parameterizations with resolution.

Another approach to the sub-grid scale cloud problem is so-called superparameterization, in which limited area fine resolution models with bulk microphysics parameterizations are run embedded in each AGCM gird box. As noted by Randall et al. (2003b) one advantage of this approach is that no adjustment in the parameterization as a function of AGCM resolution should be needed, and that in the limit of very small AGCM grid spacing this model should seamlessly evolve into a global explicit cloud-resolving model. Of course, such a seamless evolution only occurs with particular formulations of the superparameterization. Notably it requires superparameterization schemes that fill each GCM grid box with a full 3D cloud-resolving model, rather than 2D arrays (which is the approach that actually has been applied most extensively so far).

1.6 Middle Atmosphere

While the simulated zonal-mean circulation in the troposphere is only moderately sensitive to the horizontal and vertical resolution employed, it appears that the zonal-mean simulation in the middle atmosphere can be much more sensitive to numerical resolution even as the resolution becomes quite fine. Mahlman and Umschied (1987) noted that the simulation of the basic extratropical stratospheric mean temperature and wind structure in the SKYHI improved dramatically as the latitude–longitude grid resolution was enhanced from $9° \times 10°$ to $5° \times 6°$, to $3° \times 3.6°$ and $1° \times 1.2°$. The high latitude winter stratospheric temperatures were much too cold in the low-resolution versions and this "cold pole bias" problem became less severe as resolution was improved. Jones et al. (1997) and Hamilton et al. (1999) show that this improvement continues even as the resolution is reduced to $0.33° \times 0.4°$. It seems that vertical eddy transports of zonal momentum by gravity waves drive a meridional circulation that warms the winter pole in the stratosphere and reduces the strength of the westerly polar night jet (e.g., Garcia and Boville, 1994; Hamilton, 1996). Coarse-resolution models are not able to explicitly resolve all the gravity waves that are important in the real world, and this leads to a winter cold pole bias, unless the model includes a parameterization of expected gravity wave effects on the mean flow. As the resolution is improved, more of the spectrum can be explicitly resolved and the drag on the mean flow, and consequent dynamical warming at high latitudes are larger (Hayashi et al., 1989; Hamilton et al., 1995, 1999). These issues in the winter polar stratosphere have a counterpart in the summer hemisphere, where the unrealistically weak eddy

1.6 Middle Atmosphere

forcing of the mean upwelling circulation leads to a simulated polar mesopause that is too warm (Hamilton 1996; Hamilton et al., 1995, 1999).

In the tropical stratosphere it appears that the zonal-mean circulation is much less sensitive to changes in the horizontal resolution, but may depend critically on the vertical resolution employed. Historically most AGCMs have simulated winds in the tropical stratosphere that are much too steady, notably lacking the strong interannual quasibiennial oscillation (QBO). The first study to show that an AGCM could simulate a strong mean flow oscillation with the descending intense vertical shear zones that are a prominent feature of the observed QBO was that of Takahashi (1996). He took a standard T21 AGCM and reduced the vertical level spacing to about 500 m as well as reducing the subgrid-scale horizontal diffusivity. He found that an oscillation in the zonal-mean wind developed with a period of about 1.5 years. In the middle stratosphere the amplitude was comparable to that of the observed QBO, but the simulated amplitude was unrealistically weak in the lower stratosphere. Horinouchi and Yoden (1998) and Hamilton et al. (1999, 2001) found similar results, i.e., global AGCMs with fine enough vertical resolution, small enough subgrid-scale viscosity and enough resolved gravity wave flux in the tropics can produce large amplitude, low frequency mean flow oscillations in the tropical stratosphere that clearly resemble the QBO but may differ by having somewhat different periods or vertical structures (referred to now as QBO-like oscillations).

An example of how the mean flow structure in the tropical middle atmosphere changes with vertical resolution is provided by the case of the SKYHI model discussed in Hamilton et al. (2001). SKYHI when run with 40 vertical levels (L40) between the ground and the mesopause (this configuration has about 1.5 km level spacing in the lower-middle stratosphere) lacks any semblance of a QBO. The picture changes dramatically when the vertical resolution is increased to L80 or L160. In these higher-resolution versions of the model the mean flow in the stratosphere is dominated by downward propagating easterly and westerly wind regimes separated by intense shear zones in a QBO-like oscillation. In the stratosphere, the peak equatorial shears in the L40 simulation (run at about 100 km horizontal grid spacing) are $\sim 0.004\,\mathrm{s}^{-1}$. In the L160 simulation they are roughly five times as large ($\sim 0.02\,\mathrm{s}^{-1}$) and are comparable to those seen in monthly mean observations near the equator in the real stratosphere (e.g., Naujokat, 1986; Baldwin et al., 2001). The dramatic change with resolution is presumably related to the ability of fine vertical resolution models to more adequately represent the interaction of the mean flow with a broad spectrum of vertically propagating gravity waves. It is important to note that simply increasing vertical resolution does not seem to initiate QBO-like variability in the tropical stratosphere of all models, however (e.g., Boville and Randel, 1991; Hamilton and Yuan, 1992). Indeed Takahashi (1996) noted that his model produced a QBO-like oscillation only when a moist convective adjustment parameterization was employed. Apparently there needs to be enough resolved gravity wave momentum flux in the appropriate frequency and wavenumber ranges to actually generate the needed mean flow accelerations, and this is controlled to some degree by the convective parameterization employed in a given model.

None of the Takahashi (1996), Horinouchi and Yoden (1998), and Hamilton et al. (1999, 2001) models had a parameterization of nontopographic subgrid-scale gravity wave effects. The QBO has also been successfully simulated in AGCMs that do include a parameterization of the effects the nontopographic gravity waves that are thought to be important in the tropical stratosphere. For example Giorgetta et al. (2002) simulated a rather realistic mean flow QBO in a model that included a Doppler-spread parameterization of nonstationary gravity waves (Manzini et al., 1997). These authors find that their model with T42 horizontal resolution and roughly 500 m level spacing in the stratosphere did produce a nice QBO, but that the same model run with level spacings of more than 1 km did not. They found that in the version with the QBO, roughly half the forcing of mean flow accelerations came from resolved waves and half from parameterized waves (with the role of resolved waves being more important in the lowermost stratosphere).

All these studies have shown that the explicitly resolved upward gravity field emerging from the troposphere plays a critical role in the simulation even of the largest scale features of the middle atmospheric circulation. This raises the question of the validity of the gravity wave field simulation itself. A detailed comparison with observations is complicated by two problems. One is that no current instrument or analysis system can produce an instantaneous global picture of the flow on the horizontal scales of the relevant gravity waves (tens to hundreds of km). Comparisons can be made of statistics of gravity wave variances, covariances, etc. with single station observations (balloons, rockets, radars, lidars) or limited horizontal track data (from aircraft and space shuttle flights) or limited satellite swath data. Each technique has its own bias in terms of the part of the spectrum that can be efficiently detected. Unfortunately, as noted e.g., by Hamilton (1993), the most easily detected parts of the spectrum are the most energetic, which may not correspond to those with the largest eddy fluxes of momentum. Among the limited comparisons that have been published are those of variations in AGCMs with the (1) rocket soundings (Hamilton, 1989), (2) lidars (Hamilton, 1996), and (3) the Kyoto University MU radar (Sato et al., 1999). These comparisons have been fairly encouraging and suggest that current AGCMs with reasonably fine horizontal and vertical resolution may be able to reasonably simulate at least the most basic aspects of the observed gravity wave field in the middle atmosphere. Much more work on this issue needs to be performed, however.

1.7 Coupled Global Ocean–Atmosphere Model Simulations and Climate Sensitivity

How the resolution of global atmospheric AGCMs affects the simulation in coupled ocean–atmosphere global climate models is an issue of obvious importance for climate studies, but one that has thus far received only modest attention from researchers.

1.7 Coupled Global Ocean–Atmosphere Model Simulations and Climate Sensitivity

Emanuel (2001) noted that tropical cyclones induce strong vertical mixing within the upper ocean, leaving cold wakes that are restored to normal conditions to a large extent by surface fluxes from the atmosphere. This restoration is associated with net heating of the ocean column, which is balanced by oceanic heat transport out of the regions affected by the storms. The power input into the ocean from wind (which determines the strength of the vertical ocean mixing) varies as the cube of the surface wind speed. Thus the effects of very intense storms are particularly pronounced. As noted above, the intensity of the strongest storms simulated by AGCMs is a strong function of horizontal resolution, at least down to ~10 km grid spacing. So if the effect Emanuel identifies is significant for the global heat budget, then we should anticipate systematic biases in the simulated climate in a coupled model without extremely fine resolution.

Gualdi et al. (2005) performed a series of 6-month ensemble forecasts using the 19-level ECHAM4 AGCM coupled to a global ocean model, with a focus on the accuracy of forecasts of the development of the El Nino/Southern Oscillation (ENSO) phenomenon in the tropical Pacific. Many of the forecasts were repeated with both T42 and T106 versions of the ECHAM4 atmospheric component, with everything else (initial conditions, ocean model resolution, atmospheric vertical resolution) kept the same. The differences in the forecasts were considerable, particularly for the growing phase of El Nino or La Nina events. At T42 resolution the initial perturbation of the coupled system decays quickly, while the T106 model can sustain the growth of disturbances, leading to significant improvement in the forecasts.

A key application of coupled climate models is to determine the response of climate to imposed natural or anthropogenic perturbations, such as increasing concentrations of greenhouse gases. In an early study Senior (1995) examined the role of model horizontal resolution in determining the equilibrium response to a doubling of atmospheric carbon dioxide content in an AGCM coupled to a mixed layer ocean model. He found nearly identical global mean surface warming in versions of the model with roughly 500 and 250 km horizontal grid spacings. However, the latitudinal variation of warming was somewhat different, with the lower-resolution version displaying a greater intensification of the warming at high latitudes. Senior attributed this difference to the much more realistic representation of storm tracks in the higher-resolution version, which allows the eddy fluxes to respond more effectively to the reduced equator-pole temperature gradient.

The sensitivity of a model climate to large-scale perturbations is determined by the strengths of various feedback processes in the model. It is fairly clear now that the important feedback that is most uncertain in current models is the cloud feedback, particularly in the tropics and subtropics (Cess et al., 1990; Stowasser et al., 2006). Among current global models even the sign of this feedback is not consistent, and this uncertainty leads to a variation in simulated sensitivity of the global mean surface temperature of a factor of 2–3 (e.g., Stowasser et al., 2006). An interesting question is how the simulated cloud feedbacks depend on model resolution. Stowasser and Hamilton (2006) examined a related issue, namely how the simulated monthly mean cloud fields in the tropics and subtropics vary in relation to the interannual fluctuations of the large-scale circulation. In particular, they examined the connection between cloud forcing and monthly mean meteorological fields in a large number of the global coupled climate models included in the preparation of the IPCC Fourth Assessment Report. They found a very wide range of results for the various models. It was interesting that no systematic variation in the results as a function of the spatial resolution of the models was apparent. In fact, Stowasser and Hamilton (2006) examined results from two versions of the Japanese MIROC (Model for Interdisciplinary Research on Climate) model, one run at T42L20 and the other at T106L56, and the results were very similar. The implication seems to be that, at least in the range of resolutions considered, the cloud and convection parameterizations are much more important than the numerical resolution in determining how the cloud feedbacks are simulated in a global model.

Ingram (2002) investigated the water vapor feedback operating in climate change experiments in several versions of an AGCM with widely different vertical model resolutions. He found that the feedbacks were insensitive to the vertical resolution once some modest threshold was passed (i.e., results were very similar for models with 19, 38 and 100 total model levels).

1.8 Summary

A numerical AGCM is a finite numerical approximation to the continuous differential equations governing atmospheric circulation. With current resources it is generally not possible to show that our AGCM solutions have completely converged, and at least modest changes in the statistical properties AGCM simulated circulation seem to occur with improving horizontal and vertical resolution. One issue that has been fairly extensively addressed is the dependence of the zonal-mean climatological circulation on the horizontal grid spacing or (for spectral models) horizontal wavenumber truncation. For the troposphere it seems that such changes are quite important up to about T42, and still significant, if more modest, at higher resolutions. For the stratosphere and mesosphere it appears that model results for the zonal-mean simulation may depend more dramatically on the resolution. In particular, the overall structure of the extratropical middle atmosphere in coarse or moderate resolution AGCMs tends to be unrealistically close to radiative equilibrium (too cold in the high latitude winter and too warm in the high latitude summer), and this problem is progressively alleviated as horizontal resolution is improved.

Zonal-mean tropospheric simulations appear not to be strongly dependent on vertical resolution, but the zonal-mean circulation in the tropical stratosphere, in some models at least, has a very strong dependence on vertical resolution. With vertical level spacings of ~ 1 km or more in the stratosphere it appears that most (perhaps all) AGCMs will simulate nearly steady prevailing winds in the tropical stratosphere, a very unrealistic representation of the most basic aspect of the general circulation in this region of the atmosphere. When model vertical grid spacing is reduced to about

0.5 km or finer, some models display strong interannual oscillations of the zonal-mean wind in the tropical stratosphere with the alternating downward-propagating shear zones that are well-known features of the observed QBO.

In the last few years there has been considerable development in high-performance computing facilities available for atmospheric simulation, notably with the 2002 inauguration of the Earth Simulator in Japan. This has made possible global atmospheric simulations at unprecedented fine resolution. This raises issues of evaluation of the very complex and detailed simulations that result. It has been shown that at least some fine resolution models simulate a flow with realistic horizontal energy spectra. A number of studies have examined the tropical cyclone simulations within global models. These show that even modest resolution models can spontaneously simulate a climatology with a reasonable number of tropical cyclones, but that simulation of storm intensity become much more realistic as model horizontal resolution is improved.

It is understood that the model simulations will have deficiencies simply associated with the fact that components of the real circulation will not be explicitly resolved in the finite numerical approximation employed. As explicit model resolution is changed, the parameterizations used to incorporate subgrid-scale effects must also be modified. For example it is known that subgrid-scale diffusivity and viscosity coefficients must be lowered as horizontal resolution is improved. Presumably some similar scaling should apply to vertical subgrid-scale mixing, but little work on this problem seems to have been published. The effects of resolution on the performance of moist convection parameterizations are a complicated issue, and at least some published studies suggest that the performance of models with state-of-the-art convection schemes may not converge toward realistic results.

Experience with limited-area models suggests that model performance may pass a threshold when horizontal grid spacings are reduced to ~ 1 km or less. At this point the explicit dynamics along with a bulk microphysics parameterization may realistically represent many features of moist convection and clouds. Recent work with the NICAM global model run on the Earth Simulator has approached within a factor of 3 of this hypothesized horizontal resolution threshold, and initial results are encouraging that such global models can simulate realistic mesoscale organization of cloud-scale features.

Acknowledgments This work was supported by NSF Award ATM02-19120 and by JAMSTEC through its sponsorship of the International Pacific Research Center.

References

Arakawa, A. and W.H. Schubert, 1974: Interaction of cumulus cloud ensemble with the large-scale environment, Part I. *J. Atmos. Sci.*, **31**, 671–701.

Baldwin, M., L. Gray, T. Dunkerton, K. Hamilton, P. Haynes, W. Randel, J. Holton, M. Alexander, I. Hirota, T. Horinouchi, D. Jones, J. Kinnersley, C. Marquardt, K. Sato and M. Takahashi, 2001: The Quasi-biennial Oscillation. *Rev. Geophys.*, **39**, 179–229.

Bengtsson, L., M. Botzet and M. Esh, 1995: Simulation of hurricane-type vortices in a general circulation model. *Tellus*, **47A**, 175–196.

Boer, G.J. and B. Denis, 1997: Numerical convergence of the dynamics of a GCM. *Clim. Dyn.*, **13**, 359–374.

Boer, G.J. and M. Lazare, 1988: Some results concerning the effect of horizontal resolution and gravity-wave drag on simulated climate. *J. Clim.*, **1**, 789–806.

Boer, G.J. and T.G. Shepherd, 1983: Large-scale two-dimensional turbulence in the atmosphere. *J. Atmos. Sci.*, **40**, 164–184.

Boville, B.A., 1991: Sensitivity of simulated climate to model resolution. *J. Clim.*, **4**, 469–485.

Boville, B.A. and W.J. Randel, 1991: Equatorial waves in a stratospheric GCM: Effects of vertical resolution. *J. Atmos. Sci.*, **49**, 785–801.

Boyle, J.S., 1993: Sensitivity of dynamical quantities to horizontal resolution for a climate simulation using the ECMWF (cycle 33) model. *J. Clim.*, **6**, 796–815.

Broccoli, A. and S. Manabe, 1990: Can existing climate models be used to study anthropogenic changes in tropical cyclone climate. *Geophys. Res. Lett.*, **17**, 1917–1920.

Cess, R.D. et al., 1990: Intercomparison and interpretation of climate feedback processes in 19 atmospheric general circulation models. *J. Geophys. Res.*, **95**, 16601–16615.

Conaty, A.L., J.C. Jusem, L. Takacs, D. Keyser and R. Atlas, 2001: The structure and evolution of extratropical cyclones, fronts, jet streams, and the tropopause in the GEOS General Circulation Model. *Bull. Am. Meteor. Soc.*, **82**, 1853–1867.

Dunnavan, G.M. and J.W. Dierks, 1980: An analysis of Supertyphoon Tip (October 1979). *Mon. Weather Rev.*, **108**, 1915–1923.

Emanuel, K., 2001: Contribution of tropical cyclones to meridional heat transport by the oceans, *J. Geophys. Res.*, **106**, 14771–14782, doi:10.1029/2000JD900641.

Emanuel, K. and M. Zivkovic-Rothman, 1999: Development and evaluation of a convective scheme for use in climate models. *J. Atmos. Sci.*, **56**, 1766–1782.

Enomoto, T., A. Kuwano-Yoshida, N. Komori and W. Ohfuchi, 2007: Description of AFES 2: Improvements of high-resolution and coupled simulations. *High Resolution Numerical Modelling of the Atmosphere and Ocean*, (W. Ohfuchi and K. Hamilton, eds.), Springer Publications, Chapter 5.

Garcia, R.R. and B.A. Boville, 1994: "Downward control" of the mean meridional circulation and temperature distribution of the polar winter stratosphere. *J. Atmos. Sci.*, **51**, 2238–2245.

Giorgetta M.A., E. Manzini and E. Roeckner, 2002: Forcing of the quasi-biennial oscillation from a broad spectrum of atmospheric waves. *Geophys. Res. Lett.*, **29**, 1245, doi:10.1029/2002GL014756.

Griffies, S.M. and R.W. Hallberg, 2000: Biharmonic friction with a Smagorinsky-like viscosity for use in large-scale eddy-permitting ocean models. *Mon. Weather Rev.*, **128**, 2935–2946.

Gualdi, S., A. Alessandri and A. Navarra, 2005: Impact of atmospheric horizontal resolution on El Nino/Southern Oscillation forecasts. *Tellus*, **57A**, 357–374.

Hamilton, K., 1989: Evaluation of the gravity wave field in the middle atmosphere of the GFDL "SKYHI" general circulation model. *World Meteorological Organization Technical Document #273*, pp. 264–271.

Hamilton, K, 1993: What we can learn from general circulation models about the spectrum of middle atmospheric motions. *Coupling Processes in the Lower and Middle Atmosphere* (E. Thrane, T. Blix and D. Fritts, eds.), Kluwer Academic Publishers, pp. 161–174.

Hamilton, K., 1996: Comprehensive meteorological modelling of the middle atmosphere: A tutorial review. *J. Atmos. Terr. Phys.*, **58**, 1591–1628.

Hamilton, K. and R.S. Hemler, 1997: Appearance of a super-typhoon in a global climate model simulation. *Bull. Am. Meteor. Soc.*, **78**, 2874–2876.

Hamilton, K. and L. Yuan, 1992: Experiments on tropical stratospheric mean wind variations in a spectral general circulation model. *J. Atmos. Sci*, **49**, 2464–2483.

Hamilton, K., R.J. Wilson, J.D. Mahlman and L.J. Umscheid, 1995: Climatology of the SKYHI troposphere–stratosphere–mesosphere General Circulation Model. *J. Atmos. Sci.*, **52**, 5–43.

Hamilton, K., R.J. Wilson and R.S. Hemler, 1999: Middle atmosphere simulated with high vertical and horizontal resolution versions of a GCM: Improvement in the cold pole bias and generation of a QBO-like oscillation in the tropics. *J. Atmos. Sci.*, **56**, 3829–3846.

References

Hamilton, K., R.J. Wilson and R.S. Hemler. 2001: Spontaneous stratospheric QBO-like oscillations simulated by the GFDL SKYHI General Circulation Model. *J. Atmos. Sci.*, **58**, 3271–3292.

Hayashi, Y., D. G. Golder, J. D. Mahlman and S. Miyahara, 1989: The effect of horizontal resolution on gravity waves simulated by the GFDL "SKYHI" general circulation model. *Pure Appl. Geophys.*, **130**, 421–443.

Hayashi, Y., D.G. Golder and P.W. Jones, 1997: Tropical gravity waves and superclusters simulated by high-horizontal-resolution SKYHI general circulation models. *J. Meteor. Soc. Jpn.*, **75**, 1125–1139.

Held, I.M. and M.J. Suarez, 1994: A proposal for the intercomparison of the dynamical cores of atmospheric General Circulation Models. *Bull. Am. Meter. Soc.*, **75**, 1825–1830.

Horinouchi T. and S. Yoden, 1998: Wave-mean flow interaction associated with a QBO-like oscillation simulated in a simplified GCM. *J. Atmos. Sci.*, **55**, 502–526.

Ingram, W.J., 2002: On the robustness of the water vapor feedback: GCM vertical resolution and formulation. *J. Clim.*, **15**, 917–921.

Inness, P.M., J.M. Slingo, S.J. Woolnough, R.B. Neale and V.D. Pope, 2001: Organization of tropical convection in a GCM with varying vertical resolution; implications for the simulation of the Madden–Julian Oscillation. *Clim. Dyn.*, **17**, 777–793.

Janjic, Z.I., J.P. Gerrity Jr., and S. Nickovic, 2001: An alternative approach to nonhydrostatic modeling. *Mon. Weather Rev.*, **129**, 1164–1178.

Jones, P.W., K. Hamilton and R.J. Wilson, 1997: A very high-resolution general circulation model simulation of the global circulation in austral winter. *J. Atmos. Sci.*, **54**, 1107–1116.

Kang, I.-S., K. Jin, B. Wang, K.-M. Lau, J. Shukla, V. Krishnamurthy, S. Schubert, D. Wailser, W. Stern, A. Kitoh, G. Meehl, M. Kanamitsu, V. Galin, V. Satyan, C.-K. Park, and Y. Liu., 2002: Intercomparison of the climatological variations of Asian summer monsoon precipitation simulated by 10 GCMs. *Clim. Dyn.*, **19**, 383–395.

Kawatani, Y. and M. Takahashi, 2003: Simulation of the Baiu front in a high-resolution AGCM. *J. Meteor. Soc. Jpn.*, **81**, 113–126.

Kobayashi C. and M. Sugi, 2004: Impact of horizontal resolution on the simulation of the Asian summer monsoon and tropical cyclones in the JMA global model. *Clim. Dyn.*, **23**, 165–176.

Koshyk, J.N. and K. Hamilton, 2001: The horizontal kinetic energy spectrum and spectral budget simulated by a high-resolution troposphere–stratosphere–mesosphere GCM. *J. Atmos. Sci.*, **58**, 329–348.

Koshyk, J.N., B.A. Boville, K. Hamilton, E. Manzini and K. Shibata, 1999: The kinetic energy spectrum of horizontal motions in middle atmosphere models. *J. Geophys. Res.*, **104**, 27177–27190.

Lander, J. and B.J. Hoskins, 1997: Believable scales and parameterizations in a spectral model. *Mon. Weather Rev.*, **125**, 292–303.

Levy, H., J.D. Mahlman and W.J. Moxim, 1982: Tropospheric N2O variability. *J. Geophys. Res.*, **87**, 3061–3080.

Lindborg, E., 1999: Can the atmospheric kinetic energy spectrum be explained by two-dimensional turbulence? *J. Fluid Mech.*, **388**, 259–288.

Lindzen, R.S. and M. Fox-Rabinovitz, 1989: Consistent vertical and horizontal resolution. *Mon. Weather Rev.*, **117**, 2575–2583.

Mahlman, J.D. and L.J. Umscheid, 1987: Comprehensive modeling of the middle atmosphere: The influence of horizontal resolution. *Transport Processes in the Middle Atmosphere* (G. Visconti and R. Garcia, eds.), Reidel Publishing, pp. 251–266.

Manabe, S., J. Smagorinsky and R. F. Strickler, 1965: Simulated climatology of a general circulation model with a hydrologic cycle. *Mon. Weather Rev.*, **93**, 769–798.

Manzini, E., N.A. McFarlane and C. McLandress, 1997: Impact of the Doppler spread parameterization on the simulation of the middle atmosphere circulation using the MA/ECHAM4 general circulation model. *J. Geophys. Res.*, **102**, 25751–25762.

Mizuta, R., T. Uchiyama, K. Kamiguchi, A. Kitoh and A. Noda, 2005: Changes in extremes indices over Japan due to global warming projected by a global 20-km-mesh atmospheric model. *SOLA*, **1**, 153–156.

Mullen, S.L. and R. Buizza, 2002: The impact of horizontal resolution and ensemble size on probabilistic forecasts of precipitation by the ECMWF Ensemble Prediction System. *Weather Forecast.*, **17**, 173–191.

Nastrom, G.D. and K.S. Gage, 1985: A climatology of atmospheric wavenumber spectra of wind and temperature observed by commercial aircraft. *J. Atmos. Sci.*, **42**, 950–960.

Naujokat, B., 1986: An update of the observed Quasi-Biennial Oscillation of the stratospheric winds over the tropics. *J. Atmos. Sci.*, **43**, 1873–1877.

Ohfuchi, W., H. Nakamura, M.K. Yoshioka, T. Enomoto, K. Takaya, X. Peng, S. Yamane, T. Nishimura, Y. Kurihara, and K. Ninomiya, 2004: 10-km mesh meso-scale resolving simulations of the global atmosphere on the earth simulator – preliminary outcomes of AFES (AGCM for the Earth Simulator). *J. Earth Simulator*, **1**, 8–34.

Ohfuchi, W., H. Sasaki, Y. Masumoto and H. Nakamura, 2005: Mesoscale-resolving simulations of the global atmosphere and ocean on the Earth Simulator. *Eos*, 86, 45–46.

Oouchi, K., J. Yoshimura, H. Yoshimura, R. Mizuta, S. Kusunoki and A. Noda, 2006: Tropical cyclone climatology in a global-warming climate as simulated in a 20 km-mesh global atmospheric model: Frequency and wind intensity analyses. *J. Meteor. Soc. Jpn.*, **84**, 259–276.

Orlanski, I. and C. Kerr, 2007: Project TERRA: A glimpse into the future of weather and climate. *High Resolution Numerical Modelling of the Atmosphere and Ocean*, (W. Ohfuchi and K. Hamilton, eds.), Springer Publications, Chapter 3.

Palmer, T.N., 2001: A nonlinear dynamical perspective on model error: A proposal for non-local stochastic-dynamic parameterization in weather and climate prediction models. *Q. J. R. Meteor. Soc.*, **127**, 279–304.

Pope, V. and R. Stratton, 2002: The processes governing horizontal resolution sensitivity in a climate model. *Clim. Dyn.*, **19**, 211–236.

Randall, D., S. Krueger, C. Bretherton, J. Curry, P. Duynkerke, M. Moncrieff, B. Ryan, D. Starr, M. Miller, W. Rossow, G. Tselioudis and B. Wielicki, 2003a: Confronting models with data: The GEWEX Cloud Systems Study. *Bull. Am. Meter. Soc.*, **84**, 455–469.

Randall, D., M. Khairoutdinov, A. Arakawa, and W. Grabowski, 2003b: Breaking the cloud parameterization deadlock. *Bull Amer Met Soc.*, **84**, 1547–1564.

Ricciardulli, L. and R.R. Garcia, 2000: The excitation of equatorial waves by deep convection in the NCAR Community Climate Model (CCM3). *J. Atmos. Sci.*, **57**, 3461–3487.

Ricciardulli, L. and P.D. Sardeshmukh, 2002: Local time- and space scales of organized tropical deep convection. *J. Atmos. Sci.*, **59**, 2775–2790.

Roebber, P.J., D.M. Schultz, B.A. Colle and D.J. Stensrud, 2004: Towards improved prediction: High-resolution and ensemble modeling systems in operations. *Wea. Forecasting*, **19**, 936–949.

Roeckner, E., R. Brokopf, M. Esch, M. Giorgetta, S. Hagemann, L. Kornblueh, E. Manzini, U. Schlese and U. Schulzweida, 2006: Sensitivity of simulated climate to horizontal and vertical resolution in the ECHAM5 atmosphere model. *J. Clim.*, **19**, 3771–3791.

Sato, K., T. Kumakura and M. Takahashi. 1999: Gravity waves appearing in a high-resolution GCM simulation. *J. Atmos. Sci.*, **56**, 1005–1018.

Senior, C.A., 1995: The dependence of climate sensitivity on the horizontal resolution of a GCM. *J. Clim.*, **8**, 2860–2880.

Shen, B.-W., R. Atlas, J.-D. Chern, O. Reale, S.-J. Lin, T. Lee and J. Chang, 2006: The 0.125 degree finite-volume general circulation model on the NASA Columbia supercomputer: Preliminary simulations of mesoscale vortices. *Geophys. Res. Lett.*, **33**, doi:10.1029/2005GL024594.

Smagorinsky, J., 1963: General circulation experiments with the primitive equations. I. The basic experiment. *Mon. Weather Rev.*, **91**, 99–164.

Sperber, K.R., S. Hameed, G.L. Potter and J.S. Boyle, 1994: Simulation of the northern summer monsoon in the ECMWF model: sensitivity of horizontal resolution. *Mon. Weather Rev.*, **122**, 2461–2481.

Stevenson, D.B., F. Chauvin and J.-F. Royer, 1998: Simulation of the Asian summer monsoon and its dependence on model horizontal resolution. *J. Meteor. Soc. Jpn.*, **76**, 237–265.

Stowasser, M. and K. Hamilton, 2006: Relationships between cloud radiative forcing and local meteorological variables compared in observations and several global climate models. *J. Clim.*, **19**, 4344–4359.

References

Stowasser, M., K. Hamilton and G.J. Boer, 2006: Local and global climate feedbacks in models with differing climate sensitivities. *J. Clim.*, **19**, 193–209.

Sugi, M., A. Noda and N. Sato, 2002: Influence of global warming on tropical cyclone climatology: An experiment with the JMA global model. *J. Meteor. Soc. Japan*, **80**, 249–272.

Takahashi, M., 1996: Simulation of the stratospheric quasi-biennial oscillation using a general circulation model. *Geophys. Res. Lett.*, **23**, 661–664.

Takahashi, Y.O., K. Hamilton and W. Ohfuchi, 2006: Explicit global simulation of the mesoscale spectrum of atmospheric motions. *Geophys. Res. Lett.*, **33**, L12812, doi:10.1029/2006GL026429.

Tokioka, T. and I. Yagai, 1987: Atmospheric tides appearing in a global atmospheric general circulation model. *J. Meteor. Soc. Japan*, **65**, 423–438.

Tomita, H., H. Miura, S. Iga, T. Nasuno and M. Satoh, 2005: A global cloud-resolving simulation: Preliminary results from an aqua-planet experiment, *Geophys. Res. Lett.*, **32**, L08805, doi:10.1029/2005GL022459.

Tompkins, A.M and K.A. Emanuel, 2000: The vertical resolution sensitivity of simulated equilibrium tropical temperature and water vapour profiles. *Q. J. R. Meteor. Soc.*, **126**, 1219–1238.

Tsutsui, J., 2002: Implications of anthropogenic climate change for tropcial cyclone activity. A case study with the NCAR CCM2. *J. Meteor. Soc. Japan*, **80**, 45–65.

Williamson, D.L., 1999: Convergence of atmospheric simulations with increasing horizontal resolution and fixed forcing scales. *Tellus*, **51A**, 663–673.

Yamada, Y., T. Sampe, Y.O. Takahashi, M.K. Yoshioka, W. Ohfuchi, M. Ishiwatari, K. Nakajima and Y.-Y. HayashiI, 2005: A resolution dependence of equatorial precipitation activities represented in a general circulation model. *Theor. Appl. Mech. Jpn.*, **54**, 289–297.

Yoshioka, M.K., Y. Kurihara and W. Ohfuchi, 2005: Effect of the thermal tidal oscillation of the atmosphere on tropical cyclones. *Geophys. Res. Lett.*, **32**, L16802, doi:10.1029/2005GL022716.

Yu, R.C., W. Li, X. Zhang, Y.M. Liu, Y.Q. Yu, H.L. Liu and T.J. Zhou., 2000: Climatic features related to eastern China summer rainfalls in the NCAR CCM3. *Adv. Atmos. Sci.*, **17**, 503–518.

Yuan, L. and K. Hamilton, 1994: Equilibrium dynamics in a forced-dissipative f-plane shallow water model. *J. Fluid Mech.*, **280**, 369–394.

Zhou, T.J. and Z.X. Li, 2002: Simulation of the east Asian summer monsoon by using a variable resolution atmospheric GCM. *Clim. Dyn.*, **19**, 167–180.

Zwiers, F. and K. Hamilton, 1986: The Simulation of atmospheric tides in the Canadian Climate Centre general circulation model. *J. Geophys. Res.*, **91**, 11877–11898.

Chapter 2
The Rationale for Why Climate Models Should Adequately Resolve the Mesoscale

Isidoro Orlanski

Summary A review of the importance of the cyclone-frontal scale system in climate variability and the ability of present climate models to simulate them has been presented.

The analysis of three different Climate models, GISS, the NCAR community climate model CCM3, and the GFDL Finite volume AM2 (M90), have been discussed. The intention here was not to determine which one is better but rather to indicate what deficiency may be common to all of them. Evidence shows that the three models tend to be deficient in the generation of cyclone wave activity with the consequences that heat, momentum, and moisture may be deficient in the extratropical and subpolar regions. This will affect cloudiness, wind stress, and precipitation. Bauer and Del Genio (2005) have shown that the deficiency of moisture and cloudiness over the subpolar regions was due to the lack of cyclone waves to transport moisture and clouds to these regions. A discussion of complementary work done on clustering of cyclone trajectories by Gaffney et al. (2005) was also presented. Consistent with the present analysis, this study also showed that differences in trajectories between reanalysis and model simulation for each cluster of trajectories were here interpreted to be related to the lack of intense high frequency eddies of the GCM.

The previous two studies depend on the surface characteristics based on trajectories of the high frequency eddies. The present analysis on the GFDL-GCM is totally eulerian and based on the upper level eddy activities (300 mb). However, a similar conclusion has been drawn from the analysis of the band pass frequency of energy and momentum for the GFDL AM2_M90 17 year runs, where it is quite clear that the momentum and energy of the very high frequency is much lower in the model simulation than in the reanalysis. The variance of meridional velocity also shows that the deficiency of the high frequency is in the latitude area where the reanalysis shows it to be positioned in the storm track: the model displaces it south of that. There is also a suggestion that to achieve the correct intensity of the high frequency baroclinic eddies, models should have enough resolution to resolve them, since this intensity depends on the lower level circulation of the frontal circulation system. The mesoscale circulation associated with cyclones could be adequately represented in models with resolution equal or superior to $1/4°$ resolution. It is clear that to adequately resolve the mesoscale, it is necessary to not only improve the resolution but also to improve the boundary layer and surface fluxes. Clearly, at the present low resolution of climate models, this improvement is probably unattainable. However, if the cloudiness and sea ice over the subpolar regions are important to the overall climate, this should be an attainable goal because no sophistication in the moist convection or sea-ice model could correct those deficiencies due to the unresolved dynamics.

2.1 Introduction

Since the beginning of Numerical Weather Prediction (NWP) more than 50 years ago, model resolution was one of the major concerns to improve the accuracy of weather prediction. The forecast skill, measured by the correlation coefficients of the 3-day height forecasts for the extratropical northern and southern hemispheres over the last 20 years shows a constant increase in skill (from 0.5 to 0.9). This improvement is shared by the National Center for Environmental Prediction (NCEP), the ECMWF, and other prediction centers. Three main factors contributed to this improvement over these last 20 years: better observations, improved physical parameterizations and initialization schemes, and increased model resolution. Although the three improvements are probably of equal importance, the emphasis in the present discussion will be focused on the improvement of horizontal model resolution. For instance, Fig. 2.1

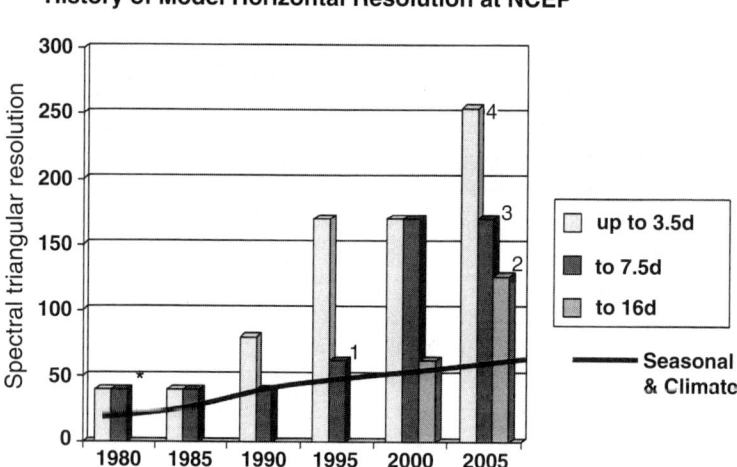

Fig. 2.1 A history of model horizontal resolution at NCEP from 1980 to 2005. The different color bars represent the model resolution for the three forecast products

shows a schematic of the model resolution changes during this period for NCEP. In 1980, they used an equivalent spectral triangular resolution of T42, whereas today they use a T254 operationally (for forecasts up to 3.5 days). Because of computational limitations, model resolution has degraded for medium range forecasts, i.e., T126 (up to 16 days).

For the same period, climate studies had a very modest increase in model resolution. The resolution in most models participating in the IPCC's fourth assessment report is of the order of T62 about 2° × 2° resolution. As we will show in this paper, the high-frequency eddy activity (weather) is of paramount importance in the quality of climate simulations. Presently most climate models are tested in a climate mode, meaning that they are run for a number of years and the simulation is verified with present climate conditions. However, there is a growing consensus among modelers (see Phillips et al. 2004) that climate models should also be tested in an NWP mode. If so, the forecast skill scores for these operational centers with model resolutions similar to T62 was around 0.55 for a 7-day forecast. This forecast would have had very modest skill, with increasing model resolution (to T170), the skills increase to 0.65 in later years. This also assumes that the other components such as data quality and physics do not change much during these few years. It should be noted that the forecast skill is a measure that has a bias to larger scales since it is the forecast of geopotential anomalies at 500 mb. Improvement in skill, although not that impressive, is mainly due to the better resolution of the shorter waves but is perhaps diminished by the fact that the error growth is faster for those scales. If we relate these results to climate model testing, in principle, the expectation to resolve the smaller scales, such as individual cyclone waves, in a climate simulation does not seem realistic. However, it could be beneficial for testing the effects of new physics in these models. But for climate variability, we might expect that at least the statistics of these high frequency waves are correctly simulated in both their behavior and intensity.

2.2 The Role of the High Frequency Wave Activity in Climate Variability

Recent studies show a convergence of conclusions regarding the role of mid latitude baroclinic cyclones in climate (see a review by Chang et al. 2002). It is well known that baroclinic eddies define the storm tracks in middle latitudes of both hemispheres; i.e., in the winter season for the North Hemisphere and both seasons for the South Hemisphere. This baroclinic eddy activity contributes substantially to momentum, heat and moisture fluxes, and clouds and wind stresses. Large differences are observed due to the interannual variability of the storm tracks; different behaviors of the baroclinic eddy life cycle are partly responsible for those changes (Lau 1985; Held et al. 1989; Orlanski 1998). Recent articles on storm track variability for the warm and cold cycles of ENSO over the eastern Pacific Ocean show that the characteristics of the high frequency baroclinic waves are mostly responsible for such changes (Orlanski 2005). It has been known for some time that baroclinic eddies have two possibilities for interacting with the larger scale flow; basically, if they break anticyclonically (LC1), the flux of momentum moves poleward pushing the jet poleward as well; if they break cyclonically (LC2), they will push the jet equatorward (Simmons and Hoskins 1980; Thorncroft et al. 1993).

Recently Orlanski (2003) clarified the role of long and short baroclinic wave breaking. The study found that short upper level waves will break either cyclonically or anticyclonically depending on the wave energy (see Fig. 2.2). Most weak waves with

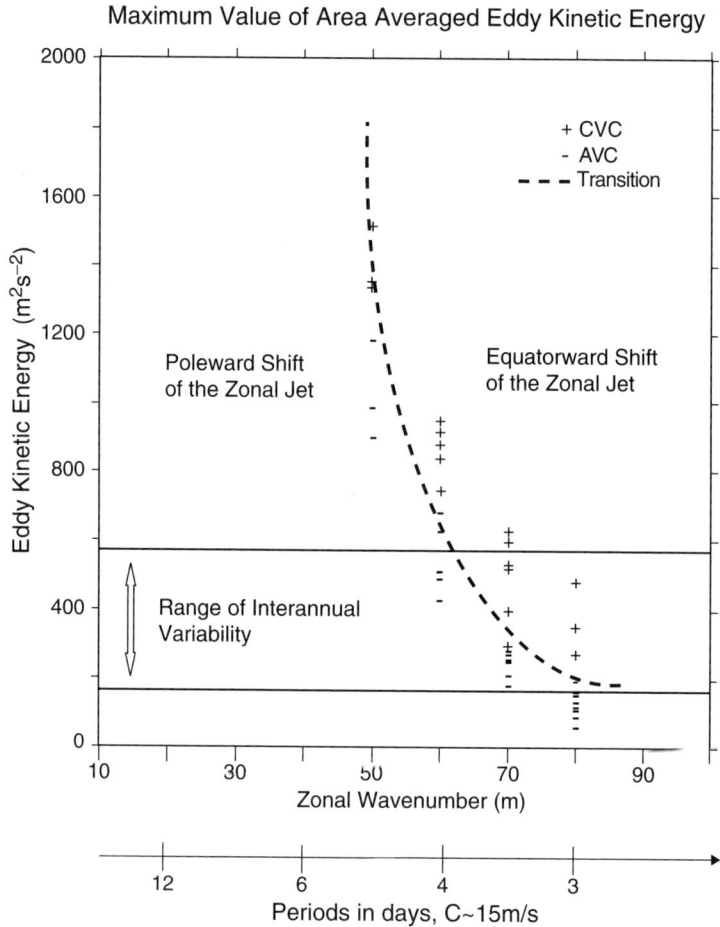

Fig. 2.2 Transition curve separating the anticyclonic wave breaking (AWB) and the cyclonic wave breaking (CWB) processes in a two – dimensional space made by eddy kinetic and wave number (from Orlanski 2003). By using a typical value for the phase velocity of synoptic waves in the atmosphere (C $\sim$ 15 m/s), the same can be viewed as EKE vs. frequencies

2.2 The Role of the High Frequency Wave Activity in Climate Variability

low energies, if they break, will be anticyclonic; as the wave energy increases the wave breaking becomes more intense and pushes the jet more poleward. However, the wave could reach a threshold energy in which case the breaking reverses with a cyclonic circulation and the jet at the end of the wave-life cycle will be positioned equatorward of the breaking wave. Using a shallow water global model, it was found that the threshold energy depends heavily on the wavelength; for shorter waves the energy threshold is low and increases rapidly as the wavelength increases. For energy levels found in our climate, wave numbers larger than 7 (shorter waves, higher frequencies) may break either way, whereas wave numbers smaller than 7 (long waves, intermediate frequencies) will only break anticyclonically. There is a bifurcation in which waves that cross a threshold energy will flux momentum equatorward; smaller energies will flux momentum poleward and the energy threshold increases with wavelength. These results underline the importance of the high frequency eddy intensity and its effect on the interannual variability of the extratropic storm track and as a consequence on climate variability.

A recent study by Riviere and Orlanski (2007) shows a very clean verification of these behaviors and is shown in Fig. 2.3. They used NCEP-NCAR reanalysis data (1979–1995) and showed behaviors that resulted if momentum fluxes were split into different frequency bands (high frequency, 1–5 days; intermediate frequencies, 5–12 days)[1]. The upper graphs of Fig. 2.3 show the meridional momentum fluxes for both bands; on the left is 5–12 days and 1–5 days is on the right. Positive values of $u'v'$ indicate poleward momentum flux, whereas negative values are equatorward. Only large values of positive (dark gray) and negative (very light gray) fluxes are in the high frequency band, whereas the intermediate band is predominantly positive.

The lower graphs in Fig. 2.3 show the kinetic energy and the E-vector $= ((v'v' - u'u')/2 - u'v')$ for the same bands. Notice the considerable energy for the high frequency band and the equatorward and poleward vectors indicate, as before, the waves flux momentum in both directions for these frequency bands. However, for the intermediate band the vectors are predominantly equatorward (momentum fluxes poleward). Figure 2.3 is a very robust verification of the results shown in Fig. 2.2. The energy level in the high frequencies is sufficient to break either way, whereas for the intermediate frequencies the predominant breaking is anticyclonic.

This result regarding the high frequency eddy intensity opens the question about the ability of low resolution climate models to resolve the baroclinic wave spectrum correctly since it seems very important for all aspects of the extratropics.

The previous results highlight the importance of the high frequency in our present climate. A valid question is how climate models could reproduce the energy partition between the high and intermediate scales as shown in Fig. 2.3. Recent analyses have been performed to evaluate characteristics of the baroclinic wave spectrum for three GCM's and its comparison with the corresponding reanalysis NCEP_NCAR and ERA40. (GISS-GCM; Bauer and Del Genio (2005); NCAR-CCM3, Gaffney et al. (2005); and in the next section of this article the GFDL-AM2).

---

[1] It was tested by changing in the boundary from 12 to 15 day without major differences.

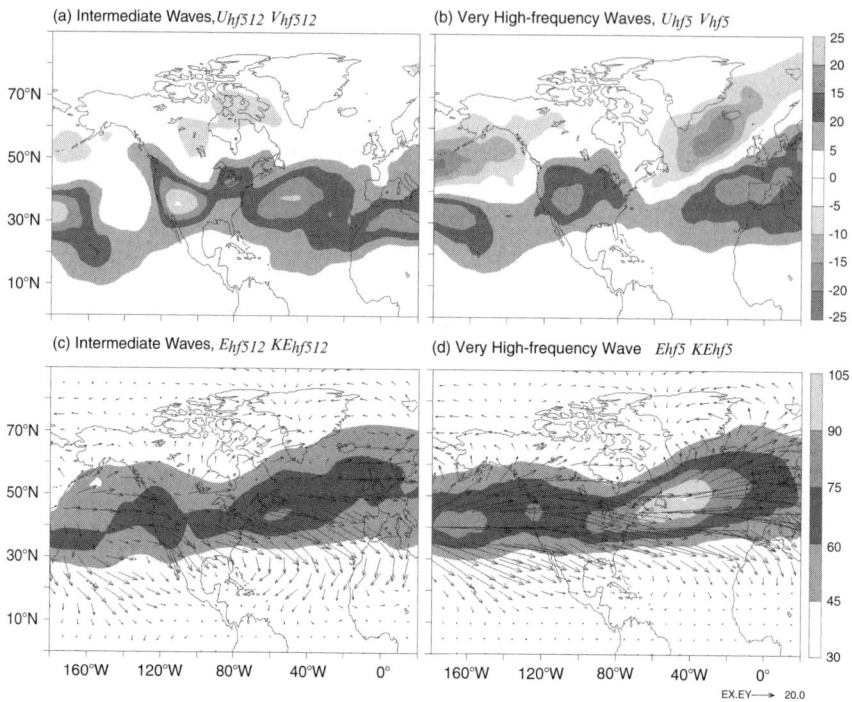

Fig. 2.3 NCEP/NCAR Reanalysis data. Period 1979–1995. (**a**) and (**b**) are respectively meridional and momentum fluxes for intermediate and very high-frequency waves. (**c**) and (**d**) are eddy kinetic energy (*gray-shadings*) and E-vector (*arrows*) for intermediate and very-high frequency. Intermediate waves correspond here to a band pass filter between 5–12 day and very-high-frequency waves to those inferior to 5-day

2.3 The Performance of the Eddy Activity in Three Climate Models

Bauer and Del Genio 2005 (BD05 hereafter), using a novel approach, have compared the characteristics, composition, and structure of 10 winters of storms simulated by the GISS-GCM and in two reanalysis products. In BD05, frequency histograms of DJF cyclones were calculated as a function of intensity index for different horizontal resolutions of the GISS-GCM, the ERA Reanalysis, and the NCEP-NCAR Reanalysis. The results are shown in Fig. 2.4. Clearly, the $4° \times 5°$ degree resolution has fewer cyclones, which have low intensity compared to the two reanalyses.

The $2° \times 2.5°$ degree resolution shows a tendency for more intense cyclones but they are still very few in number (should be noted that there were the lower resolution data was taking 6 h apart, the $2° \times 2.5°$ degree resolution was taking 12 h

2.3 The Performance of the Eddy Activity in Three Climate Models

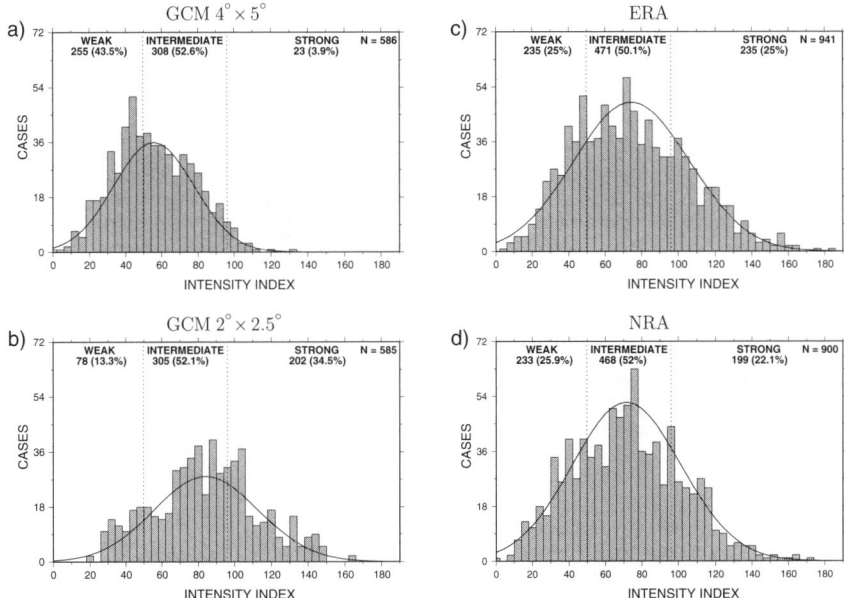

Fig. 2.4 Frequency histograms of DJF cyclone occurrence as a function of intensity index in the GISS GCM at (**a**) 4° × 5° × 23 L resolution, and (**b**) 2° × 2.5° × 32 L resolution GCM come from only 4 winters and have been projected to the 10 winters to facilitate comparison with the other histograms (from Bauer and Del Genio 2005)

apart). Looking to the trajectories and other characteristics of the GCM cyclones they concluded that "The GISS-GCM produced fewer, generally weaker and slower moving cyclones, the simulated cyclones are shallower and drier aloft than those in the reanalysis and the frontal regions are less tilted due to a weaker ageostrophic circulation." Their findings show that the reanalysis cyclones are the major dynamical source of water vapor over the extratropical oceans; the GCM, however, does not have the moisture sources from the lower more humid areas southwest of the cyclone. They concluded that these shortcomings may be common to most climate GCMs, which do not resolve the mesoscale structure of frontal systems and this may account for some universal problems in climate GCM midlatitude cloud properties. They also concluded that the GCM's underprediction of mid latitude cirrus was more dynamical rather than parameterization inadequacies (see more of cloud feedback in BD05).

Complementary new results that support BD05 were shown by Gaffney et al. (2005). Actually this study examines a different aspect regarding the deficiency of the very high frequency of baroclinic eddies in climate models. Using the NCAR-coupled model, they use 15 winter GCM simulations to compare with 44 years of

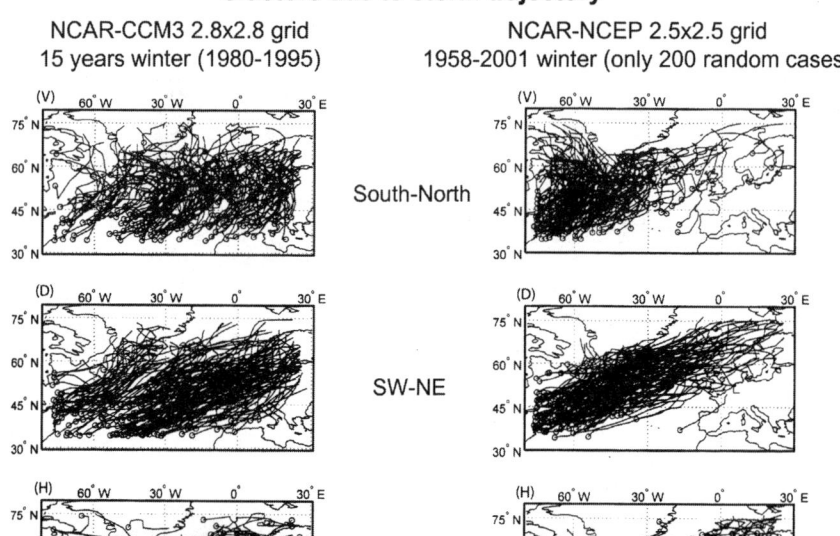

Fig. 2.5 Three trajectories clusters are shown for the CCM3 (*left*) and NCEP/NCAR (*right*) (see text, figures are from Gaffney et al. 2005)

reanalysis. The novel technique here was based on clustering the trajectories of the surface low pressure centers. Focusing over the region of the Atlantic Ocean, the trajectories were divided into three clusters: south-to-north, southwest-to-northeast, and west-to-east; the three NCEP-NCAR reanalysis trajectory clusters are shown on the right side of Fig. 2.5. The trajectories appear to be reasonably well clustered. The first two seem very homogeneous in the group of trajectories; in fact, the first cluster may represent the typical "Northeastern storms" and the middle one is representative of the positive phase of the NAO (North Atlantic Oscillation) signal. The corresponding clusters from the GCM simulation are shown on the left side of Fig. 2.5. Note that there is far more scatter in each cluster. Moreover, the lack of a concentrated region on the western Atlantic of most baroclinic cyclones that are generated for the GCM seems to be a deficiency related to the intensity of the cyclone scale simulation. By comparison, the reanalysis suggests that cyclones normally reaching the ocean become more explosive, Class B cyclogenesis (an upper level wave as it reaches the warm ocean can produce a secondary development), as discussed below; this seems to be absent in the GCM (Fig. 2.5, left side). We can derive two conclusions from this analysis. These trajectories suggest that the model generates less poleward trajectories and a number of weak storms over the Atlantic Ocean rather than more

2.3 The Performance of the Eddy Activity in Three Climate Models

intense storms over the entrance of the Atlantic storm track. It is well known that intense cyclones translate more poleward; this is consistent with the strong poleward trajectories shown in the first two clusters of the reanalysis data. Moreover, the fewer trajectories of the eastern Atlantic shown in the first two clusters of the reanalysis (Fig. 2.5, right) is an indication, again as it is well known, that as larger waves in the early winter season deplete the surface baroclinicity over the ocean, is why the winter averaged maximum baroclinicity is encountered at the entrance of the storm track and is very weak over the eastern ocean (see the baroclinic conversion in Chang et al. 2002, their Fig. 9). Only a weak array of waves will be spread over the entire ocean (as seen in left side of Fig. 2.5). Since weak waves minimally mix the ocean baroclinicity, they have a baroclinic source all over the ocean rather than concentrated over the western region. This is only a supposition; the other possibility is that the model exerts a very weak surface mixing in the ocean.

In order to complement the two previous studies, we present the statistical analysis of the upper troposphere eddy activity in the GFDL AM2-M90 runs. This analysis differs from the other two; the previous studies analyses are primarily based on trajectory analysis of surface low pressure centers to track cyclonic activity. We will use results from the GFDL AM2 at M90 resolution (The GFDL Global Atmospheric Model Development Team, 2004) (hereafter GFDL04) to confirm the suggestions by BD05 that most GCMs have a similar deficiency. This model at M90 resolution (Lin 2004) is a finite volume $\sim 1° \times 1°$ degree resolution. The atmospheric model runs were done for the period 1979–1995 and used prescribed SST's. This period is similar to the reanalysis data used in Fig. 2.3. Figure 2.6 shows the results from the GCM, and to simplify the comparison, the results are presented in the same format as Fig. 2.3. Although the distribution for the intermediate frequencies are similar to the reanalysis data, large differences are noticed in the high frequency momentum fluxes and kinetic energy; they obviously are much weaker in the GCM (Fig. 2.6) when compared to the reanalysis (Fig. 2.3).

To further understand the possible effects from the reduction in high frequency eddy activity, we can examine three variances over the winter hemisphere. The first is the variance of the monthly mean meridional velocity deviation at 300 mb ($V_p = V - V_{month}$); the second is the variance of $V_s = (V_{month} - V_{climo})$ and the third is $V_{d24} = (V(t) - V(t-24\,h))$. The variance of $\langle V_p \rangle$ contains most of the eddy activity; $\langle V_s \rangle$ can be interpreted as containing the very low-frequency (seasonal and interannual) and $\langle V_{d24} \rangle$ has been used by many authors to reflect the very high frequency variability due to individual cyclones, according to the discussion in Sect. 2.4 (see Chang et al. 2002).

Figure 2.7 shows the variance of these three quantities using the ECMWF ERA-40 reanalysis ($1.3° \times 1.3°$ degree resolution) (left side) and the differences between the GFDL AM2-M90 and the ERA-40 (right side). The first graph on the left of Fig. 2.7 shows $\langle V_p \rangle$ for the reanalysis; this basically defines the storm track for the Northern Hemisphere. When we look at the difference between the GFDL-M90 and the reanalysis (upper right), it appears to be a very good simulation on the overall eddy

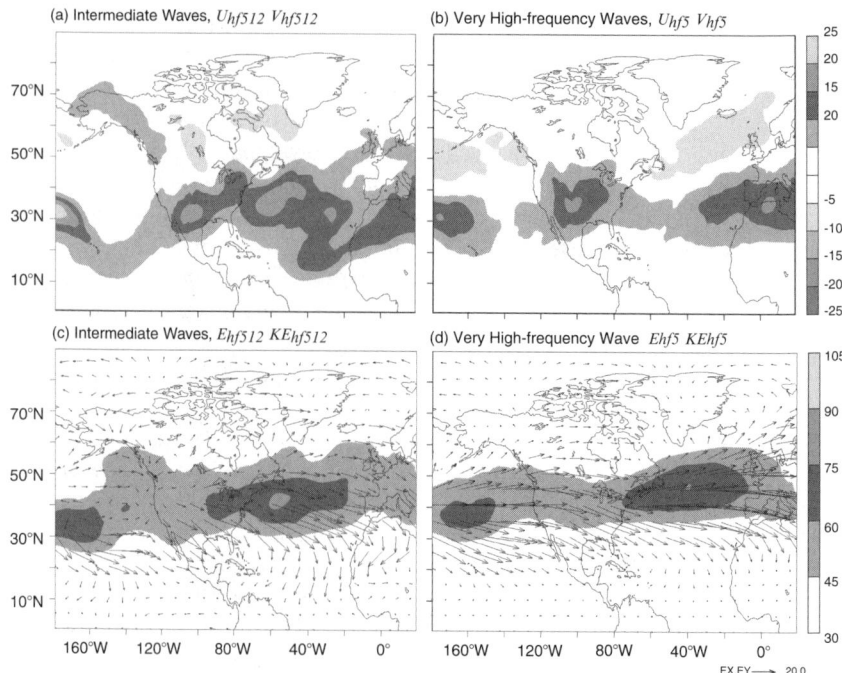

Fig. 2.6 The same as Fig. 2.3, but for the GFDL-GCM-M90

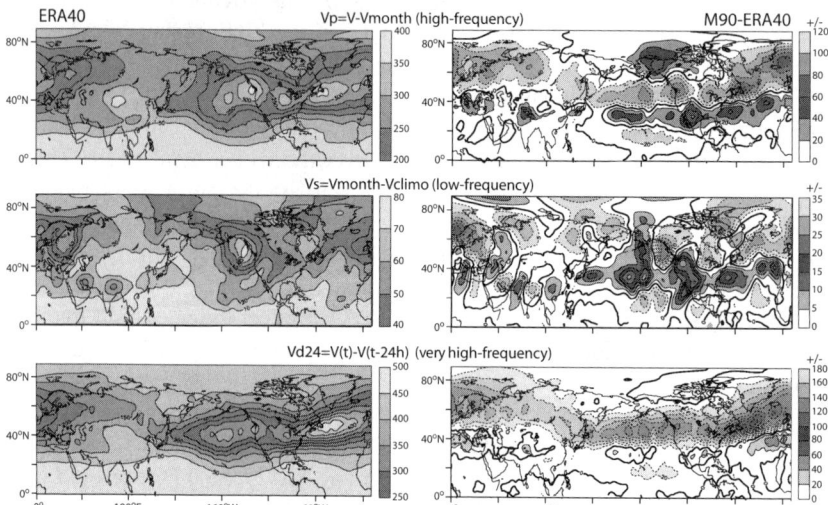

Fig. 2.7 Variance of meridional velocity from ERA40 reanalysis (left column) and GFDL-GCM-M90-ERA 40 (right column). The upper figures are the variance of the daily minus monthly meridional velocity. The middle graphs are the variance of the monthly minus climatological month and the lower figures correspond to the 24 h difference of meridional velocity

activity. The differences are basically that the storm track seems displaced southward for the model simulation, a well known result from these runs where the large scale circulation and eddy activity are slightly displaced southward from the reanalysis (as found in previous simulations with the B_grid model scheme, GFDL_GCM04). A complementary result is shown for the low frequency variance (middle graphs). The middle left graph shows the ERA-40 in which the larger variation over the eastern Pacific and eastern Atlantic are an indication of the major variability due to the PNA and NAO. Again, the difference between model and reanalysis is, as previously mentioned, the displacement to the south of the major variabilities.

The variability of the very high frequency $V_{d24\,h}$ (lower left) presents the most interesting result. It shows a more extended storm track, but notice that the maximum is over the western Atlantic (as previously discussed for the cluster trajectories in Fig. 2.5). The most revealing results can be interpreted from the differences (lower right) in which the model seems to have produced very little of the very high frequency activity. Notice that there is not, as before, a displacement of the eddy activity but rather a net negative anomaly from the GCM minus the ERA analysis. Even more interesting is where the missing activity of the model simulation is; it is in the area of the deficiency in eddy activity for the total variance of high frequency $\langle V_p \rangle$. It should be noted that at this model resolution the deficiency is probably half the amount of variance of the very high frequency eddy activity. However, in the low resolution the same variable in a $4° \times 5°$ degree resolution described in BD05 is much less (no more than 25% of the coarse reanalysis). To confirm the previous results on the reduction of high frequency activity shown in Fig. 2.7c. We have also analyzed the eddy activity at 850 mb, by calculating the zonal gradient of meridional velocity (a proxy for low level relative vorticity), and calculating the variance for the entire period (Jan 1979–Dec 1995). As before, it shows the model (AM2-M90) has a lower variance intensity of the zonal gradient of meridional velocity as compared with the ERA-40 reanalysis for the entire northern hemisphere storm track (the differences are about 20%). It should be pointed out that the AM2-M90 shows a large intensity very close to all the major topographic features of the globe. Although the fact that the model seems to have weaker vorticity in the lower levels is in agreement with all the results discussed previously, the caveat is that the data in the model was only saved for daily averaged values of meridional velocity, the daily average values used for the reanalysis was calculated using only four times daily values. It is well known that the cyclone-frontal system requires more instantaneous values than does that derived from daily averages.

2.4 The Cyclone-Frontal System

The simulation of cyclone-frontal systems has been extensively studied (see Orlanski et al. 1985; Orlanski and Katzfey 1987; Uccellini et al. 1983, among others). Fronts provide the main supply of surface heat and moisture to the cyclone. As the cyclone evolves from a weak disturbance, it tends to produce strong surface temperature gradients that we define as fronts; the cold front is the region where the cold air advances producing an edge for the warm moist air ahead of it. The warm front is where the warm air advances and produces a statically unstable region. As the surface cyclone intensifies, the warm air that is drawn into the cyclone contracts to the point that it collapses (occluded system); at this point cyclone intensification stops. While the surface cyclone intensifies, there is considerable vorticity stretching that produces upper level vorticity and the wave becomes deeper aloft. Basically, this is the evolution of the very high frequency eddy. It is of paramount importance then that the frontal regions are well resolved in order to generate intense cyclones because it is the only way that heat and moisture will feed the cyclone intensification. Clouds are an integral part of the cyclone-frontal system and their characteristic heavily depends on its position along the synoptic system. As described in Lau and Crane (1995) satellite data analysis shows that cloud types with different cloud tops and thickness are found over the cyclone-frontal system.

There are suggestions in the literature on what numerical resolution is necessary to accurately resolve the frontal systems. Case studies of intense cyclones, such as the Presidents Day Storm (Orlanski and Katzfey 1987; Uccellini et al. 1983, among others), indicate that the scale should be better than $1/2°$ resolution, but it is probably more likely that $1/3°$ to $1/4°$ resolution is needed to capture the characteristics and intensity of strong cyclones. Klein and Jakob (1999) did a comparison analysis of the forecast of cloud amounts. Their study analyzed the characteristics of cloud water paths for different model resolutions (T63 and T213) in the ECMWF operational forecast and did not find any significant difference over the extratropical latitudes. Assuming, as previously stated, that fronts are a major source of moisture through very thin areas along the frontal system and the fact that simulating the dynamics requires a high resolution model, we should expect larger differences for these two model resolutions. However, the authors suggested that the overabundance of high-top thin and medium clouds in the ERA composite as compared with the ISCCP composite would be diminished if the ERA were performed at horizontal resolution of T213.

In recent papers, the role of horizontal resolution have been revisited (Lin 2004; Orlanski 2005). In the testing of a new vertically Lagrangian scheme for global models (Lin 2004) shows the life cycle of baroclinic eddies in a dry idealized atmosphere for three models resolutions ($2° \times 2.5°$, $1° \times 1.25°$, and $0.5° \times 0.625°$). His conclusion was that although the solutions of the baroclinic waves seems very close, it seems not to converge yet, the frontal system in the very high resolution seems much sharper than the rest, and it seems the cyclonic areas more intensively developed. These result may have a larger differences if the environment would have moisture. In a different setting than the previous study. Orlanski (2005) used a very high resolution non-hydrostatic model to study the response of eddy activity over the Pacific Ocean to an ENSO (warm event). That study found that the cyclone-frontal-system characteristics did not change much between two simulations with horizontal resolutions

2.4 The Cyclone-Frontal System

**Pacific Storm Track at Different Horizontal Resolutions
Snap-shot of Column Liquid Water Content**

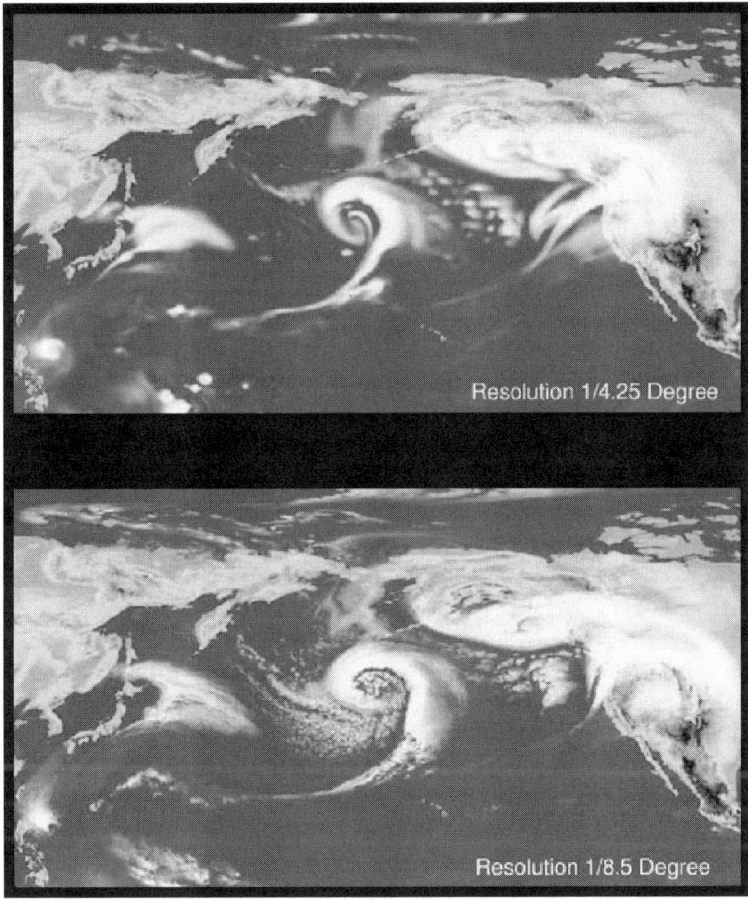

Fig. 2.8 A snapshot of the cyclone frontal-system by a nonhydrostatic model run with two very high horizontal resolutions

of $1/(4.25)°$ and $1/(8.5)°$. Figure 2.8 shows a snapshot of the total column liquid water distribution for both model resolutions. The surface relative vorticity analysis has shown that aside from the small scale, the two fronts are very comparable. Although it is not suggested that a model should go to this high resolution, it is clear that the simulation of the cyclone-frontal-system seems to convergence at about a $1/4°$ degrees.

2.5 Summary and Conclusions

We have reviewed the analysis of three different Climate models, GISS (Bauer and Del Genio 2005), the NCAR community climate model CCM3 (Gaffney et al. 2005), and the GFDL Finite volume AM2-M90 (The GFDL Global Atmospheric Model Development Team 2004; Lin 2004). The intention here is not to show which one is better but rather to determine what deficiency may be common among them. Evidence shows that the three models tend to be deficient in the generation of cyclone wave activity with the consequences that heat, momentum, and moisture may be deficient in the extratropical and subpolar regions. This will affect cloudiness, wind stress, and precipitation.

Bauer and Del Genio (2005) have shown that the deficiency of moisture and cloudiness over the subpolar regions was due to the lack of cyclone waves transport of moisture and clouds to these regions. They showed that the number of strong cyclones over ten winter years is small compared to observations. They also suggested that it was not the moist parameterization scheme but rather the low resolution that could be the culprit for inadequately resolving the mesoscale features along the fronts, moist advection, and cloudiness.

A discussion of a complementary work done on clustering of cyclone trajectories by Gaffney et al. (2005) was also presented. Consistent with the present analysis, this study also showed that differences in trajectories between reanalysis and model simulation for each cluster; these differences interpreted here to be related to the lack of intense high-frequency eddies. The model cyclones are weak and have more zonal trajectories than the classical strong storms shown in the reanalysis that have poleward trajectories. Moreover, the lack of asymmetry exhibited in the reanalysis trajectories shows that they definitely have a greater concentration of genesis in the western Atlantic Ocean. This concentration is due to two effects: the production of Class B cyclogenesis and that the ocean baroclinicity is stronger in the western Atlantic. This is due to the fact that the meridional SST's over the middle Atlantic become much weaker due to the mixing by strong waves early in the season at the middle of the ocean. The GCM solutions do not show that asymmetry; a reasonable conclusion is that models do not produce cyclones intense enough for ocean mixing.

The previous two studies depend on the surface characteristics based on trajectories of the high frequency eddies. Our analysis of the GFDL-GCM is totally eulerian and based on the upper level eddy activities (300 mb). However, a similar conclusion has been drawn from the analysis of the band pass frequency of energy and momentum for the GFDL-AM2 (M90) 17 year runs, where it is quite clear that the momentum and energy of the very high frequency is much lower in the model simulation than in the reanalysis. The variance of meridional velocity also shows that the deficiency of the high frequency is in the latitude area where the reanalysis shows the position of the storm track, whereas the model displaces it south of that. The results suggest that if the model would have better high frequency levels, the position of the storm track as measured by the variance of the meridional velocity may be positioned more correctly. There is also a suggestion that to achieve the correct intensity of the high frequency baroclinic eddies, models should have enough resolution to resolve the lower level circulation of the frontal circulation system. The mesoscale circulation associated with cyclones could be achieved with resolution equal or superior to 1/4° resolution. But it is clear that to adequately resolve the mesoscale, it is necessary to not only improve the resolution but also to improve the boundary layer and surface fluxes. Clearly, at the present low resolution of climate models, it is a considerable and probably unattainable resolution. But, if the cloudiness and sea ice over the subpolar regions are important to the overall climate, this should be an attainable goal in the very near future because no sophistication in the moist convection or sea-ice model could correct those deficiencies.

Acknowledgments The author is indebted to Drs. M. Bauer and R. Del Genio for providing Fig. 2.4 and to Drs. S. Gaffney, A. Robertson, P. Smyth, S. Camargo, M. Ghil for providing Fig. 2.5 of this publication. The author also extend the appreciation to Drs. Brian Gross, S. J. Lin, Mr Larry Polinsky, and Dr. Chris kerr for reading the manuscript and for suggestions that clarified the paper. The author also thanks Mr. Bill Stern and Bruce Wyman for facilitating the data for the analysis of the GFDL AM2-M90 runs.

References

Bauer, M., and A. Del Genio. 2005: Composite Analysis of Winter Cyclones in a GCM: Influence on Climatological Humidity, *Journal of Climate*: Vol. 19, pp. 1652–1672.

Chang, E. K. M., S. Lee and K. L. Swanson. 2002: Storm Track Dynamics. *Journal of Climate*: Vol. 15, No. 16, pp. 2163–2183.

Gaffney, S., A. Robertson, P. Smyth, S. Camargo, M. Ghil. 2006: Probabilistic Clustering of Extratropical Cyclones Using Regression Mixture Models, *Technical Report UCS-ICS 06-02,* Bren School of Information and Computer Sciences, University of California, Irvine.

The GFDL Global Atmospheric Model Development Team. 2004: The New GFDL Global Atmosphere and Land Model AM2-LM2: Evaluation with Prescribed SST Simulations. *Journal of Climate*: Vol. 17, No. 24, pp. 4641–4673.

Held, I. M., S. W. Lyons and S. Nigam. 1989: Transients and the Extratropical Response to El Niño. *Journal of the Atmospheric Sciences*: Vol. 46, No. 1, pp. 163–174.

Klein, S. A. and Jakob, C. 1999: Validation and Sensitivities of Frontal Clouds Simulated by the ECMWF Model. *Monthly Weather Review*: Vol. 127, No. 10, pp. 2514–2531.

Lau, N. 1985: Modeling the Seasonal Dependence of the Atmospheric Response to Observed El Niños in 1962–76. *Monthly Weather Review*: Vol. 113, No. 11, pp. 1970–1996.

Lau, N. and M. W. Crane. 1995: A Satellite View of the Synoptic-Scale Organization of Cloud Properties in Midlatitude and Tropical Circulation Systems. *Monthly Weather Review*: Vol. 123, No. 7, pp. 1984–2006.

Lin, S. J. 2004: "Vertically Lagrangian" finite-volume dynamical core for global models. *Monthly Weather Review*: Vol. 132, No.10, pp. 2293–2307.

Orlanski, I. 2005: A New Look at the Pacific Storm Track Variability: Sensitivity to Tropical SSTs and to Upstream Seeding. *Journal of the Atmospheric Sciences*: Vol. 62, No. 5, pp. 1367–1390.

—. 2003: Bifurcation in Eddy Life Cycles: Implications for Storm Track Variability. *Journal of the Atmospheric Sciences*: Vol. 60, No. 8, pp. 993–1023.

—. 1998: Poleward Deflection of Storm Tracks. *Journal of the Atmospheric Sciences*: Vol. 55, No. 3, pp. 2577–2602.

— and J. J. Katzfey. 1987: Sensitivity of Model Simulations for a Coastal Cyclone. *Monthly Weather Review*: Vol. 115, No. 11, pp. 2792–2821.

—, B. Ross, L. Polinsky and R. Shaginaw, 1985: Advances in the Theory of Atmospheric Fronts. *Advances in Geophysics*: Vol. 28B, pp. 223–252.

Phillips et al. 2004: Evaluating Parameterizations in General Circulation Models: Climate Simulations Meets Weather Prediction. *Bulletin of the American Meteorological Society*: Vol. 85, pp. 1903–1915.

Riviere, G. and I. Orlanski, 2007: Characteristics of the Atlantic storm track eddy activity and its relationship with the North Atlantic Oscillation. *Journal of the Atmospheric Sciences*: Vol. 64, pp. 241–266.

Simmons, A. J. and B. J. Hoskins. 1980: Barotropic Influences on the Growth and Decay of Nonlinear Baroclinic Waves. *Journal of the Atmospheric Sciences*: Vol. 37, No. 8, pp. 1679–1684.

Thorncroft, C. D, B. J. Hoskins and M. E. McIntyre. 1993: Two Paradigms of Baroclinic Wave Life-Cycle Behavior. *Quarterly Journal of the Royal Meteorological Society*: Vol. 119, pp. 17–55.

Uccellini, I., R. Petersen, P. Kocin, M. Kaplan, J. Zack and W. C. Wang. 1983: Mesoscale Numerical Simulation of the President Day Cyclone: Impact of Sensible and Latent Heat on the Pre-Cyclogenetic Environment. Preprint 6th Conf. Numerical weather Prediction. Omaha. *American Meteorological Society*: pp. 45–52.

Chapter 3
Project TERRA: A Glimpse into the Future of Weather and Climate Modeling

Isidoro Orlanski and Christopher Kerr

Summary One major challenge in obtaining useful numerical simulations of weather and climate is addressing the sensitivity of these simulations to the characteristics and distribution of clouds in the model(s). Latent heat release produced in clouds as a consequence of moist convection can dramatically affect the dynamics that govern the development of larger scale weather systems and storm tracks. Also, given the profound effects of cloud distribution on the radiative characteristics of the atmosphere, these interactions critically affect the models' climate and thus our conclusions regarding climate change.

A very high resolution global model has recently been run at NOAA's Geophysical Fluid Dynamics Laboratory (GFDL) to investigate the potential value of cloud-resolving numerical models to weather forecasts and climate simulations. Dubbed "Project TERRA", this experiment was conceived as an experimental 1-day simulation with GFDL's ZETAC model.

3.1 Introduction

The ZETAC model (Paulius et al. 2006) is the first Global Mesoscale Circulation Model (GMCM) to be run at GFDL and perhaps the first global, cloud-resolving model run anywhere that uses a grid resolution of 10–12 km. Models with similar horizontal resolution have been run at Japan's Earth simulator but, like most models run at the world's meteorological centers, these models employ the hydrostatic approximation. Such models are simpler and run faster, but they are unable to simulate cloud evolution explicitly. Instead, they depend on physical parameterizations to represent the structure of the clouds and their effects on the surrounding environment.

The ZETAC model used in Project TERRA is of a different breed. Nonhydrostatic and employing explicit moist convection (cloud resolving), ZETAC can simulate phenomena on a wide range of scales, from individual clouds to the entire global circulation. It is presently in regular use at GFDL in a limited area mode (1/6 of the globe) to study the impact of sea surface temperatures (SSTs) in storm track dynamics (Orlanski 2003, 2005). TERRA marks its first use in a global domain. The ZETAC model (Pauluis et al. 2006) was written by Dr. Steve Garner at the GFDL using the GFDL's Flexible Modeling System (GFDL 2004).

3.2 High Resolution Results

Project TERRA included simulations with different horizontal resolutions. They range from a low resolution simulation of $1° \times 1°$ (361 × 165 longitude × latitude points), to the highest resolution of $1/8° \times 1/8°$ (2 880 × 1 400 points). The experiment used a terrain-following vertical coordinate with 56 vertical levels up to an altitude of 25 km. The simulations were initialized using the results of the coarser ($1° \times 1°$) model run for 40 days, which was started from zonal initial conditions in the atmosphere and idealized SSTs, similar to that used in the storm track experiments. The successively coarser resolution simulations provided the initial fields for the progressively finer resolution runs. For instance, the simulation of the $1/4°$ case was initiated from the $1/2°$ lower-resolution result and, after 2 days, the $1/8°$ simulation was initiated from the $1/4°$ results.

3.3 The Versatility Offered by Nonhydrostatic GCM's

At present, all of the operational medium-range prediction centers around the world employ a variety of hydrostatic models with a wide range of model resolutions to generate global and limited-area forecasts. In some cases, these hydrostatic models are used even for resolutions at which the classical assumption is quite questionable (Orlanski 1981; Daley 1988).

Over the past several years, however, several international centers and research institutions have been engaged in efforts to develop unified models that could be used for all scales from global to regional (e.g., British Meteorological Office, Canadian Meteorological Center, Japan Meteorological Agency, etc). This approach has a number of very attractive advantages. Perhaps the most important of these is that, because the global and limited area models share a common dynamic core and physics, the boundary conditions passed on from the global model to drive the limited area model possess more compatible characteristics. Because regional models can employ resolutions of only a few kilometers, and therefore are firmly in the nonhydrostatic regime, it is preferable that the boundary conditions be as compatible as possible, i.e., also nonhydrostatic.

Another benefit of using nonhydrostatic models is its ability to handle convective instability. In a hydrostatic system the vertical acceleration is unbound because of the nature of the approximation; gravitational or convective instability for dry or moist atmospheres degenerates to the finer horizontal scales and exhibits its larger growth rates. Nonhydrostatic systems do not have such limitations. Convective instability is not very uncommon even in larger mesoscale circulations (e.g., warm fronts in developing baroclinic waves). Hydrostatic models require considerable vertical mixing (dry convective adjustment) to control such singularities. Nonhydrostatic models handle such instabilities explicitly, thereby avoiding the need for large artificial vertical diffusion.

In the case of moist convection, hydrostatic General Circulation Models (GCMs) divide the process into two steps. First, they calculate moist variables as tracers that are advected by the flow resolved in the model. Second, they check areas of moist instability and apply some form of cloud parameterization to eliminate the instability, thus modifying the thermodynamic and radiative outputs that affect a variety of feedbacks, including: cloud distribution, latent heating, cloud radiative feedback, etc. In contrast, nonhydrostatic models explicitly generate vertical momentum and heat fluxes consistent with the dynamic resolution of the model.

In order to demonstrate the versatility of nonhydrostatic models, the ZETAC model was run at several different grid resolutions, using bulk microphysics in place of any moist convective parameterizations. It should be stressed that these runs even with the highest resolution of $1/8° \times 1/8°$ were done only to show the versatility of the nonhydrostatic system to handle large scale convection and not necessarily to suggest that the coarser resolution simulations should be done without further cloud parameterizations. More detailed studies should be performed to compare the explicit cloud simulations with those produced by cloud parameterizations schemes in order to evaluate their merits.

Inspecting (not shown here) a snapshot of the global simulations centered on the Pacific storm track, using the vertically averaged liquid water content, for four different resolutions of the model, it was found that the model handles the large scale convection without any deep convective parameterization at even $1°$ resolution. It also shows (Fig. 3.1) the realism of the cloud features at $1/8°$ resolution compared to the cloud distribution at $1/4°$ resolution. However, as reassuring as it may be that the structure and evolution of the weather systems are very similar, this is no guarantee that the cloud distribution and cloud spectrum are correct, or even superior to analogous results from a hydrostatic model. Figure 3.2 shows a snapshot of the global solution. Specifically it displays the cloud distribution (column integrated liquid water) after 24 h integration with 12 km resolution. Work is currently underway to compare the quality of the cloud distributions for a moderate resolution (say, $1/4°$) hydrostatic model using parameterized convection with the nonhydrostatic model employing only bulk microphysics. A necessary future work should be to develop more appropriate new bulk microphysics, or "moist grid parameterization" that could be used with mesoscale resolution (coarser cloud resolving models $\sim$10 to 20 km).

3.4 Computational Requirements

In addition to being an ambitious project in numerical meteorological modeling, TERRA simultaneously tested the limits of GFDL's current High Performance Computing System (HPCS). The model was run on one SGI 3900 node consisting

Pacific Storm Track at Different Horizontal Resolutions
Snap-shot of Column Liquid Water Content

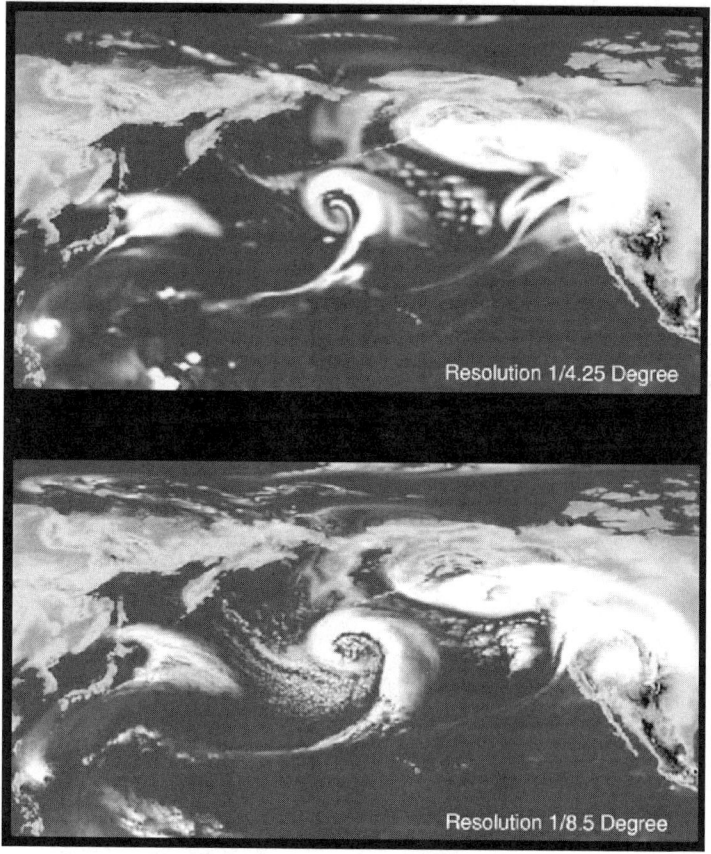

Fig. 3.1 A snapshots of the cyclone frontal-system by a nonhydrostatic model with two very high horizontal resolutions

of 512 MIPS R14000 processors. The present speed of the global ZETAC model on this computer system (or likely any current scalable computer system) makes the use of such models impractical for current operational weather prediction or climate research.

This is why, historically, hydrostatic models have been favored by numerical modelers – they are significantly faster and therefore cheaper to run, in terms of computational resource requirements. However, this advantage is limited primarily to the dynamical core. As models have become more complex, the fraction of the total computational time represented by the core has fallen to 20–30%. If the nonhydrostatic core takes 50% more computational time than its hydrostatic counterpart, this still means only a 10–15% increase in the total computing time for the model as a whole.

3.5 Summary and Conclusion

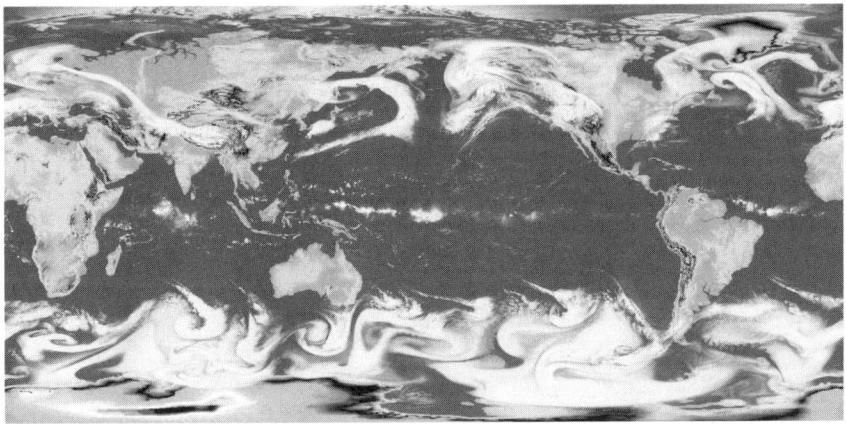

Fig. 3.2 A global view of the cloud distribution (column integrated liquid water) after 24 h integration using 12 km resolution

In fact, one might expect this increase to be reduced somewhat due to the reduction of necessary parameterization calculations.

An estimate of the computational power required to produce operational forecasts using a model such as that used in Project TERRA can be obtained as follows. Suppose that there is an operational requirement for high-resolution weather forecasts model with a $1/8°$ resolution, or a $1/4°$ resolution GCM for seasonal prediction to run: the first one, 24 h forecast in 15 min of wall clock time, and the second one, 4 days in 15 min wall-clock, equivalent to 1 model year per day of wall-clock time.

The current ZETAC model, using a 3 s time step and 504 CPU's, will complete 1 h of model time in 5 h of wall-clock time. Assuming that implementation of a polar filter will permit an increase in the time-step to 30 s, this will reduce the cost of 1 h of model time to 30 min wall clock time, or, equivalently, 1 model day per 12 h wall clock time.

Increasing the number of CPU's to 5 000 (roughly the size of Japan's Earth Simulator) and assuming scalability, the cost of 1 model day is reduced to 1.2 h (72 min) wall clock time (roughly the performance of the Earth Simulator running hydrostatic models of comparable resolution). If the individual processors are only a factor of 5 faster, this brings performance to within our goal of 15 min wall clock time per model day. This combination of increased processor count and faster CPU's seems well within the realm of possibility over the next 3–5 years.

3.5 Summary and Conclusion

Project TERRA achieved its goal of demonstrating that current nonhydrostatic models are already quite robust and are limited only by the computational resources available to run them. It is envisioned that, within a few years and with a concerted effort and adequate resources, new bulk microphysics, or a "moist grid parameterization", that could be used with mesoscale resolution (coarser cloud resolving models ~10 to 20 km), model efficiencies and advances in computer performance could lead to a major breakthrough in weather forecasting and climate prediction through the use of cloud resolving models like ZETAC.

Acknowledgments The authors are indebted to Drs. Ants Leetmaa and Brian Gross for their support to the Project TERRA, Mr. Larry Polinsky for reading the manuscript and for suggestions that clarified the chapter. The authors also extend their appreciation to the computer operations support staff for the efficiently running such a large model.

References

Daley, R., 1988: The normal modes of the spherical non-hydrostatic equations with applications to the filtering of acoustic modes. *Tellus*, 40A, 96–106.

The GFDL Global Atmospheric Model Development Team, 2004: The new GFDL global atmosphere and land model AM2-LM2: Evaluation with prescribed SST simulations. *Journal of Climate*, Vol. 17, No. 24, pp. 4641–4673.

Orlanski, I., 2005: A new look at the pacific storm track variability: Sensitivity to tropical SSTs and to upstream seeding. *Journal of the Atmospheric Sciences*, Vol. 62, No. 5, pp. 1367–1390.

—, 2003: Bifurcation in eddy life cycles: Implications for storm track variability. *Journal of the Atmospheric Sciences*, Vol. 60, No. 8, pp. 993–1023.

—, 1981: The quasi-hydrostatic approximation. *Journal of the Atmospheric Sciences*, Vol. 38, No. 16, pp. 572–582.

Paulius, O., Frierson, D., Garner, S., Held, I., and Vallis, G. (2006) The hypohydrostatic rescaling and its impact on atmospheric convection. *Theoretical and Computational Fluid Dynamics*, Vol. 20, pp. 485–499.

Chapter 4
An Updated Description of the Conformal-Cubic Atmospheric Model

John L. McGregor and Martin R. Dix

Summary An updated description is presented for the quasi-uniform Conformal-Cubic Atmospheric Model. The model achieves high efficiency as a result of using semi-Lagrangian, semi-implicit time differencing. A reversible staggering treatment for the wind components provides very good dispersion characteristics. An MPI methodology is employed that allows the model to run efficiently on multiple processors. The physical parameterizations for the model are briefly described, and results are shown for the Held-Suarez test, the Aqua-Planet Experiment and an AMIP simulation having 125 km resolution. Antarctic snow accumulation is also shown from a shorter simulation having 50 km resolution.

4.1 Introduction

During the past decade the Conformal-Cubic Atmospheric Model (CCAM) has been developed at CSIRO, Australia. CCAM is formulated on a quasi-uniform grid, derived by projecting the panels of a cube onto the surface of the Earth. The conformal-cubic grid was devised on these panels by Rancic et al. (1996), and is isotropic except at the eight singular vertices themselves. An example of a C48 grid is shown in Fig. 4.1, having 48×48 grid points on each panel and a quasi-uniform resolution of 208 km. The dynamical formulation of CCAM includes some distinctive features. The model is hydrostatic, with two-time-level semi-implicit time differencing. It employs semi-Lagrangian advection with bicubic horizontal interpolation (McGregor 1993, 1996). The grid is unstaggered, but the winds are transformed reversibly to/from C-grid staggered locations before/after the gravity wave calculations following McGregor (2005b), providing improved dispersion characteristics. Three-dimensional Cartesian representation is used during the calculation of departure points, and also for the advection or diffusion of vector quantities. The model uses deformation-based horizontal diffusion. As with most semi-Lagrangian models, the time differencing is made weakly implicit by off-centering (Rivest et al. 1994) in order to avoid resonances near steep orography for large Courant numbers. The model employs global *a posteriori* conservation for mass and moisture. A short description of CCAM was provided by McGregor and Dix (2001). The present article provides extra details of the updated model formulation, as well as the results of several simulations.

Fig. 4.1 A view of the CCAM C48 grid, having quasi-uniform resolution of 208 km

Section 4.2 describes the model formulation. Section 4.3 provides an overview of the physical parameterizations. Section 4.4 discusses the message passing configuration and compares the model efficiency on several computers. Section 4.5 presents several examples of CCAM simulations. Section 4.6 provides some concluding comments.

4.2 Dynamical Formulation of CCAM

4.2.1 Primitive Equations

The primitive equations for the conformal-cubic model are the same as those for a polar-stereographic or Lambert conformal grid, except that the map factor m is that of the conformal-cubic projection (derived by the power-series procedure of

4.2 Dynamical Formulation of CCAM

Rancic et al. (1996)). First, the material time derivative is denoted by

$$\frac{d}{dt} \equiv \frac{\partial}{\partial t} + mu\frac{\partial}{\partial x} + mv\frac{\partial}{\partial y} + \dot{\sigma}\frac{\partial}{\partial \sigma}, \quad (4.1)$$

where p_s is the surface pressure, u and v are the velocity components, and the terrain-following vertical coordinate is $\sigma \equiv p/p_s$. The horizontal momentum equations are

$$\frac{du}{dt} + m\frac{\partial \phi_v}{\partial x} + mR_d T_v \frac{\partial \ln p_s}{\partial x} = (f + f_m)v \quad (4.2)$$

$$\frac{dv}{dt} + m\frac{\partial \phi_v}{\partial y} + mR_d T_v \frac{\partial \ln p_s}{\partial y} = -(f + f_m)u, \quad (4.3)$$

where ϕ_v is the geopotential (including virtual temperature contributions) and f is the Coriolis parameter; R_d is the gas constant for dry air. The virtual temperature, T_v, is defined by

$$T_v = T\left[1 + \left(\frac{R_v}{R_d} - 1\right)q\right] \quad (4.4)$$

where T is temperature, R_v is the gas constant for water vapour, and q is the mixing ratio of water vapour. Contributions from physical parameterizations are not included in these equations, but are treated in a split manner each time step after the "dynamics" computations. The extra map projection terms f_m are small and are given by

$$f_m = u\frac{\partial m}{\partial y} - v\frac{\partial m}{\partial x}. \quad (4.5)$$

In fact these metric terms are omitted during the solution procedure, as they are completely subsumed into the process of semi-Lagrangian trajectory estimation and interpolation (Bates et al. 1990).

The temperature equation is

$$\frac{dT}{dt} - \frac{R_d T}{c_p \sigma}\frac{\omega}{p_s} = 0, \quad (4.6)$$

where c_p is the specific heat of dry air at constant pressure and ω is the pressure vertical velocity. The equation for water vapour mixing ratio is

$$\frac{dq}{dt} = 0, \quad (4.7)$$

whilst the continuity equation is given by

$$\frac{d \ln p_s}{dt} + D + \frac{\partial \dot{\sigma}}{\partial \sigma} = 0, \quad (4.8)$$

where the divergence D is given by

$$D = m^2\left\{\frac{\partial(u/m)}{\partial x} + \frac{\partial(v/m)}{\partial y}\right\}. \quad (4.9)$$

As will be seen during the semi-Lagrangian discretization, the vertical velocities $\dot{\sigma}$ and ω may be diagnosed from the vertically integrated continuity equation. The hydrostatic equation,

$$\frac{\partial \phi_v}{\partial \sigma} = -\frac{R_d T_v}{\sigma}, \tag{4.10}$$

may be integrated assuming a simple average layer virtual temperature between levels, to give

$$\phi_{v_k} = \phi_{v_{k-1}} - \frac{R_d}{2}\left(T_{v_k} + T_{v_{k-1}}\right) \ln \frac{\sigma_k}{\sigma_{k-1}} \quad \text{for } k = 2, 3, \ldots, K. \tag{4.11}$$

For the lowest level, a standard temperature lapse from the surface of 6.5° km^{-1} gives

$$\phi_{v_1} = \phi_s - \left\{\sigma_1^{-0.0065 R_d/g} - 1\right\} \frac{R_d T_{v_1}}{0.0065}, \tag{4.12}$$

where g is the gravity constant and ϕ_s is the surface geopotential. The final expression for the geopotential is thus of the form

$$\phi_{v_k} = \phi_s + \sum_{j=1}^{k} B_{k,j} T_{v_j} \quad \text{for } k = 1, 2, \ldots, K, \tag{4.13}$$

where

$$B_{1,1} = -\left\{\sigma_1^{-0.0065 R_d/g} - 1\right\} \frac{R_d}{0.0065}$$

$$B_{k,k} = -\frac{R_d}{2} \ln \frac{\sigma_k}{\sigma_{k-1}} \quad \text{for } k = 2, \ldots, K$$

$$B_{k,j} = B_{j-1,j-1} + B_{j,j} \quad \text{for } k = 2, \ldots, K; \ j = 2, \ldots, k-1$$

$$B_{k,j} = 0 \quad \text{for } k = 1, 2, \ldots, K; \ j = k+1, \ldots, K.$$

It is noted here that CCAM may also be used in variable-resolution mode, by employing the Schmidt (1977) transformation. With this transformation, the primitive equations are unchanged, but the map factors have altered values; details are provided by McGregor (2005a).

4.2.2 Semi-Lagrangian Discretization

The primitive equations are solved by two-time-level semi-Lagrangian discretization. Values at the current time level are denoted with superscript τ, those at the new time level by $\tau + 1$, and those at the departure points at time τ (having arrival positions at

4.2 Dynamical Formulation of CCAM

the $\tau + 1$ grid points) by τ_*. The method for calculating the τ_* departure values will be described in the following subsection. To avoid mountain resonances, Rivest et al. (1994) advocated off-centring of the time-averaged terms. This is included below in terms of ε, a small constant having a typical value of 0.1.

Evaluating the above primitive equations near the mid-points of the fluid trajectories, the continuity equation (4.8) becomes

$$\left\{\ln p_s + (1+\varepsilon)\frac{\Delta t}{2}\left(D + \frac{\partial \dot{\sigma}}{\partial \sigma}\right)\right\}^{\tau+1} = \frac{\Delta t}{2} M_{p_s} \qquad (4.14)$$

where Δt is the time increment and

$$\frac{\Delta t}{2} M_{p_s} = \left\{\ln p_s - (1-\varepsilon)\frac{\Delta t}{2}\left(D + \frac{\partial \dot{\sigma}}{\partial \sigma}\right)\right\}^{\tau_*}.$$

Introducing the vertical integral notation

$$\overline{(\ )}^\sigma = \int_0^\sigma (\ )\, d\sigma, \qquad (4.15)$$

(4.14) may be vertically integrated throughout the depth of the atmosphere to provide

$$\left\{\ln p_s + (1+\varepsilon)\frac{\Delta t}{2}\overline{D}^1\right\}^{\tau+1} = \frac{\Delta t}{2}\overline{M_{p_s}}^1. \qquad (4.16)$$

Substituting $\ln p_s$ back into (4.14), and now integrating from 0 to σ yields

$$\dot{\sigma}^{\tau+1} = \left(\sigma \overline{D}^1 - \overline{D}^\sigma\right)^{\tau+1} - \left(\sigma \overline{M_{p_s}}^1 - \overline{M_{p_s}}^\sigma\right)/(1+\varepsilon). \qquad (4.17)$$

Note that the pressure vertical velocity, ω, is defined by

$$\frac{\omega}{p_s} = \frac{1}{p_s}\frac{dp}{dt} = \dot{\sigma} + \sigma \frac{d\ln p_s}{dt}. \qquad (4.18)$$

Substituting $d\ln p_s/dt$ from (4.8) into (4.18), then $\dot{\sigma}$ from (4.17), all at time step $\tau + 1$, yields

$$\left(\frac{\omega}{p_s}\right)^{\tau+1} = -\left(\overline{D}^\sigma\right)^{\tau+1} + \left(\overline{M_{p_s}}^\sigma - \sigma M_{p_s}\right)/(1+\varepsilon). \qquad (4.19)$$

The temperature equation (4.6) is rearranged in terms of a reference temperature $\bar{T}$ as

$$\frac{dT}{dt} - \frac{R_d \bar{T}}{c_p \sigma}\frac{\omega}{p_s} = N_T \qquad (4.20)$$

where

$$N_T = \frac{R_d(T - \bar{T})}{c_p \sigma}\frac{\omega}{p_s}.$$

Ideally, the N_T term would be evaluated midway along the trajectory between the τ_* and $\tau + 1$ positions. A time-extrapolated value could be used for N_T, but usually in CCAM it is assumed that these terms change slowly in time and can be adequately approximated by τ values at the appropriate spatial locations. Durran and Reinecke (2004) have suggested that such terms be treated by an Adams-Bashforth procedure; this has been implemented as an option in CCAM, but no added benefit has so far been observed from its use. Using the simpler time treatment for evaluating N_T, semi-Lagrangian discretization of (4.20) gives

$$\left\{ T - (1+\varepsilon) \frac{\Delta t}{2} \frac{R_d \bar{T}}{c_p \sigma} \frac{\omega}{p_s} \right\}^{\tau+1} = A_T \qquad (4.21)$$

where

$$A_T = G^{\tau_*} + \frac{\Delta t}{2} N_T^{\tau}$$

$$G^{\tau_*} = \left\{ T + (1-\varepsilon) \frac{\Delta t}{2} \frac{R_d \bar{T}}{c_p \sigma} \frac{\omega}{p_s} + \frac{\Delta t}{2} N_T \right\}^{\tau_*}.$$

Note that in CCAM, $\bar{T}$ is prescribed each time step; it is allowed to vary horizontally, but not vertically. Presently we choose $\bar{T} = T^{\tau}(\sigma = 0.78)$ for each grid point; there seems to be little sensitivity to the choice of vertical level. Equations (4.17) and (4.19) provide a separation of ω/p_s into further "linear" and "nonlinear" components (with respect to divergence, D). Substituting (4.19) into (4.21) gives the final semi-Lagrangian version of the temperature equation,

$$\left\{ T + (1+\varepsilon) \frac{\Delta t}{2} \frac{R_d \bar{T}}{c_p \sigma} \overline{D}^\sigma \right\}^{\tau+1} = X_T \qquad (4.22)$$

where

$$X_T = A_T + \frac{\Delta t}{2} \frac{R_d \bar{T}}{c_p \sigma} \left(\overline{M_{p_s}}^\sigma - \sigma M_{p_s} \right).$$

4.2.2.1 Calculation of the Semi-Lagrangian Departure Values

The departure values of the temperature quantity G in (4.21) are calculated by a split procedure. First, note that G^{τ_*} is simply the value of G proceeding from time τ to time $\tau + 1$, after purely advective processes. For this advection purpose, we use time extrapolated ($\tau + 1/2$) values for u and v, as advocated by Temperton and Staniforth (1987), written in vector notation as

$$\mathbf{u}^{\tau+1/2} = (15\mathbf{u}^\tau - 10\mathbf{u}^{\tau-1} + 3\mathbf{u}^{\tau-2})/8, \qquad (4.23)$$

together with the current (τ) values for $\dot{\sigma}$. It is possible that some form of time-extrapolated $\dot{\sigma}$ may be beneficial, but this has not been investigated for CCAM.

4.2 Dynamical Formulation of CCAM

The advection of G is split into three parts: vertical advection for $\Delta t/2$, horizontal advection for Δt, and vertical advection for $\Delta t/2$. The vertical advection parts are calculated by the total variation diminishing (TVD) method, as advocated by Thuburn (1993); alternative flux limiters are available, but we usually use the "MC" (monotonized centred) flux-limiter of van Leer (1977). During experimentation, we have found that the TVD vertical advection of temperature leads to a sharper tropopause than using semi-Lagrangian vertical advection. The horizontal advection is performed using semi-Lagrangian advection. The departure points are evaluated in three-dimensional Cartesian space either by iteration or by the method of McGregor (1993), then bicubic spatial interpolation on the panels is used to evaluate G at those departure points. Departure values of the $\ln p_s$, u, and v terms (to be defined in the following subsection) are evaluated by a similar procedure. Note, however, that the interpolated values of the u and v terms are calculated via interpolation of their corresponding three-dimensional Cartesian components, as these vary smoothly even near the vertices; by this procedure, there is no special treatment required near the eight vertices of the grid. For moisture and trace gases, the monotonic limiter of Bermejo and Staniforth (1992) is also applied.

4.2.2.2 Semi-Lagrangian Momentum Equations

First, define an "augmented" geopotential

$$P = \phi + R_d \bar{T} \ln p_s, \tag{4.24}$$

where ϕ is calculated using temperature rather than virtual temperature, and where the surface pressure term is multiplied by the reference temperature $\bar{T}$ (as defined in the previous section). Equations (4.2) and (4.3) may be expressed as

$$\frac{du}{dt} + m\frac{\partial P}{\partial x} - fv = N_u \tag{4.25}$$

$$\frac{dv}{dt} + m\frac{\partial P}{\partial y} + fu = N_v, \tag{4.26}$$

where

$$N_u = f_m v + m\frac{\partial}{\partial x}(\phi - \phi_v + R_d \bar{T} \ln p_s) - mR_d T_v \frac{\partial \ln p_s}{\partial x}$$

$$N_v = -f_m u + m\frac{\partial}{\partial y}(\phi - \phi_v + R_d \bar{T} \ln p_s) - mR_d T_v \frac{\partial \ln p_s}{\partial y}.$$

Semi-Lagrangian discretization then produces

$$\left\{ u + (1+\varepsilon)\frac{\Delta t}{2}\left(m\frac{\partial P}{\partial x} - fv\right)\right\}^{\tau+1} = A_u \tag{4.27}$$

$$\left\{v + (1+\varepsilon)\frac{\Delta t}{2}\left(m\frac{\partial P}{\partial y} + fu\right)\right\}^{\tau+1} = A_v, \qquad (4.28)$$

where

$$A_u = \left\{u - (1-\varepsilon)\frac{\Delta t}{2}\left(m\frac{\partial P}{\partial x} - fv\right) + \frac{\Delta t}{2}N_u\right\}^{\tau_*} + \frac{\Delta t}{2}N_u^{\tau}$$

$$A_v = \left\{v - (1-\varepsilon)\frac{\Delta t}{2}\left(m\frac{\partial P}{\partial y} + fu\right) + \frac{\Delta t}{2}N_v\right\}^{\tau_*} + \frac{\Delta t}{2}N_v^{\tau}.$$

There are subtleties in the finite-differencing treatment of the above pressure gradient terms, as a result of the availability of reversible staggering. Note first that the eventual semi-implicit solver will be using staggered values of $u^{\tau+1}$ and $v^{\tau+1}$. The arrangement of the staggered grid points is illustrated in Fig. 4.2. To provide eventual consistency between the pressure gradient terms (catering for situations in which there has been no change at a grid point due to advection), all terms involving horizontal derivatives on the right-hand-sides of (4.27) and (4.28) are first evaluated at the staggered locations, then transformed reversibly to unstaggered positions. Note that for alternate time steps, alternately "left" and "right" pivot points are used in the

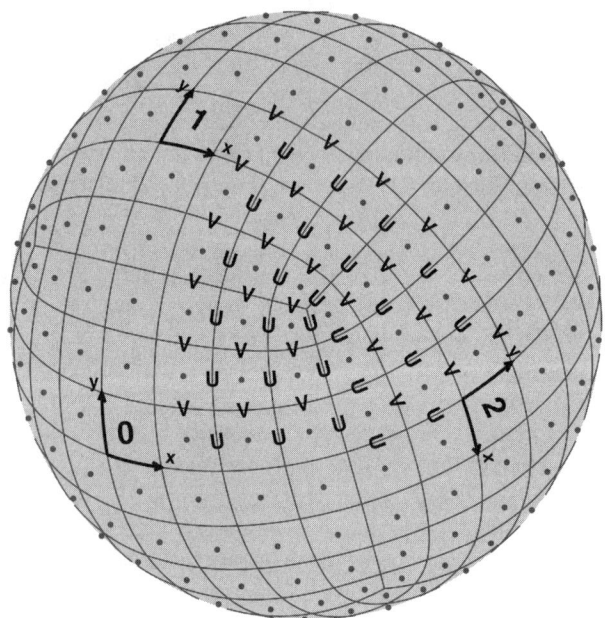

Fig. 4.2 View of the CCAM grid illustrating the winds in their staggered locations. Nonstaggered variables are located at the "dot" positions. Three of the conformal-cubic panels, with their local grid orientation, are indicated by the symbols 0, 1, and 2

4.2 Dynamical Formulation of CCAM

reversible transformations, in order to achieve the excellent dispersion characteristics outlined by McGregor (2005b). This technique is an important ingredient in the CCAM formulation, and is a technique not readily available to spectral models, or models formulated on noncyclic grids.

4.2.2.3 Reorganization of the Momentum Equations

Solving the simultaneous equations (4.27) and (4.28) in terms of (nonstaggered) $u^{\tau+1}$ and $v^{\tau+1}$ yields

$$\left\{u(1+F^2) + (1+\varepsilon)\frac{\Delta t}{2}m\left(\frac{\partial P}{\partial x} + F\frac{\partial P}{\partial y}\right)\right\}^{\tau+1} = X_u \quad (4.29)$$

$$\left\{v(1+F^2) + (1+\varepsilon)\frac{\Delta t}{2}m\left(\frac{\partial P}{\partial y} - F\frac{\partial P}{\partial x}\right)\right\}^{\tau+1} = X_v, \quad (4.30)$$

where

$$X_u = A_u + FA_v$$

$$X_v = A_v - FA_u$$

$$F = (1+\varepsilon)f\Delta t/2.$$

Note that the reversible staggering procedure avoids the averaging truncation errors of the Coriolis terms normally associated with a C-grid. Division of each of (4.29) and (4.30) by $m(1+F^2)$ gives

$$\left\{\frac{u}{m} + \frac{\Delta t}{2}\alpha\left(\frac{\partial P}{\partial x} + F\frac{\partial P}{\partial y}\right)\right\}^{\tau+1} = \frac{\alpha X_u}{m(1+\varepsilon)} \quad (4.31)$$

$$\left\{\frac{v}{m} + \frac{\Delta t}{2}\alpha\left(\frac{\partial P}{\partial y} - F\frac{\partial P}{\partial x}\right)\right\}^{\tau+1} = \frac{\alpha X_v}{m(1+\varepsilon)}, \quad (4.32)$$

where

$$\alpha = (1+\varepsilon)/(1+F^2).$$

By substituting $\ln p_s$ from (4.16) and T from (4.22) into (4.24), $P^{\tau+1}$ may be written in terms of $D^{\tau+1}$ and known quantities. By horizontally differentiating (4.31) and (4.32), $D^{\tau+1}$ may be written in terms of horizontal derivatives of $P^{\tau+1}$ and then substituted into (4.24), whence a Helmholtz equation arises in terms of $P^{\tau+1}$. Further details of the derivation, including the eigenvector decomposition, are provided by

McGregor (2005a). After solving the Helmholtz equation (see Sect. 4.3), back substitution into (4.31) and (4.32) gives $u^{\tau+1}$ and $v^{\tau+1}$, then substitution into (4.16) and (4.22) gives $\ln p_s^{\tau+1}$ and $T^{\tau+1}$.

4.3 Physical Parameterizations

CCAM includes a fairly comprehensive set of physical parameterizations. The diurnally varying GFDL parameterization for long-wave and short-wave radiation (Fels and Schwarzkopf 1975; Schwarzkopf and Fels 1991) is employed. Cloud distributions are derived in conjunction with the liquid and ice-water microphysics scheme of Rotstayn (1997). The gravity wave drag scheme of Chouinard et al. (1986) is included. The model employs a stability-dependent boundary layer scheme based on Monin-Obukhov similarity theory (McGregor et al. 1993), together with nonlocal vertical mixing (Holtslag and Boville 1993). The model employs a canopy scheme as described by Gordon et al. (2002), having six layers for soil temperatures, six layers for soil moisture (solving Richard's equation), and three layers for snow. It also includes a simple parameterization to enhance sea-surface temperatures under conditions of low wind speed and large downward solar radiation, for the purpose of calculating surface fluxes. Horizontal diffusion is not required; however, weak horizontal diffusion is usually included following Smagorinsky et al. (1965), but with the diffusivities derived from deformation of the three-dimensional Cartesian representation of the horizontal wind field; in the case of winds, the diffusion is also applied via the three-dimensional Cartesian components.

CCAM employs the mass-flux cumulus convection scheme described by McGregor (2003), and includes evaporative downdrafts. Cloud base is determined by proceeding downwards to find the first moist adiabatically unstable layer; the mixing ratio of the air parcel at the potential cloud base is enhanced by 5% over sea and 15% over land (the greater value over land represents the greater spatial variability in the boundary layer expected over land). Convection is permitted up to the uppermost model layer for which the parcel is moist adiabatically unstable. The closure is that convection continues to exhaustion in a 20-min convective timescale (the 20-min timescale only applies if the model time step is smaller than 20-min). Exhaustion occurs for the smallest possible mass flux such that the modified environment no longer provides the given cloud base or convective stability criteria for all cloud layers. In any time step, three passes of the convection scheme are performed, to avoid the possibility that the cloud-base or cloud-top layers are only marginally satisfying the convective stability criteria. Although the cloud base condition is simple, it permits an estimate of the new cloud-base conditions as a result of the modified environment, and hence provides a simple and natural cumulus closure. A simple detrainment scheme is used for cumulus convection, in which 30% of the condensed moisture is used to moisten the upper half of the environment surrounding the convective tower. Shallow convection is treated following Smith (1990), in which vertical turbulent mixing is enhanced in regions of cloudy air.

4.4 Parallel Aspects

This section describes the parallel version of CCAM, outlining the parallelization strategy and the performance achieved. The parallelization was done using the Message-Passing Interface (MPI). This is the most portable approach possible at present, it works on purely distributed machines but also on shared memory machines like the SGI Altix and NEC SX-6.

4.4.1 Grid Decomposition

There are two simple strategies for the grid decomposition, shown in Fig. 4.3. The first, "face" decomposition, keeps the cube faces together where possible; for example with 6 processors each gets a complete face. The second, "uniform" decomposition, partitions each face amongst all processors. Majewski et al. (2002) use a uniform decomposition in their icosahedral model so that each processor has a region on each of the 10 faces. The uniform decomposition should minimize load balance problems. With the face scheme, one face (processor) will have the high N latitudes while another has the high S latitudes. For example, one of these will have a lot more sunlit points than the other except near the time of the equinoxes. However, the uniform scheme will have a larger message volume. Suppose each face has $n \times n$ points and there are N_f faces and N_p processors. In the face approach, the processor boundary region length is $n/\sqrt{N_p/N_f}$. In the uniform scheme it is $N_f n/\sqrt{N_p}$. This means that the message volume for uniform decomposition is $\sqrt{N_f}$ times larger, about 2.5 for a conformal cubic grid and 3.2 for an icosahedral grid.

Whether the overhead in message volume is worth the improvement in the load balance can only really be decided experimentally. In the parallel CCAM, the use of indirect addressing means that most of the model code is independent of the decomposition and it is easy to switch from one to the other. An advantage of the uniform scheme is that it can use any number of processors, while the face scheme is limited to a factor or a multiple of 6 (e.g., on a 16 processor machine it could only use 12). In the standard CCAM code, indirect addressing arrays are used to get the indices of neighbouring points. It is possible to use explicit indexing within a face but this is not generally done. For simplicity, consider the 6 processor parallel version, where each face is on a different processor. Figure 4.4 shows the neighbours of face 0 for a C4 grid (reduced size just for clarity).

In the standard code, global arrays are dimensioned $i_{full}=n \times n \times 6$. In the MPI version, i_{full} is redefined as the number of points on the processor, in this example $i_{full}=n \times n$. The required neighbouring points are held in an extra region at the end of the array, i.e., a global array is now dimensioned $i_{full}+i_{extra}$, where i_{extra} is the size of the buffer region (as illustrated by the shaded region in Fig. 4.4), which depends on the number of processors. MPI send and receive operations are used to send the neighbouring regions between processors. Within each processor, the index arrays are remapped to point into this end-region where necessary. This only has to be

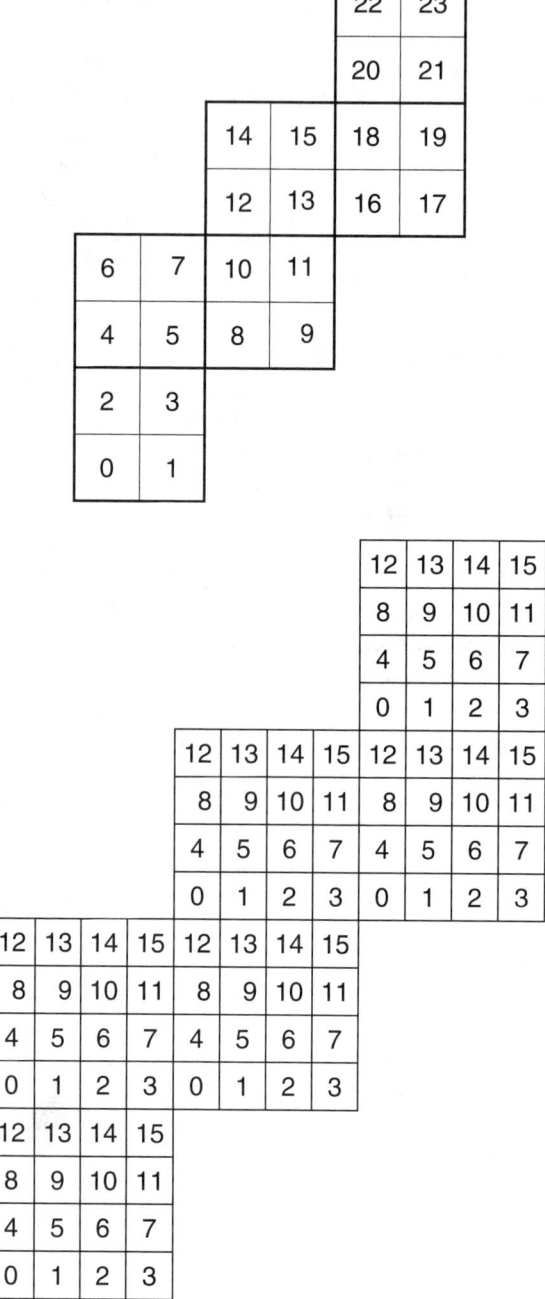

Fig. 4.3 Grid decomposition showing face decomposition with 24 processors (*top*) and uniform decomposition with 16 (*bottom*)

4.4 Parallel Aspects

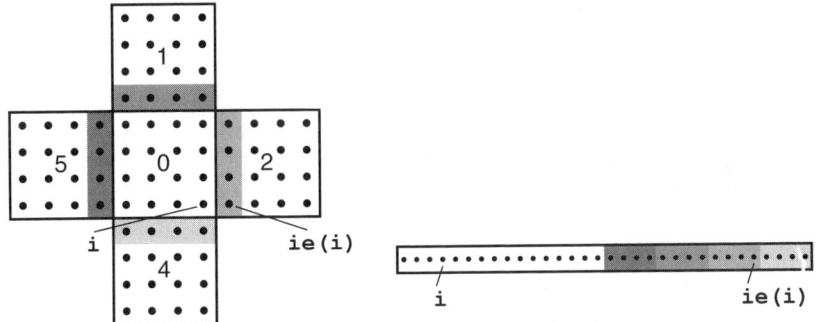

Fig. 4.4 Remapping of off-processor neighbour indices to buffer region. The east neighbour of point i, defined by the indirect addressing array ie(i) is on a different processor

done once in a setup routine. With wind variables the situation is more complicated due to the changes in face orientation. Over a panel boundary the required neighbour of a u component could actually be a v component, as can be seen from Fig. 4.2. In the MPI code, the wind rotation is taken care of when setting up the neighbour data to be transferred, the appropriate u or v values are used as required. The extra memory usage of the buffer region is reasonably small, with 6 processors it is of order 20% and with 24 processors about 40% (note that the total memory usage per processor decreases with increasing number of processors; this is just the relative contribution). Figure 4.4 here shows just one row of neighbour points on each processor, but in fact two rows are needed for the cubic interpolation.

4.4.2 Treatment of the Semi-Lagrangian Interpolations

The semi-Lagrangian interpolations of Sect. 4.4.2 require bicubic interpolation about the departure points. If the departure point is in another processor's region, this could require a larger halo region and more data transfer. Some models cope with this by limiting the time step so that the departure point is never more than one row away from the processor's region or by only transferring the extra halo region in the upwind direction. With the conformal cubic grid, getting the neighbour indexing correct for a larger halo region is very complicated near the vertices. It is simpler to use what has been called "computation on demand" (e.g., Majewski et al. (2002)). With this method, each processor does the interpolation for all departure points that fall in its region. The result is then sent to the processors that require it. This increases the number of messages, because processors have to first send a list of the points they require and later send the results. However it decreases the overall message volume and it imposes no artificial limits on the time step.

4.4.3 Helmholtz Solver

The successive over-relaxation (SOR) solver used in the non-MPI code would require three bounds synchronizations per iteration (one per "color"). A conjugate gradient (CG) solver is much preferred, requiring only one bounds synchronization and two global sums, using the formulation of D'Azevedo et al. (1992) so that the sums can be done together. Another advantage is that it requires only simple indirect addressing rather than the multiply-indirect addressing of the SOR scheme. A simple non-preconditioned CG solver takes about the same number of iterations as SOR with optimal acceleration, but is faster in practice because of the simpler indexing. The solver starts off working on all modes at once, with higher modes dropping out as they converge. With even moderate numbers of processors it is an advantage to use a local preconditioner to reduce the number of iterations (e.g., Thomas and Loft 2000). There is an option to apply an incomplete LU decomposition without fill-in, ILU(0), preconditioner on selected modes.

4.4.4 Performance

The model has been run successfully on NEC SX-6, Compaq AlphaServer SC and SGI Altix computers, and on various Linux clusters. The MPI code on a single processor runs at almost exactly the same speed as the original code. The only fundamental difference is that even the single processor version has to do some copying to account for the wind rotation. Table 4.1 shows the performance of a C48 resolution model on an SGI Altix. There is excellent parallel scaling, particularly given the modest horizontal resolution of this test case.

Table 4.1 Model performance on the SGI Altix, using face decomposition for a C48 version of CCAM having 18 vertical levels, where nproc denotes the number of processors

Nproc	Time (s/simulated day)	Speed-up	Speed-up Excluding Load Balance
1	239.9		
2	124.0	1.9	2.0
4	59.7	4.0	4.1
6	36.9	6.5	6.8
16	15.7	15.3	15.8
24	10.0	24.0	26.3
48	5.66	42.3	47.1

To allow an estimate of the overhead of the parallel communications, the Table also includes the speed-up calculated excluding the time lost due to load imbalance in the model physics (due to differing numbers of sunlit points, land points, points with convection etc. in each processor's region). For nproc = 6 the speed-up is super-linear, presumably due to better cache use.

4.5 Examples of CCAM Simulations

In order to illustrate the performance of CCAM, several quasi-uniform simulations are shown in this section. Note that CCAM is also being used extensively in variable-resolution mode, by employing the Schmidt (1977) transformation mentioned in Sect. 4.2. All the simulations shown here use 18 vertical levels.

4.5.1 Held-Suarez Test

This test was proposed by Held and Suarez (1994) to allow comparison of different dynamical cores. For this CCAM simulation a C48 grid was employed having about 208 km grid spacing. The simulation starts from rest, with no orography and uses Newtonian relaxation to a prescribed zonally (and hemispherically) symmetric temperature field; it also includes Rayleigh damping of low-level winds to represent boundary-layer friction. The simulation excludes moist processes, and includes no physical parameterizations, apart from weak horizontal diffusion. The CCAM simulation was run for 600 days, and Fig. 4.5 shows zonal averages from days 201–600 for temperature and the zonal wind component. These plots are extremely similar to the corresponding fields shown by Held and Suarez (1994) for their averages from days 201–1200. The only noticeable differences are that the CCAM equatorial stratospheric temperature minimum is a little weaker, and the stratospheric equatorial winds are also a little weaker.

4.5.2 Aquaplanet Simulation

A suite of simulations was performed for the Aqua-Planet Experiment (APE) Project, also using a C48 version of CCAM. For this intercomparison project, eight different sea surface temperature (SST) profiles are specified according to Neale et al (2000) for a version of the model without land or orography. For these simulations the usual model physical parameterizations are included, except that the solar forcing is fixed to be overhead at the equator throughout the whole experiment period. The simulations were performed for 3 years, following a 6 month spin-up period. Figure 4.6 shows the zonally averaged rainfall after spin-up and the convective component for the "qobs" SST experiment, which was designed to resemble the observed annual mean SST distribution. The peak equatorial rainfall of 14 mm day^{-1} is similar to observed climatological values and to most of the other models in the intercomparison. The proportion and distribution of convective rainfall also appears to be quite reasonable for a GCM having a resolution of about 200 km.

4.5.3 AMIP Simulation

A series of CCAM simulations was performed for the Stretched Grid Model Intercomparison Project (SGMIP) Phase 2. Phase 1 has been described by Fox-Rabinovitz

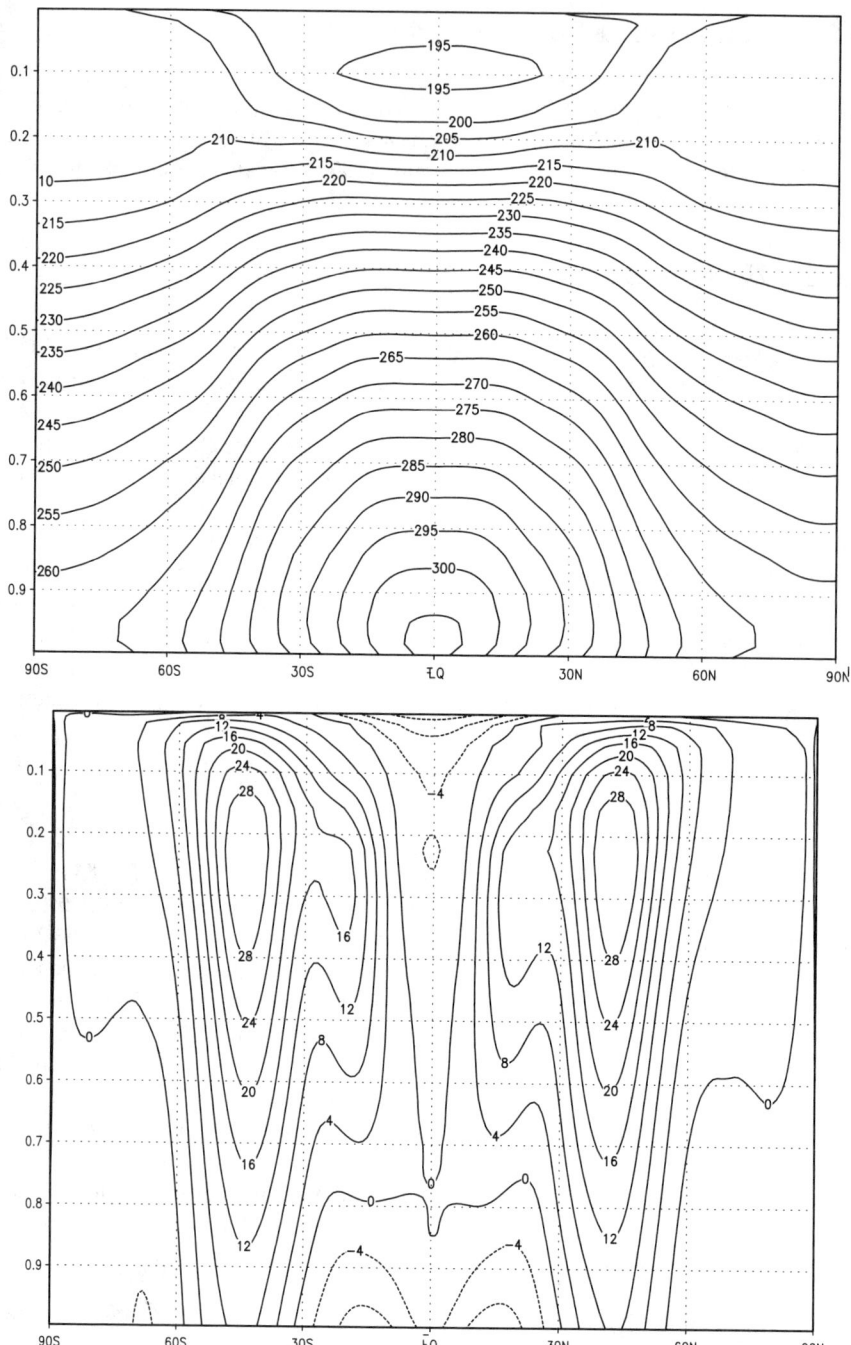

Fig. 4.5 Vertical cross sections of zonal averages of temperature profile (*top*, k), and zonal wind component (*bottom*, ms^{-1}). These are for a CCAM simulation of the Held-Suarez test using a C48 grid

4.5 Examples of CCAM Simulations

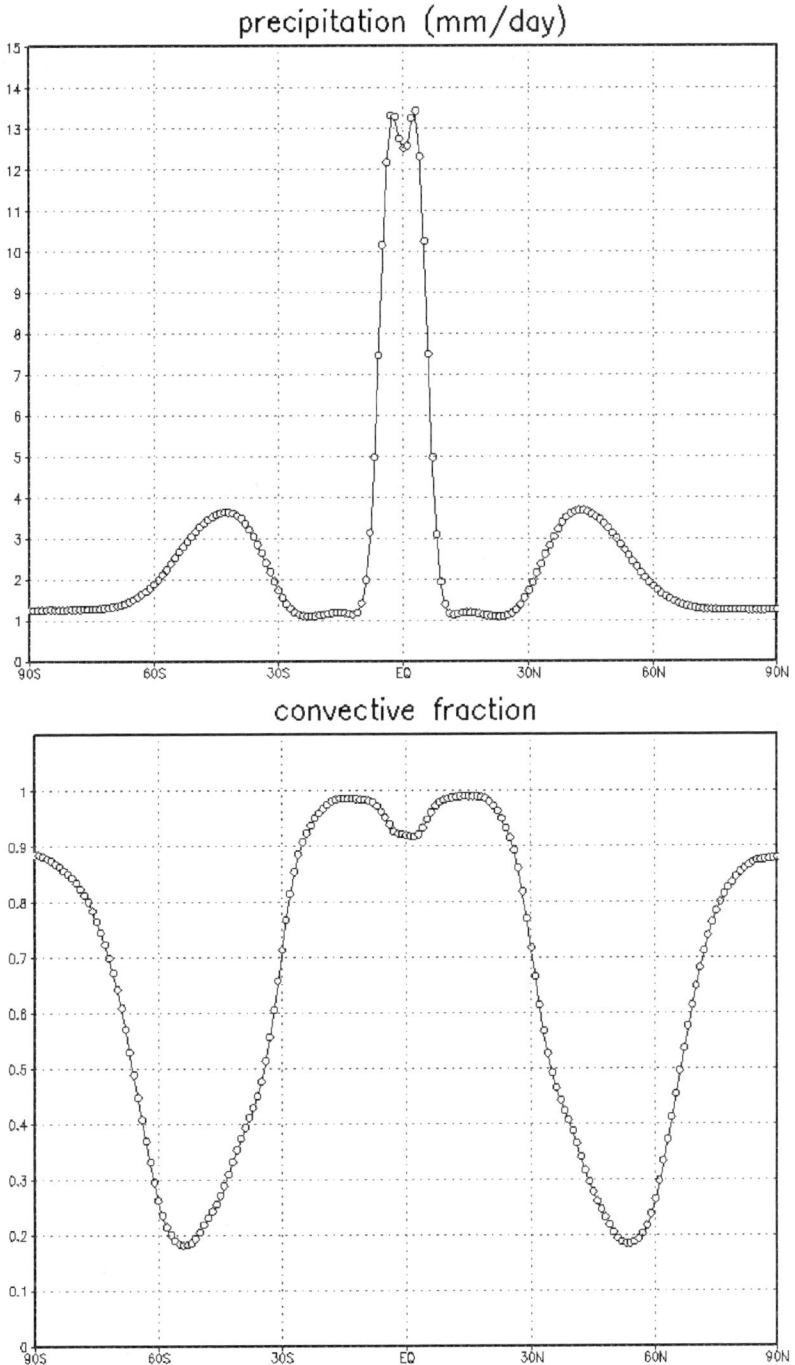

Fig. 4.6 Cross section of the mean (*top*) and convective (*bottom*) precipitation from the aquaplanet experiment on a C48 grid, for the "qobs" SST distribution

et al. (2006), and compared four variable-resolution global models (including CCAM), all the models having their finest resolution of about 50 km over North America. As part of Phase 2, we performed Atmospheric Model Intercomparison Project (AMIP) style simulations with a C80 version of CCAM having quasi-uniform grid spacing of 125 km, and using monthly varying observed SST and sea-ice distributions. Figure 4.7 shows the average DJF mean sea level pressure (MSLP) patterns for 1979–1998 for observed values as determined by NCEP reanalyses. Figure 4.7

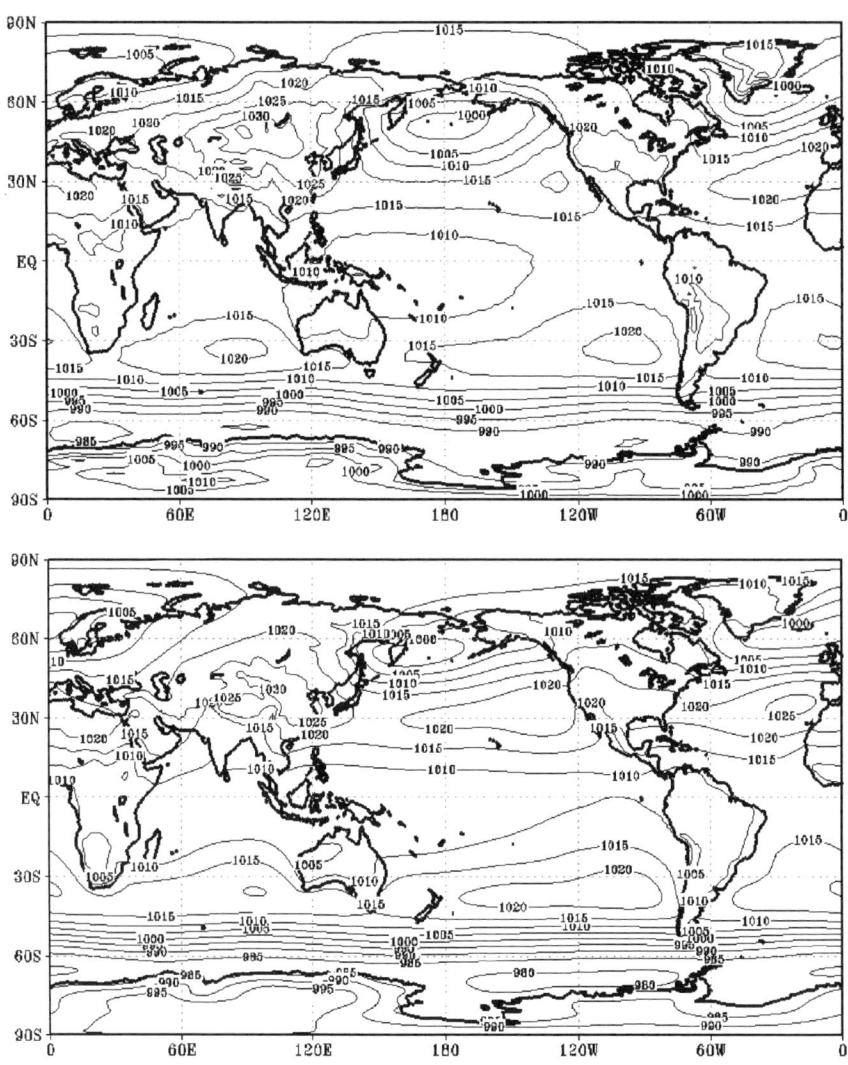

Fig. 4.7 Climatology of 1979–1998 DJF MSLP (hPa) from observations (*top*), and the C80 simulation (*bottom*) using AMIP distributions of SST and sea-ice

4.5 Examples of CCAM Simulations 69

also shows the corresponding MSLP patterns from the CCAM simulation. The simulation captures the main patterns, including the strong Southern Ocean pressure gradient, the oceanic high-pressure regions, the Aleutian low, and a weak Icelandic low. Figure 4.8 shows the similar MSLP fields but for JJA. Again the MSLP patterns are well captured, although the Southern Ocean pressure gradient is a little stronger for this season.

Figure 4.9 shows the average DJF observed precipitation patterns for 1979–1998, as analysed by Xie and Arkin (1997) and from the CCAM simulation. Figure 4.10 shows the similar precipitation patterns but for JJA. For both seasons the precipitation

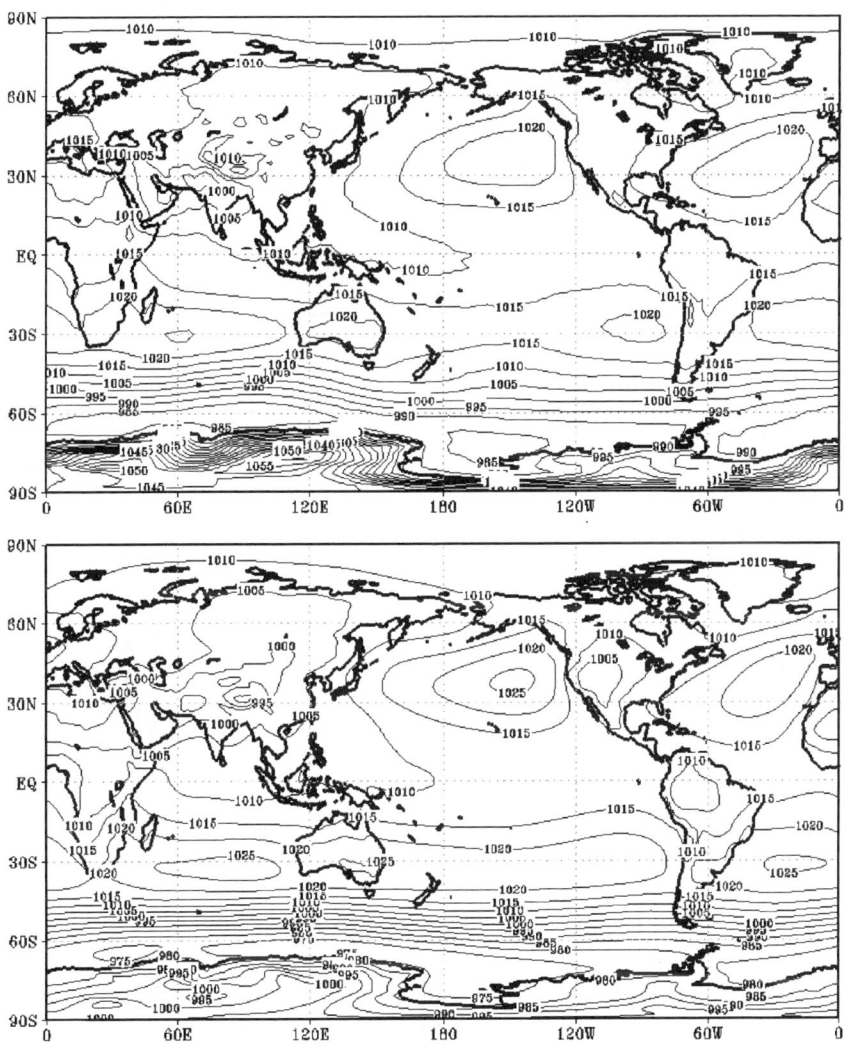

Fig. 4.8 Climatology of 1979–1998 MSLP as for Fig. 4.7, but for JJA

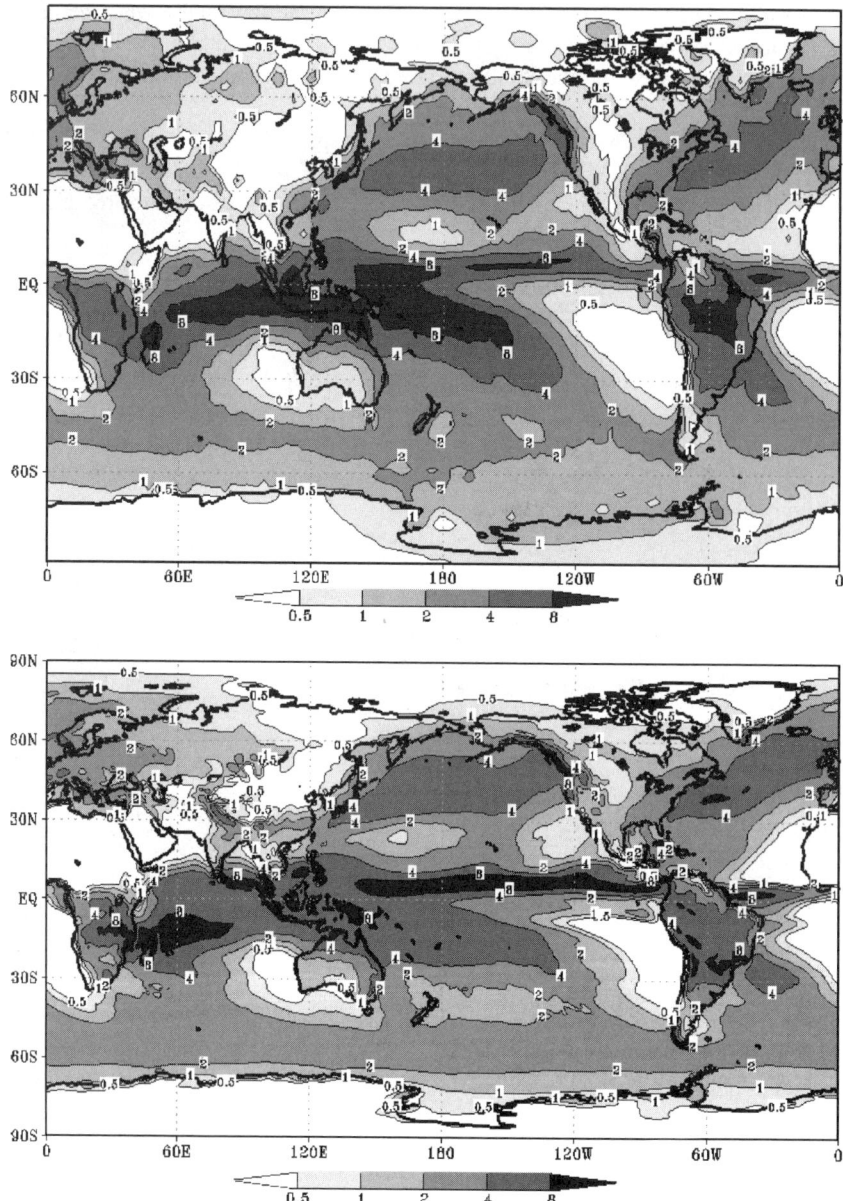

Fig. 4.9 Climatology of 1979–1998 DJF precipitation (mm day^{-1}) from observations (*top*), and the C80 simulation (*bottom*) using AMIP distributions of SST and sea-ice

4.5 Examples of CCAM Simulations

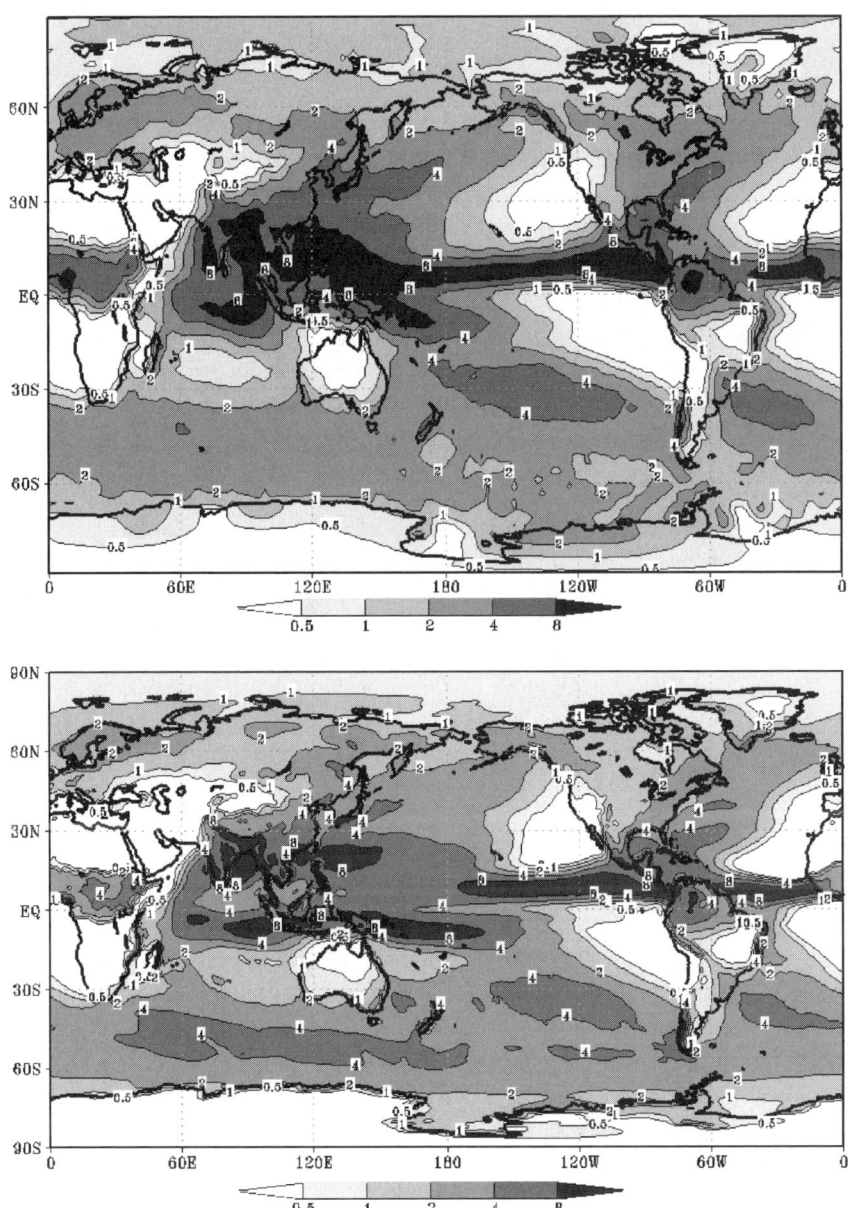

Fig. 4.10 Climatology of 1979–1998 precipitation as for Fig. 4.9, but for JJA

distributions are generally well captured over most of the globe. The main deficiency is that the simulated rainfall over the maritime continent appears to be less intense than that analysed from the observed data. The precipitation corresponding to both the Asian and Australian monsoons appears to be well captured.

4.5.4 Simulation of Antarctic Snow Accumulation

As part of the SGMIP-2 project, extra simulations were performed for 1997–2003 using a high-resolution C200 version of CCAM, having quasiuniform global resolution of 50 km. In Fig. 4.11 we show the simulated snow accumulation over Antarctica for 1997 plotted on the native CCAM grid. By plotting on the native grid, better detail is preserved than plotting the field on an interpolated latitude–longitude grid. The accumulation pattern is qualitatively very similar to the observed long-term annual accumulation of Vaughan et al. (1999), with a dry interior and much larger values near the coastline. However, the very dry interior region (less than 50 mm year^{-1}, water equivalent) seems to be too extensive in the CCAM simulation. In the Antarctic Peninsula region, the model achieves accumulation rates greater than 1000 mm year^{-1}, which is comparable to observations. In the aforementioned 125 km C80 CCAM simulation, significantly smaller peak values were obtained. Clearly, a grid spacing of about 50 km or finer is needed to resolve the precipitation characteristics of the narrow Antarctic Peninsula. This result is in accord with the experience of van Lipzig et al. (2004), who used regional models with 14 km and 55 km grid spacing to model snow accumulation on the Antarctic Peninsula; they found the finer resolution model produced larger snow accumulation by better resolving the orographic barrier of the Peninsula.

4.6 Concluding Comments

This paper has presented an updated description of CCAM. The model is formulated on the quasi-uniform conformal-cubic grid and has a number of attractive features. In particular the semi-Lagrangian semi-implicit scheme performs efficiently and accurately as a result of being able to use reversible staggering treatment for the wind components. The grid possesses eight vertices which have the characteristic of weak singularities. However, no special numerical treatment is required near the vertices; this is because all sensitive operations involving the winds (namely, calculation of departure points, interpolation of wind components, horizontal diffusion) are performed in terms of the corresponding three-dimensional Cartesian components.

The model possesses a fairly comprehensive set of physical parameterizations, enabling it to perform well as an atmospheric GCM. The model has now been coded to run efficiently on multiple processors using the MPI methodology, and timing results were presented for several computers. Very acceptable simulation results were

4.6 Concluding Comments

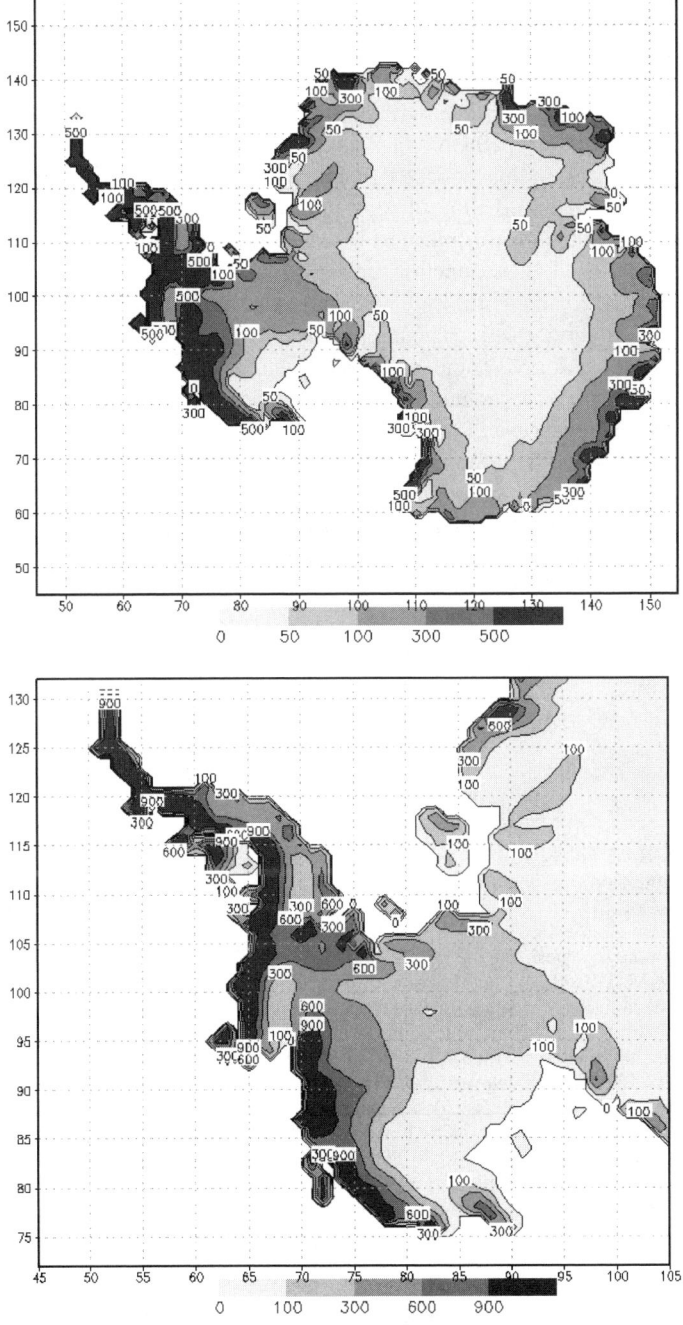

Fig. 4.11 Snow accumulation (mm year^{-1}, as water equivalent) over Antarctica, displayed on the native CCAM C200 grid, as simulated for 1997. The bottom panel shows a closer view for the Antarctic Peninsula. The axes show the grid-point numbering on the southernmost panel

shown from the Held-Suarez test, the Aqua-Planet Experiment and an AMIP simulation at 125 km resolution. Antarctic snow accumulation was also shown from one year of a 7-year simulation having a global resolution of 50 km.

The model is being used for a variety of purposes. By using the Schmidt (1977) transformation, CCAM is being run in stretched-grid mode to provide high resolution over selected regions. In this mode, CCAM is being used for both numerical weather prediction and for the modelling of regional climate change. CCAM is also being used for modelling the global transport of trace gases, and for the determination of their sources and sinks in conjunction with inverse methods.

Finally, we make some comments regarding alternative grid systems. An alternative quasi-uniform grid, that is currently popular, is the icosahedral grid. This is more uniform than the conformal-cubic grid, which becomes a little crowded near the vertices; this is not a practical problem in CCAM because the semi-Lagrangian numerics avoid any associated time step penalty. A unique attraction of the cubic grid geometry is its suitability for the reversible staggering treatment, which provides very good dispersion properties. This staggering treatment is also applicable to gnomonic-cubic grids, which can moreover be made as uniform as icosahedral grids. Our future plans include investigations of atmospheric modelling on the gnomonic-cubic grid.

Acknowledgments The authors are grateful to Jack Katzfey, Eva Kowalczyk, Marcus Thatcher, and Kim Nguyen for their continuing assistance with evaluation and development of the model. This project has received partial support from the Antarctic Climate and Ecosystems Cooperative Research Centre.

References

Bates, J. R., F. H. M. Semazzi, R. W. Higgins, and S. R. M. Barros, 1990: Integration of the shallow water equations on the sphere using a vector semi-Lagrangian scheme with a multigrid solver. *Mon. Wea. Rev.*, **118**, 1615–1627.

Bermejo, R., and A. Staniforth, 1992: The conversion of semi-Lagrangian advection schemes to quasi-monotone schemes. *Mon. Wea. Rev.*, **120**, 2622–2632.

Chouinard, C., M. Béland, and N. McFarlane, 1986: A simple gravity wave drag parametrization for use in medium-range weather forecast models. *Atmos. Ocean*, **24**, 91–110.

D'Azevedo, E. F., V. L. Eijkhout, and C. H. Romine, 1992: Conjugate gradient algorithms with reduced synchronization overhead on distributed memory multiprocessors. Technical Report LAPACK working note 56, University of Tennessee.

Durran, D. R., and P. A. Reinecke, 2004: Instability in explicit two-time-level semi-Lagrangian schemes. *Q. J. R. Meteorol. Soc.*, **130**, 365–369.

Fels, S. B., and M. D. Schwarzkopf, 1975: The simplified exchange approximation: A new method for radiative transfer calculations. *J. Atmos. Sci.*, **32**, 1475–1488.

Fox-Rabinovitz, M., J. Coté, B. Dugas, M. Déqué, and J. L. McGregor, 2006: Variable-resolution GCMs: Stretched-Grid Model Intercomparison Project (SGMIP). *J. Geophys. Res.*, 111, D16104, doi: 10.1029/2005JD006520.

Gordon, H. B., L. D. Rotstayn, J. L. McGregor, M. R. Dix, E. A. Kowalczyk, S. P. O'Farrell, L. J. Waterman, A. C. Hirst, S. G. Wilson, M. A. Collier, I. G. Watterson, and T. I. Elliott, 2002: The CSIRO Mk3 climate system model. Technical Report 60, CSIRO Atmospheric Research, 130 pp.

Held, I. M., and M. J. Suarez, 1994: A proposal for the intercomparison of the dynamical cores of atmospheric general circulation models. *Bull. Amer. Meteor. Soc.*, **75**, 1825–1830.

References

Holtslag, A. A. M., and B. A. Boville, 1993: Local versus non-local boundary layer diffusion in a global climate model. *J. Climate*, **6**, 1825–1842.

Majewski, D., D. Liermann, P. Prohl, B. Ritter, M. Buchhold, T. Hanisch, G. Paul, and W. Wergen, 2002: The operational global icosahedral-hexagonal gridpoint model GME: Description and high-resolution tests. *Mon. Wea. Rev.*, **130**, 319338.

McGregor, J. L., 1993: Economical determination of departure points for semi-Lagrangian models. *Mon. Wea. Rev.*, **121**, 221–230.

McGregor, J. L., 1996: Semi-Lagrangian advection on conformal-cubic grids. *Mon. Wea. Rev.*, **124**, 1311–1322.

McGregor, J. L., 2003: A new convection scheme using a simple closure. *BMRC Res. Rep. 93*, 33–36.

McGregor, J. L., 2005a: C-CAM: Geometric aspects and dynamical formulation [electronic publication]. Technical Report 70, CSIRO Atmospheric Research, 43 pp.

McGregor, J. L., 2005b: Geostrophic adjustment for reversibly staggered grids. *Mon. Wea. Rev.*, **133**, 1119–1128.

McGregor, J. L., and M. R. Dix, 2001: The CSIRO Conformal-Cubic Atmospheric GCM. *IUTAM Symposium on Advances in Mathematical Modelling of Atmosphere and Ocean Dynamics*, P. F. Hodnett, Ed., Kluwer: Dordrecht, 197–202.

McGregor, J. L., H. B. Gordon, I. G. Watterson, M. R. Dix, and L. D. Rotstayn, 1993: The CSIRO 9-level atmospheric general circulation model. Technical Report 26, CSIRO Atmospheric Research, 89 pp.

Neale, R. B., and B. J. Hoskins, 2000: A standard test for AGCMs and their physical parameterizations. I: The proposal. *Atmos. Sci. Letters*, **1**, 101–107.

Rancic, M., R. J. Purser, and F. Mesinger, 1996: A global shallow-water model using an expanded spherical cube: Gnomonic versus conformal coordinates. *Quart. J. Roy. Meteor. Soc.*, **122**, 959–982.

Rivest, C., A. Staniforth, and A. Robert, 1994: Spurious resonant response of semi-Lagrangian discretizations to orographic forcing: Diagnosis and solution. *Mon. Wea. Rev.*, **122**, 366–376.

Rotstayn, L. D., 1997: A physically based scheme for the treatment of stratiform clouds and precipitation in large-scale models. I: Description and evaluation of the microphysical processes. *Quart. J. Roy. Meteor. Soc.*, **123**, 1227–1282.

Schmidt, F., 1977: Variable fine mesh in spectral global model. *Beitr. Phys. Atmos.*, **50**, 211–217.

Schwarzkopf, M. D., and S. B. Fels, 1991: The simplified exchange method revisited: An accurate, rapid method for computation of infrared cooling rates and fluxes. *J. Geophys. Res.*, **96**, 9075–9096.

Smagorinsky, J., S. Manabe, and J. L. Holloway, 1965: Numerical results from a nine-level general circulation model of the atmosphere. *Mon. Wea. Rev.*, **93**, 727–768.

Smith, R. N. B., 1990: A scheme for predicting layer clouds and their water content in a general circulation model. *Quart. J. Roy. Meteor. Soc.*, **116**, 435–460.

Temperton, C., and A. Staniforth, 1987: An efficient two-time-level semi-Lagrangian semi-implicit scheme. *Quart. J. Roy. Meteor. Soc.*, **113**, 1025–1039.

Thomas, S. J., and R. D. Loft, 2000: Parallel semi-implicit spectral element methods for atmospheric general circulation models. *J. Sci. Comput.*, **15**, 499–518.

Thuburn, J., 1993: Use of a flux-limited scheme for vertical advection in a GCM. *Quart. J. Roy. Meteor. Soc.*, **119**, 469–487.

van Leer, B., 1977: Towards the ultimate conservative difference scheme IV. A new approach to numerical convection. *J. Comput. Phys.*, **23**, 276–299.

van Lipzig, N. P. M., J. C. King, T. A. Lachlan-Cope, and M. R. van den Broeke, 2004: Precipitation, sublimation, and snow drift in the Antarctic Peninsula region from a regional atmospheric model. *J. Geophys. Res.*, **109**, D24106, doi:10.1029/2004JD004701.

Vaughan, D. G., J. L. Bamber, M. Giovinetto, J. Russell, and A. P. R. Cooper, 1999: Reassessment of net surface mass balance in Antarctica. *J. Climate*, **12**, 933–946.

Xie, P., and P. A. Arkin, 1997: Global precipitation: A 17-year monthly analysis based on gauge observations, satellite estimates, and numerical model outputs. *Bull. Amer. Meteor. Soc.*, **78**, 2539–2558.

Chapter 5
Description of AFES 2: Improvements for High-Resolution and Coupled Simulations

Takeshi Enomoto, Akira Kuwano-Yoshida, Nobumasa Komori, and Wataru Ohfuchi

Summary This chapter describes the updated version of Atmospheric General Circulation Model for the Earth Simulator (AFES 2). Modifications are intended (1) to increase the accuracy and efficiency of the Legendre transform at high resolutions and (2) to improve the physical performance. In particular, the Emanuel scheme replaces a simplified version of the Arakawa-Schubert scheme for the parametrization of cumulus convection. The Emanuel scheme parametrizes O(100 m) drafts within subgrid-scale cumuli and does not have explicit dependency upon the grid size. Therefore the cloud model of the Emanuel scheme allows us to use it at high resolutions of O(10 km) where the validity of the ensemble cloud model of the Arakawa-Schubert scheme is questionable. Moreover, 10-year test runs indicate that the use of the Emanuel scheme improve the physical performance at a moderate resolution as well. Anomalies of the geopotential height and zonal winds in the middle to upper troposphere are reduced, although the improvements in terms of the distributions of precipitation and sea-level pressure are not significant. Improvements are attributable to a better vertical structure of temperature in the tropics due to more realistic estimation of mixing of the momentum, temperature, and moisture by the Emanuel scheme.

5.1 Introduction

Using AFES (Atmospheric General Circulation Model for the Earth Simulator; Numaguti et al. 1997; Shingu et al. 2002; Shingu et al. 2003), simulations at a landmarking resolution of T1297L96 (truncation wave number of 1 279, corresponding to 10-km horizontal grid-spacing and 96 levels) were conducted in August 2002 and successfully reproduced meso-scale features of three high-impact weather systems under January, June, and September conditions (Ohfuchi et al. 2004). Most notably tropical storms were simulated quite realistically. The fine resolution and large-scale forcing contribute to reproduce tropical cyclogenesis without artificial perturbations or forcing in the tropics (as known as bogus). At the same time the T1279L96 simulations left us with challenges that motivate us to improve the dynamical and physical schemes as well as input/output capabilities.

First of all, each high-resolution integration can produce tera bytes of data. In order to reduce the size by a factor of 2/9, the spectral output format is implemented as an option in addition to sequential- and direct-access formats. Data at a preferred resolution can be retrieved by the spherical synthesis. Another way to limit the data size is to save only a limited area and/or interpolate model levels to selected pressure levels. Spectral and limited-area ouputs are possible simultaneously. This is useful to investigate a local phenomena at the full resolution and large-scale flow at a lower resolution at the same.

The accuracy and computational cost of the Legendre transform are also serious issues at high resolutions. It was found that the Legendre Transform becomes less accurate as the truncation wave number N increases. The transform error grows to an unbearable degree as N becomes closer to 2000. Moreover, there were redundant transforms, although the Legendre transform code of AFES is highly optimized. Modifications to the Legendre transform are discussed in Subsect. 5.2.2.

In T1279L96 simulations, convective precipitation is less than large-scale precipitation even in the tropics. In these simulations, the Arakawa-Schubert type scheme was used to represent cumulus convection (Arakawa and Schubert 1974; Numaguti et al. 1997). Resolution dependency tests were conducted to find that convective precipitation from cumulus parametrization decreases markedly at T639 with the Arakawa-Schubert scheme. The Emanuel scheme showed very small sensitivity to horizontal resolution (Y.O. Takahashi, *pers. comm.*). This result motivated us to look for an alternative scheme and it was decided to implement the Emanuel scheme (Emanuel 1991; Emanuel and Živković-Rothman 1999; Peng et al. 2004).

Another major motivation came from the need to improve physical processes of AFES as the atmospheric component of CFES (Coupled Atmosphere-Ocean General Circulation Model for the Earth Simulator; Komori et al. 2007). The Emanuel scheme found to be useful in improving the model climate. The comparison are made by the 10-year runs using the two schemes in Subsect. 5.3.3. Modern schemes such as mstrnX (Sekiguchi et al. 2003; Sekiguchi 2004) for radiation, MATSIRO (Minimal Advanced Treatments of Surface Interactions and Run-Off; Takata et al. 2003) for land-surface process have been also introduced. Cumulus cloud cover is calculated by the simple scheme by Teixeira and Hogan (2002). A new grid condensation scheme has been also developed based on statistical methods proposed by Sommeria and Deardroff (1977) and Mellor (1977).

Use of AFES is expanded to weather research by adding an ability to use the analysis in the Japan Meteorological Agency Grid Point Value (GPV) dataset as the initial conditions. In order to avoid unwanted waves due to the difference in the topography, the surface pressure in GPV is not used and calculated from the temperature and geopotential height in GPV and the topography used in AFES. The winds, temperature, and humidity in GPV are interpolated spectrally in the horizontal and linearly in the vertical.

This article is aimed at describing modifications made after the T1279L96 experiments reported by Ohfuchi et al. (2004). AFES 2 is the new version with those modifications. Section 5.2 describes modifications to dynamical processes. Descriptions of recently introduced physical schemes follow in Sect. 5.3. Concluding remarks are found in Sect. 5.4.

5.2 Dynamical Processes

5.2.1 Formulation

The mathematical formulation of the dynamical core is briefly reviewed here.

AFES is a spectral, Eulerian and primitive-equation AGCM based on CCSR/NIES AGCM 5.4.02 (Numaguti et al. 1997). It was rewritten to run most efficiently on the Earth Simulator (Shingu et al. 2002, 2003). Its dynamical framework basically follows that of Hoskins and Simmons (1975). The vertical coordinates are in $\sigma = p/p_s$, where p and p_s are the pressure and surface pressure, respectively. The vertical finite difference scheme by Arakawa and Suarez (1983) is adopted for a better representation of the hydrostatic equation, the pressure gradient force and the conservation of energy and potential temperature. The leap-frog scheme with the Robert-Asselin filter and semi-implicit scheme are used to integrate forward in time.

The prognostic equations for the vorticity (ζ), divergence (δ) and temperature (T), specific humidity (q) and surface pressure (p_s) are as follows:

$$\frac{\partial \zeta}{\partial t} = \frac{1}{a(1-\mu^2)} \frac{\partial N_v}{\partial \lambda} - \frac{1}{a} \frac{\partial N_u}{\partial \mu} - \mathcal{D}(\zeta), \tag{5.1}$$

$$\frac{\partial \delta}{\partial t} = \frac{1}{a(1-\mu^2)} \frac{\partial N_u}{\partial \lambda} - \frac{1}{a} \frac{\partial N_v}{\partial \mu} - \nabla^2 (E + \phi + RT_0 \ln p_s) - \mathcal{D}(\delta), \tag{5.2}$$

$$\frac{\partial T'}{\partial t} = -\frac{1}{a(1-\mu^2)} \frac{\partial (UT')}{\partial \lambda} - \frac{1}{a} \frac{\partial (VT')}{\partial \mu} + T'\delta$$

$$- \dot{\sigma} \frac{\partial T'}{\partial \sigma} + \frac{RT_v \omega}{c_p \sigma p_s} + \frac{Q_{\text{diff}}}{c_p} - \mathcal{D}(T'), \tag{5.3}$$

$$\frac{\partial q}{\partial t} = -\frac{1}{a(1-\mu^2)} \frac{\partial (Uq)}{\partial \lambda} - \frac{1}{a} \frac{\partial (Vq)}{\partial \mu} + q\delta - \dot{\sigma} \frac{\partial q}{\partial \sigma} - \mathcal{D}(q), \tag{5.4}$$

$$\frac{\partial \ln p_s}{\partial t} = -\int_0^1 \left(\frac{U}{a(1-\mu^2)} \frac{\partial \ln p_s}{\partial \lambda} + \frac{V}{a} \frac{\partial \ln p_s}{\partial \mu} + \delta \right) d\sigma, \tag{5.5}$$

where λ is the longitude, ϕ the latitude $\mu = \sin(\phi)$, t time, a the planet radius, u, v the eastward and northward winds, $U = u\cos(\phi)$, $V = v\cos(\phi)$, N_u and N_v the nonlinear terms, $\mathcal{D}(\zeta)$, $\mathcal{D}(\delta)$, $\mathcal{D}(T')$, $\mathcal{D}(q)$ the horizontal dissipation, E the kinetic energy, ϕ the geopotential, $T_0(\sigma)$ the reference temperature, $T' = T - T_0$, R the gas constant for dry air, c_p the specific heat at constant pressure, T_v the virtual temperature, and Q_{diff} the heating from horizontal mechanical dissipation. The kinetic energy is written as

$$E = \frac{U^2 + V^2}{2(1 - \mu^2)}. \tag{5.6}$$

The nonlinear terms are

$$N_u = (f + \zeta)V - \frac{RT'_v}{a}\frac{\partial \ln p_s}{\partial \lambda} - \dot{\sigma}\frac{\partial U}{\partial \sigma}, \tag{5.7}$$

$$N_v = -(f + \zeta)U - RT'_v\frac{1 - \mu^2}{a}\frac{\partial \ln p_s}{\partial \mu} - \dot{\sigma}\frac{\partial V}{\partial \sigma}, \tag{5.8}$$

where f is the Coriolis parameter and $T'_v = T_v - T_0$.

Horizontal dissipation $\mathcal{D}(q)$ is composed of the Rayleigh friction and horizontal diffusion. The Rayleigh friction is an artificial forcing to reduce the wind speed and wave reflection preferentially near the upper boundary. The vertical profile of the Rayleigh friction coefficient takes the form

$$K_R = \frac{1}{\tau_R}\left[1 + \tanh\left(\frac{z - z_R}{H_R}\right)\right] \tag{5.9}$$

where τ_R is the e-folding time for the Rayleigh friction at height $z = -H \ln(\sigma) \to \infty$ and z_R the reference height $z_R = -H \ln(\sigma_{\text{top}})$, σ_{top} the first full sigma level from the top of the atmosphere, and H and H_R the scale heights. The default values are $H = 8\,000$ m and $H_R = 7\,000$ m. τ_R depends upon σ_{top}. An empirical relation is

$$(\tau_R)^{-1} = 15\sqrt{p_{\text{top}}}(\text{day}), \tag{5.10}$$

where p_{top} is the pressure at the first full sigma level. AFES 2 provides an option to exclude the zonally symmetric components from the Rayleigh friction.

The horizontal diffusion is imposed to remove energy from the smallest scales. For $X = T$ and q, the horizontal diffusion takes the form

$$\mathcal{D}(X) = K(-1)^{N_D/2}\nabla^{N_D} X, \tag{5.11}$$

where N_D is the degree of the diffusion. Usually $N_D = 4$. In order to exclude the solid body rotation, for $X = \zeta$ and δ, the horizontal diffusion takes the form

$$\mathcal{D}(X) = K\left[(-1)^{N_D/2}\nabla^{N_D} - \left(\frac{2}{a^2}\right)\right]X. \tag{5.12}$$

The pseudo pressure coordinate correction term of the form

$$\mathcal{D}_p(X) = \mathcal{D}(X) - \mathcal{D}(\ln p_s)\sigma\frac{\partial T}{\partial \sigma} \tag{5.13}$$

is added to avoid excessive vertical transport near steep orography. Horizontal dissipation $\mathcal{D}(X)$ has a significant influence upon the atmospheric spectra. In AFES 2,

5.2 Dynamical Processes

relatively small values are used for the horizontal diffusion coefficient in order to be consistent with the observed $-5/3$ curve. An empirically obtained relation is

$$K^{-1} = 19.9N^{-0.831}(\text{day}^{-1}) \tag{5.14}$$

where N is the truncation wave number (Y.O. Takahashi, *pers. comm.*) as opposed to the previously used relation

$$K^{-1} = 200N^{-1.4}(\text{day}^{-1}). \tag{5.15}$$

Atmospheric spectra of AFES are further discussed by Takahashi et al. (2006).

5.2.2 Modifications to the Legendre Transform

The spherical transform method has been used extensively in many global atmospheric models since it has advantages in the effective resolution and wave propagation properties. It should be reminded that the perfectly uniform resolution in the spherical geometry is not possible with any finite difference methods beyond icosahydron but with the spectral method using the triangular truncation. The cost of the spherical transform theoretically increases as $O(N^3 \log N)$, where N is the truncation wave number. However, Shingu et al. (2002, 2003) showed that AFES achieved a slower increase of the cost of the spherical transform than $O(N^3 \log N)$ through the optimization to the vector-parallel architecture of the Earth Simulator. Untch (*pers. comm.*) showed that the cost of the Legendre transform is affordable with 17% at T2047 using the linear grid in the ECMWF (European Centre for Medium-Range Weather Forecasts) IFS (Integrated Forecast System) using semi-Lagrangian advection with the minimal number of Legendre transforms. With a larger truncation wave number, where the nonhydrostatic dynamical core is appropriate, it is also necessary with more detailed or explicit physical schemes such as cloud microphysics. Those schemes are much more computationally demanding, thus the cost of the Legendre transform may not be a serious problem although the spectral transform method eventually become less efficient than finite difference methods.

The accuracy of the Legendre transform is essential to that of spectral models. The Legendre transform becomes less accurate as the truncation wave number increases in AFES. The Legendre transform only has single precision at T1279 and finally the error becomes unacceptably large near T1800. Although the cause of this problem is not yet known, it is considered that errors are accumulated in the three-point recurrence formula

$$\tilde{P}_n^m(\theta) = \frac{1}{\epsilon_n^m}\left[\cos\theta\, \tilde{P}_{n-1}^m(\theta) - \epsilon_{n-1}^m \tilde{P}_{n-2}^m\right], \tag{5.16}$$

where $\theta = \pi/2 - \phi$ is the colatitude and $\tilde{P}_n^m(\theta)$ is the normalized associated Legendre function

$$\tilde{P}_n^m(\theta) = \sqrt{\frac{2n+1}{2} \frac{(n-m)!}{(n+m)!}} P_n^m(\theta) \tag{5.17}$$

with

$$\tilde{P}_0^0(\theta) = \frac{1}{\sqrt{2}},$$

$$\tilde{P}_1^0(\theta) = \sqrt{\frac{3}{2}} \cos\theta,$$

$$\tilde{P}_m^m(\theta) = \sqrt{\frac{2m+1}{2m}} \sin\theta \, \tilde{P}_{m-1}^{m-1},$$

$$\tilde{P}_{m+1}^m(\theta) = \sqrt{2m+3} \cos\theta \, \tilde{P}_m^m \tag{5.18}$$

and

$$\epsilon_n^m = \sqrt{\frac{n^2 - m^2}{4n^2 - 1}}. \tag{5.19}$$

The values on the diagonal lines ($n = m, n = m + 1$, Fig. 5.1a) are obtained by (5.18) to calculate the value at $n+1$ those at n and $n-1$ for each m. Errors accumulate as m increases along $n = m$ and $n = m + 1$ and as n increases for each m. Thus it is anticipated that the largest error appears in a region with large n moderate m.

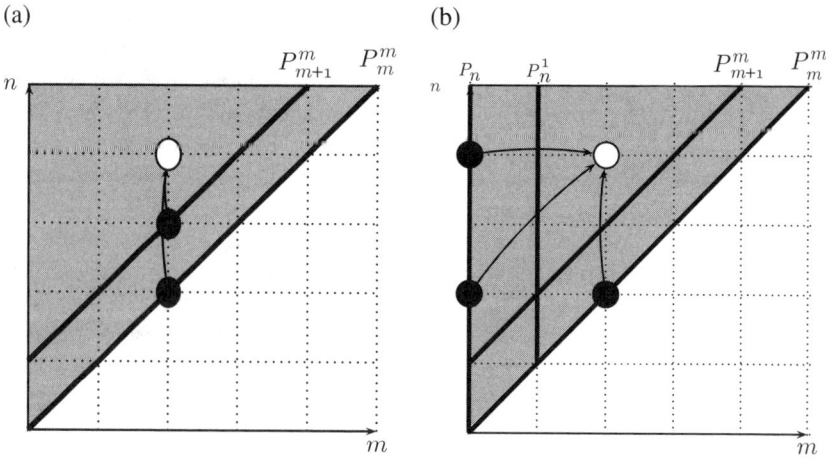

Fig. 5.1 Schematic representation of the (**a**) three- and (**b**) four-point recurrence relations. Horizontal and vertical axes represent zonal wave number (m) and total wave number (n), respectively. The value at white dot is calculated from those at *black dots*. Values at the points on the *yellow lines* are precalculated to obtain values at other points. *Grey shades* denote the region with $n \geq m$

5.2 Dynamical Processes

Swartztraber (1993) suggested an alternative four-point recurrence equation. Its normalized form is

$$\tilde{P}_n^m = \frac{1}{\sqrt{(n+m-1)(n+m)}} \left\{ \sqrt{\frac{2n+1}{2n-3}} \times \right.$$

$$\left[\sqrt{(n+m-2)(n+m-3)} \tilde{P}_{n-2}^{m-2} + \sqrt{(n-m)(n-m-1)} \tilde{P}_{n-2}^m \right]$$

$$\left. - \sqrt{(n-m+1)(n-m+2)} \tilde{P}_n^{m-2} \right\}. \tag{5.20}$$

The values of $\tilde{P}_n^0$ are calculated using the Fourier expansion found by Laplace and Legendre (Hobson 1931)

$$P_n(\theta) = \sum_{k=0}^{n} a_{n,k} \cos k\theta$$

$$= 2 \frac{1 \cdot 3 \cdot 5 \cdots (2n-1)}{2 \cdot 4 \cdot 6 \cdots 2n} \left[\cos n\theta + \frac{1 \cdot n}{1 \cdot (2n-1)} \cos(n-2)\theta \right.$$

$$\left. + \frac{1 \cdot 3}{1 \cdot 2} n(n-1)(2n-1)(2n-3) \cos(n-4)\theta + \cdots \right]. \tag{5.21}$$

With the normalizing factor, the coefficients $a_{n,k}$ can be obtained by

$$a_{n,n} = \sqrt{1 - \frac{1}{4n^2}} a_{n-1,n-1}, \quad a_{0,0} = \frac{\sqrt{2}}{2} \tag{5.22}$$

and

$$a_{n,n-l} = \frac{(l-1)(2n-l+2)}{l(2n-l+1)} a_{n,n-l+2} \tag{5.23}$$

where $l \equiv n - k$. The values of $\tilde{P}_n^1$ are found immediately from (5.17) and (5.21) (Belousov 1962)

$$P_n^1(\theta) = \frac{1}{\sqrt{n(n+1)}} \sin\theta \frac{dP_n(\theta)}{d(\cos\theta)} = -\frac{1}{\sqrt{n(n+1)}} \frac{dP_n(\theta)}{d\theta}$$

$$= \frac{1}{\sqrt{n(n+1)}} \sum_{k=0}^{n} k a_{n,k} \sin k\theta. \tag{5.24}$$

P_n and P_n^1 are the starting values to calculate the values at even and odd m, respectively.

The recurrence relation (5.20) is schematically illustrated in Fig. 5.1b. This method avoids accumulation of error as n increases since $\tilde{P}_n$ are calculated directly from its Fourier representation.

Figure 5.2 show the distribution of the maximum root-mean square error in a test with T2159. Each harmonics is initialized with the value of 1 for both real and imaginary components. As anticipated large errors are found in a region with

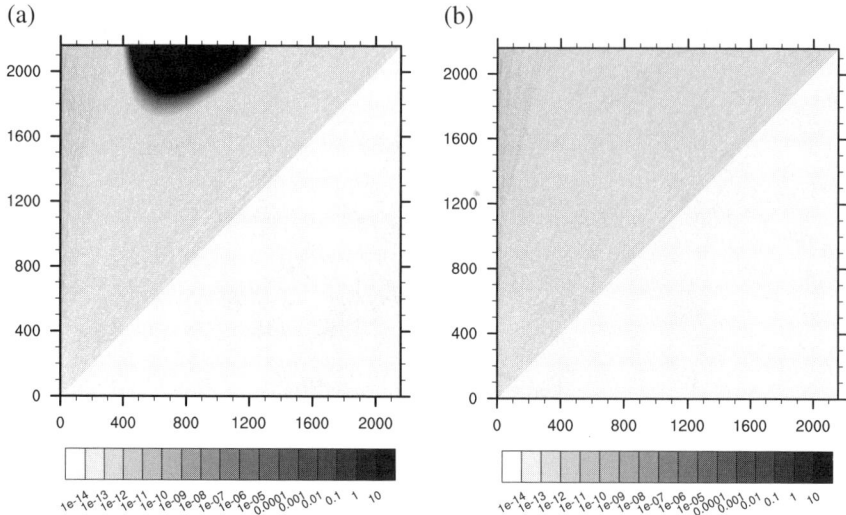

Fig. 5.2 Distribution of error caused from a pair of the Legendre synthesis and analysis using (**a**) three- and (**b**) four-point recurrence equations

$400 < m < 1\,400$ and $n > 1\,700$ in the test with the three-point recurrence [(5.16), Fig. 5.2a] in that with four-point recurrence [(5.20), Fig. 5.2b]. The transform using associated Legendre functions calculated with (5.16) is inaccurate for $N > 1\,700$. The error from the transform using that with (5.20) is still negligible at T2559 (not shown). The difference between (5.16) and (5.20) is negligible at for small truncation wave numbers ($\leq$ T639) and (5.20) is recommended for T1279 and required for truncation wave numbers larger than about 1 700.

Following Swartztrauber (2002) the Gaussian latitudes and weights are calculated in terms of θ in place of $\cos\theta$ to avoid errors near the poles with (5.21). Tests with the sum of the Gaussian weights show that the error reduced by a few digits at all resolutions tested (Enomoto et al. 2004). AFES 2 adopted this method as default.

The number of Legendre transforms were reduced by 2/3 in AFES 2 by evaluating meridional derivatives in the spectral space (Temperton 1991) as follows. Noting the identity

$$\tilde{H}_n^m(\theta) = \cos\phi \frac{\mathrm{d}}{\mathrm{d}\phi}\left[\tilde{P}_n^m(\phi)\right]$$
$$= -n\epsilon_{n+1}^m \tilde{P}_{n+1}^m(\theta) + (n+1)\epsilon_n^m \tilde{P}_{n-1}^m(\theta) \quad (5.25)$$

(U and V) are obtained from the synthesis of

$$U_n^m = \frac{1}{a}\left[im\chi_n^m + (n-1)\epsilon_n^m \psi_{n-1}^m - (n+2)\epsilon_{n+1}^m\right]$$
$$V_n^m = \frac{1}{a}\left[im\psi_n^m - (n-1)\epsilon_n^m \chi_{n-1}^m + (n+2)\epsilon_{n+1}^m \zeta_{n+1}^m\right] \quad (5.26)$$

where $\psi = a^2 \nabla^{-2} \zeta$ and $\chi = a^2 \nabla^{-2} \delta$ are stream function and divergence, respectively. Similarly ζ and δ are obtained as

$$\zeta_n^m = \frac{1}{a}\left[im\tilde{V}_n^m - n\epsilon_n^m \tilde{U}_{n+1}^m + (n+1)\epsilon_n^m \tilde{U}_{n-1}^m\right]$$

$$\delta_n^m = \frac{1}{a}\left[im\tilde{U}_n^m + n\epsilon_{n+1}^m \tilde{V}_{n+1}^m - (n+1)\epsilon_n^m \tilde{V}_{n-1}^m\right] \quad (5.27)$$

where $\tilde{U}_n^m, \tilde{V}_n^m$ are the Legendre transforms of winds scaled with $\cos^2 \phi$ in the Fourier space.

5.3 Physical Processes

Following the advection by the dynamical core, the tendencies from the physical schemes are successively added. First the cumulus parametrization is called except when the Emanuel dry convective adjustment precedes it. The grid condensation, radiation, vertical diffusion, and surface processes are then called. Finally the original dry convective adjustment is called except when it is turned off. In this section, the effects of the revision of those physical schemes are documented. Especially comparisons are made between the runs with different convective parametrization schemes.

5.3.1 Radiation Scheme mstrnX

AFES has used mstrn8 (Nakajima et al. 2000). In AFES 2 the improved version called mstrnX (Sekiguchi et al. 2003; Sekiguchi 2004) is also made available as an option. It was developed to decrease heating errors found in mstrn8 of ~ 1 K day^{-1} in the troposphere and lower stratosphere against the results using the line-by-line method. A cold bias in the lower stratosphere is typically found in models using mstrn8. The updated version, mstrnX improves parameters of gaseous absorption and optimization of the correlated k-distribution using the up-to-date database. These modifications contributed to reduce heating errors to typical values of less than 0.1 K day^{-1}.

As expected, the cold bias in the lower stratosphere is mostly reduced and other local anomalies stand out instead. Figure 5.3 shows the annual average of the zonal-mean temperature in 1-year runs at T79L48 with a climatological sea-surface temperature (SST) and anomalies from that in ECMWF Re-analysis 15 dataset (ERA-15; Gibson et al. 1999). The anomalies in the tropical lower-stratosphere seem to represent error from the convective scheme. Those over the poles are related to orographically induced gravity waves.

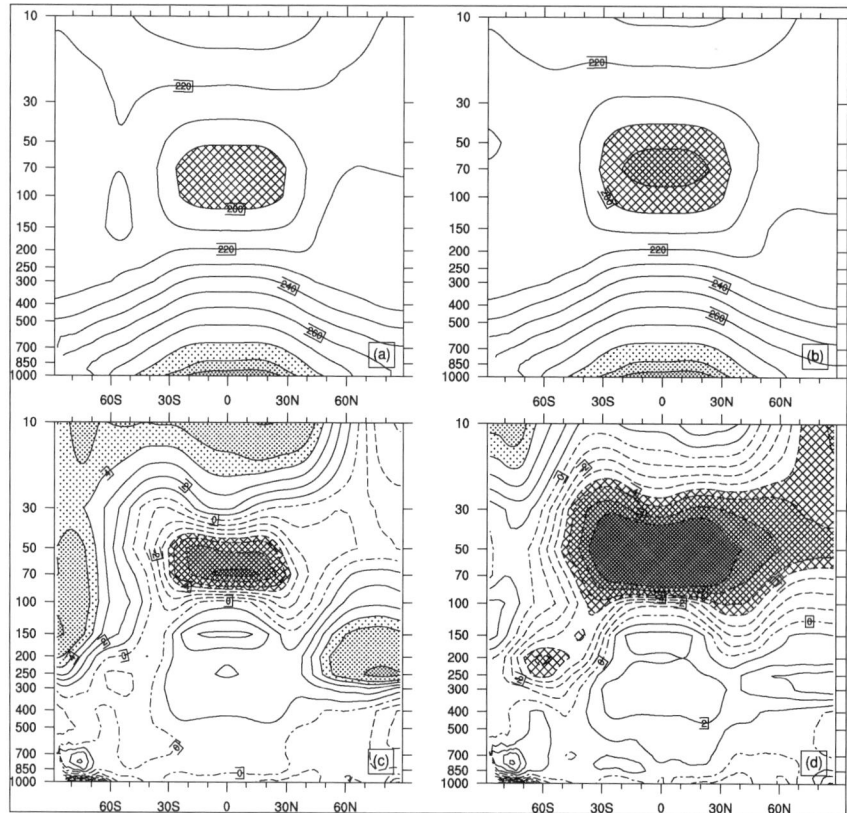

Fig. 5.3 *Panels* (**a**) and (**b**) show the annual average of the zonal-mean temperature in 1-year runs with a climatological SST using mstrnX (new) and using mstrn8 (old), respectively. The contour interval is 10 K between 180 and 300 K with 185 and 295 K contours added. Values smaller than 200 and larger than 280 K are *hatched* and *dotted*, respectively. *Panels* (**c**) and (**d**) show temperature anomalies from ERA-15 of in the runs using mstrnX and using mstrn8, respectively. Negative and naught lines are *dotted* and *line-dotted*, respectively. The contour interval is 1 K between −9 and 9 K with ±6 and ±8 K contours omitted. Values smaller than −4 and larger than 4 K are *hatched* and *dotted*, respectively

5.3.2 Dry Convective Adjustment

The original dry convective adjustment scheme was introduced to stabilize the stratosphere. The behavior of this scheme, however, is different from its original intention. The layers near the surface rather than those in the middle atmosphere are most actively adjusted to supplement the turbulent vertical diffusion. Since the adjustment takes place at the end of the physics schemes and just before the dynamics, it often "cleans up" odd tendencies of preceding physical schemes. A more serious problem with this scheme is that it sometimes causes moisture to be transported upward very rapidly upon the adjustment, probably due to the lack of removal of supersaturation.

An alternative adjustment scheme taken from the Emanuel convective parametrization has been introduced to AFES. It is intended to remove unconditional instability to ensure stable calculation of the following moist convection. In this scheme, the horizontal momentum as well as temperature and humidity are mixed vertically preserving the mass-weighted enthalpy. Supersaturation is removed and added to convective precipitation. It is possible to use one or both of the two dry adjustment schemes in AFES 2.

5.3.3 Emanuel Convective Parametrization

The Arakawa-Schubert scheme (Arakawa and Schubert 1974) in AFES assumes ensemble of cumuli at different stages of development in a rather large domain. Ascent covers in a small portion of the grid and compensating descent elsewhere. Thus it is expected that this assumption becomes less appropriated as the grid size is reduced. The convective energy is assumed to be in the quasiequilibrium with the destabilization by large-scale flow. The mass flux of each plume is assumed to grow with height as it mix with environmental air. The intensity of convection is determined by the cloud work function, vertically integrated buoyancy reduced by entrainment. The environmental air is moistened by the detrainment of cloudy air at the cloud top and subsequent subsidence. The implementation of this scheme in AFES (Numaguti et al. 1997) includes effects of downdraft but mixing of momentum.

The Emanuel scheme (Emanuel 1991; Emanuel and Živković-Rothman 1999; Peng et al. 2004) parametrizes collective effects of drafts of $O(100\,m)$ within the cumuli in a grid box. When there is upward cloud-base mass flux, a parcel of air at a single level at the maximum enthalpy below its minimum in the mid troposphere. Subparcels rise to levels between the cloud base and the level of neutral buoyancy as undilute ascents. A fraction of condensed water is converted to precipitation and downdraft is calculated in proportional to the ratio of precipitation falling outside the cloud. Subparcels then mix with environmental air and move to levels of the equal liquid potential temperature and detrain there after precipitation is removed. This procedure is called episodic mixing. The cloud-base mass flux increases if a parcel is buoyant at the lifting condensation level and damps with time toward the subcloud-layer quasiequilibrium.

The Emanuel scheme was chosen to be used at horizontal resolutions $O(10-100\,km)$. With our implementation of the Akarawa-Schubert scheme, the convective precipitation dramatically is reduced at resolutions higher than T639 (not shown). This probably implies that an assumption of ascent in a small area and descent elsewhere is no longer valid at high resolutions. By contrast, the Emanuel scheme shows only a small sensitivity to horizontal resolution in terms of ratio of precipitation from convective and grid-condensation processes (Y.O. Takahashi, *pers. comm.*). Although it is preferable that convective precipitation smoothly decreases as more convective systems are resolved in a nonhydrostatic model using even higher resolutions of O(100 m–1 km), convection in a grid box of O(10-100 km) needs to be parametrized in a hydrostatic model. It was also important to incorporate effects of mixing of temperature, moisture, and momentum with environments in a consistent manner rather than adding such a process to a current scheme. The Emanuel scheme is rather computationally intensive but the computational cost can be significantly reduced to be comparable with that of the Arakawaka-Schubert scheme by balancing loads among MPI processes.

The Emanuel scheme does not have built-in cloud fraction diagnostics. The cloud water is calculated as the weighted average of the cloud water associated with undilute updraft and the average of the cloud water of the mixture (Bony and Emanuel 2001). The cloud fraction is estimated by the method for convective clouds suggested by Teixeira and Hogan (2002).

Now the physical performance of the Emanuel scheme is compared against that of the Arakawa-Schubert scheme. Both schemes have several tuning parameters; therefore, the comparison below may not be general. However, some differences can be attributable to the formulation of these schemes. Ten-year test runs were conducted at for each convective parametrization scheme at a moderate resolution of T79L48. At this resolution, long integration is feasible and both schemes work properly. Other specifications are identical: the original dry adjustment and mstrnX were used in both tests. In each case the model was initialized with winds, temperature, humidity, and ground water on Jan 1, 1982 from ERA-40 dataset (Uppala et al. 2005) and forced with the weekly climatological SST between 1982 and 2004 produced from the NOAA Optimum Interpolation (OI) SST v2 (Reynolds et al. 2002).

Figures 5.4 and 5.5 show the distributions of the simulated sea-level pressure and model bias relative to the National Centers for Environmental Prediction/National Center for Atmospheric Research Reanalysis (CDAS) data (Kistler et al., 1999). The top panels show the simulated distribution with the Emanuel (left) and Arakawa-Schubert (right) schemes. The bottom panels compare model bias from the CDAS climatology in runs with the Emanuel (left) and Arakawa-Schubert (right) schemes. Although the amplitude of model bias is in general smaller with the Emanuel scheme, the differences are not quite significant in either seasons. Accordingly, there is not much difference in the distribution of precipitation bias (Figs. 5.6 and 5.7) relative to the Climate Prediction Center Merged Analysis of Precipitation (CMAP; Xie and Arkin 1997).

However there are improvements in vertical profiles of temperature, humidity, and momentum. Figure 5.8 compares the out-going long wave radiation (OLR). Although the distribution of bias is similar with both schemes, the amplitude is much smaller with the Emanuel scheme. The underestimation of OLR relative to the Earth Radiation Budget Experiment (ERBE) data (ERBE Science team 1986) is caused by over estimation of convective clouds in the tropics. The Arakawa-Schubert scheme adopts

5.3 Physical Processes

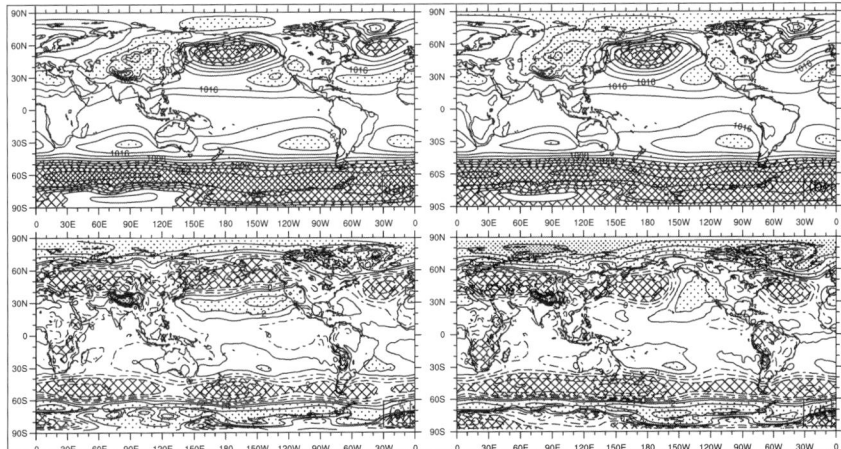

Fig. 5.4 *Panels* (**a**) and (**b**) show the December–January–February (DJF) average of the sea-level pressure (slp, hPa) in the 10-year run using the Emanuel and Arakawa-Schubert schemes, respectively. The contour interval is 4 hPa and values smaller than 1 004 and larger than 1 020 hPa are *hatched* and *dotted*, respectively. *Panels* (**c**) and (**d**) show the model bias of the runs with the Emanuel and Arakawa-Schubert schemes from CDAS, respectively. The contour interval is 2 hPa with ±1 and ± 15 hPa contours added and negative contours are *dotted*. Values smaller than −2 and larger than 2 hPa are *hatched* and *dotted*, respectively

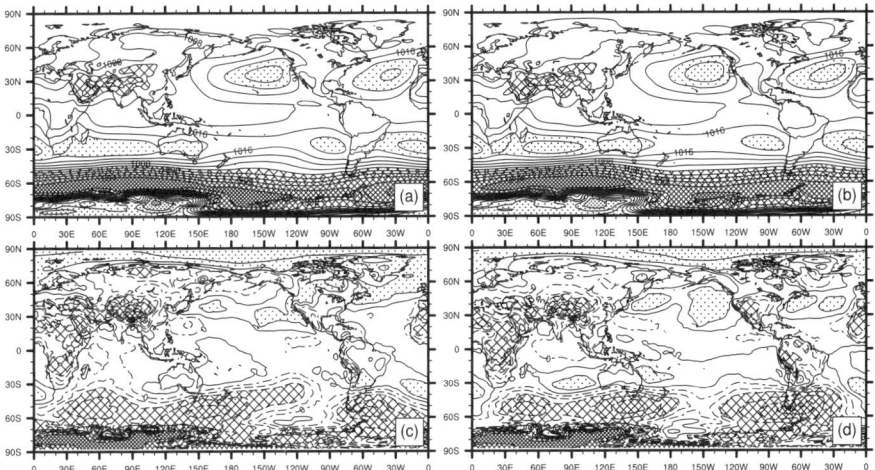

Fig. 5.5 As in Fig. 5.4 but for the June–July–August averages

the plume cloud model and assumes the increase of entrainment with height. This assumption can lead to overestimation of cloud fraction or penetration. The cloud diagnostic scheme by Teixiera and Hogan (2002) is designed to reduce overestimation of cloud fraction in the tropics. This scheme seems to contributes to the smaller

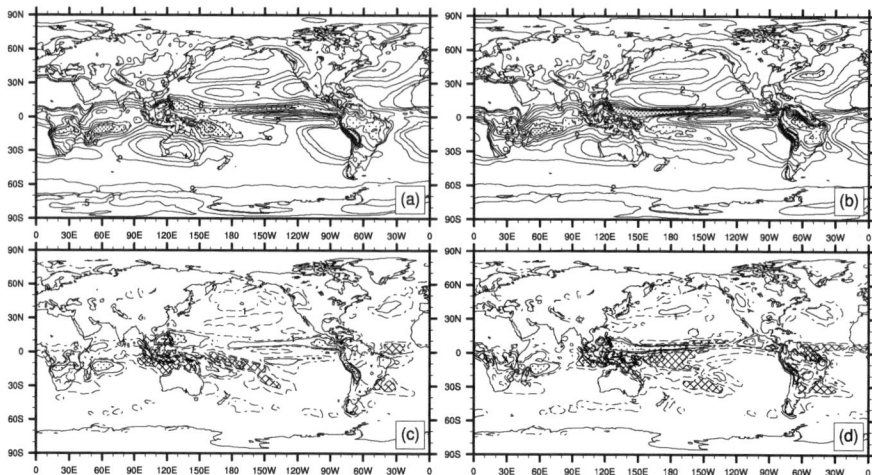

Fig. 5.6 Panels (**a**) and (**b**) show the December–January–February (DJF) average of the precipitation (mm day^{-1}) in the 10-year runs using the Emanuel and Arakawa-Schubert schemes, respectively. The contour levels are 0.5, 1, 2, 4, 6, 8, 10, 12, 17 mm day^{-1} and the values larger than 8 mm day^{-1} are *dotted*. Panels (**c**) and (**d**) show the model bias with the Emanuel and Arakawa-Schubert schemes from CMAP, respectively. The contour levels are $\pm 1, \pm 2, \pm 4, \pm 8$ mm day^{-1} and negative contours are *dotted*. Values smaller than -4 and larger than 4 mm day^{-1} are *hatched* and *dotted*, respectively

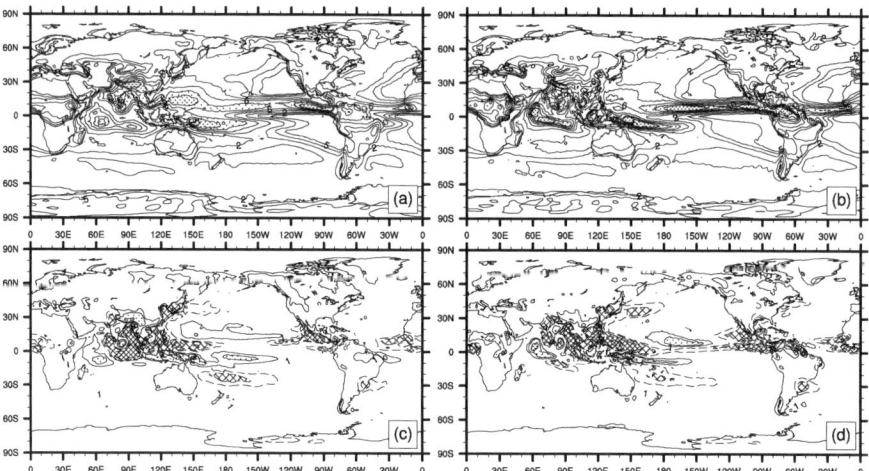

Fig. 5.7 As in Fig. 5.6 but for the June–July–August averages

anomalies together with the episodic-mixing cloud model of the Emanuel scheme that does not assume entraining plumes.

The mid tropospheric height anomalies are significantly smaller with the Emanuel scheme especially in the tropics and subtropics (Fig. 5.9). Improvements are found

5.3 Physical Processes

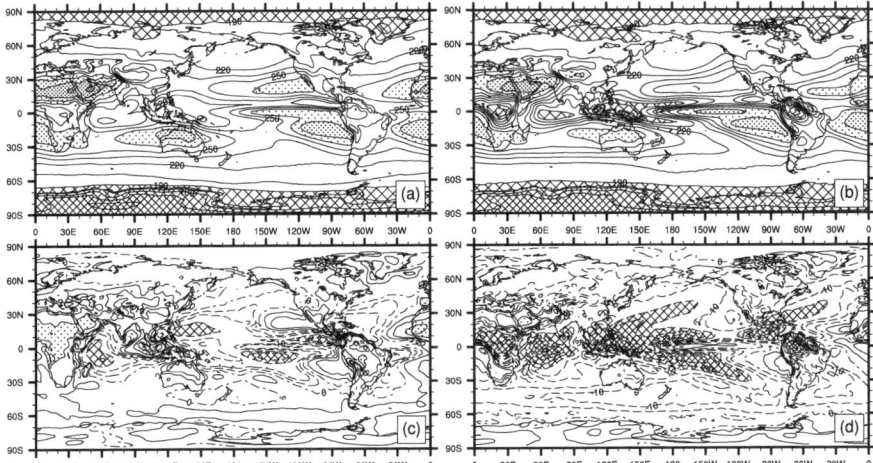

Fig. 5.8 *Panels* (**a**) and (**b**) show the annual average of the out-going long wave radiation (W m^{-2}) in the 10-year runs using the Emanuel and Arakawa-Schubert schemes, respectively. The contour interval is 15 W m^{-2} between 100 and 310 W m^{-2}. Values smaller than 190 and larger than 265 W m^{-2} are *hatched* and *dotted*, respectively. *Panels* (**c**) and (**d**) show the model bias with the Emanuel and Arakawa-Schubert schemes from ERBE, respectively. The contour interval is 10 W m^{-2} between -60 and 60 W m^{-2} with ± 5 W m^{-2} contours added and negative and naught contours are *dotted* and *dash-dotted*, respectively. Values smaller than -20 and larger than 20 W m^{-2} are *hatched* and *dotted*, respectively

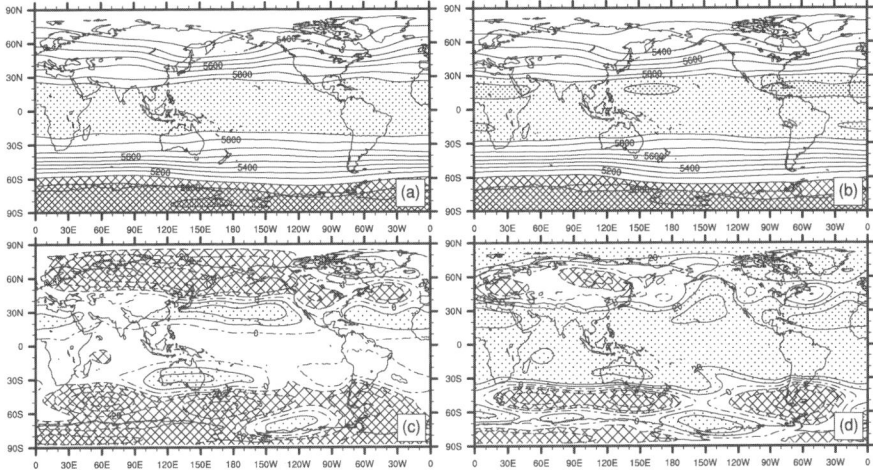

Fig. 5.9 *Panels* (**a**) and (**b**) show the annual average of the geopotential height (gpm) in the 10-year runs using the Emanuel and Arakawa-Schubert schemes, respectively. The contour interval is 100 m with 5 850 and 5 950 m contours added. Values smaller than 5 100 and larger than 5 850 m are *hatched* and *dotted*, respectively. *Panels* (**c**) and (**d**) show the model bias in the runs with the Emanuel and Arakawa-Schubert schemes from ERA-15, respectively. The contour interval is 20 m with ± 10 m contour added and negative and naught contours are *dotted* and *dash-dotted*, respectively. Values smaller than -20 and larger than 20 m are *hatched* and *dotted*, respectively

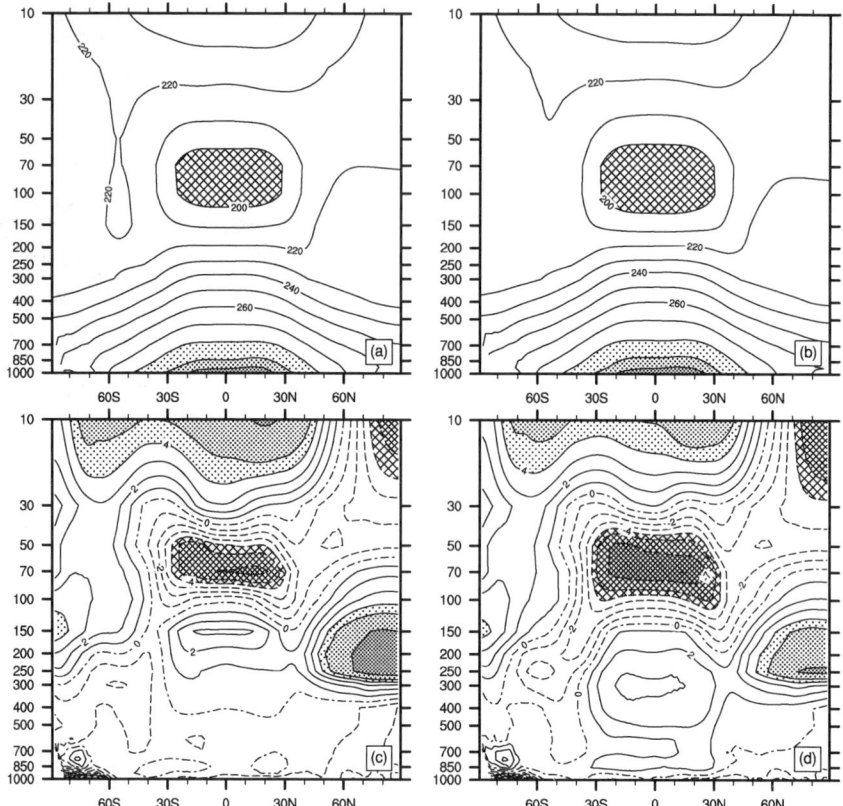

Fig. 5.10 *Panels* (**a**) and (**b**) show the annual average of the zonal-mean temperature (K) in the 10-year runs using the Emanuel and Arakawa-Schubert schemes, respectively. The contour interval is 10 K between 180 and 300 K with 185 and 295 K contours added. Values smaller than 200 and larger than 280 K are *hatched* and *dotted*, respectively. *Panels* (**c**) and (**d**) show the model bias in the runs with with the Emanuel and Arakawa-Schubert schemes from from ERA-15, respectively. Negative and naught lines are *dotted* and *line-dotted*, respectively. The contour interval is 1 K between −9 and 9 K with ±6 and ±8 K contours omitted. Values smaller than −4 and larger than 4 K are *hatched* and *dotted*, respectively

in DJF and JJA although the annual mean is shown. The lower-tropospheric temperature is also improved (not shown). These can be attributable to the improvements of vertical profile of the temperature (Fig. 5.10). The Emanuel scheme rigourously conserves the liquid water potential temperature. Such a formulation contributes to a better representation of the vertical temperature and humidity profile. In fact the humidity profile is also improved (Fig. 5.11). It is interesting to note that excess of humidity near the surface is reduced with the Emanuel scheme. The cloud model of the Emanuel scheme does not assume that of the deep convection. Any subgrid scale buoyant motions with mixing with environment can be represented. As shown

5.3 Physical Processes

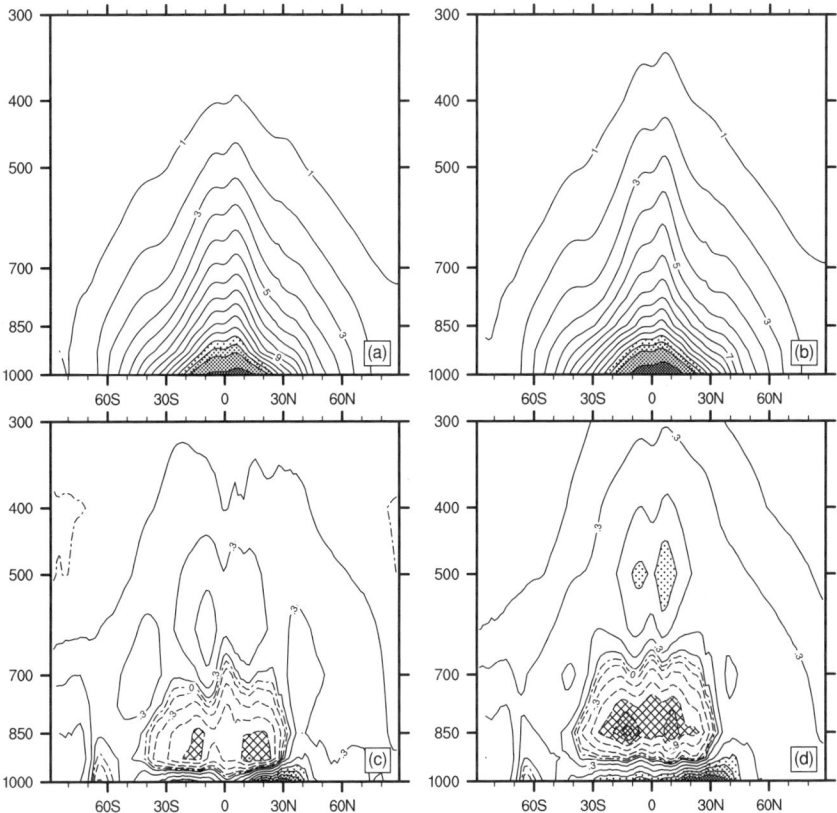

Fig. 5.11 *Panels* (**a**) and (**b**) show the annual average of the zonal-mean specific humidity (g kg^{-1}) in the 10-year runs using the Emanuel and Arakawa-Schubert schemes, respectively. The contour interval is 1 g kg^{-1} between 1 and 16 g kg^{-1} with 15 g kg^{-1} contours omitted. Values larger than 12 g kg^{-1} are *dotted*. *Panels* (**c**) and (**d**) show the model bias in the runs with with the Emanuel and Arakawa-Schubert schemes from from ERA-15, respectively. Negative and naught lines are *dotted* and *line-dotted*, respectively. The contour interval is 0.3 g kg^{-1} between −1.8 and 1.8 g kg^{-1} with ±0.1 g kg^{-1} contours added. Values smaller than −0.9 and larger than 0.9 g kg^{-1} are *hatched* and *dotted*, respectively

in Peng et al. (2004), shallow convection seems to be represented in our model using the Emanuel scheme, at least partly. Another improvement is found in the upper-troposphere. The mixing of the momentum included in the Emanuel scheme seems to remove the overestimation of the zonal wind speed of mid-latitude jets (Fig. 5.12).

Although the distribution of the precipitation is not improved much, the use of the Emanuel scheme improves the overall physical performance not only at high resolutions, where the assumption of Arakawa and Schubert (1974) is questionable, but also at moderate resolutions. Since the vertical profile seems be quite accurate, it is rather the large-scale forcing that needs to be improved.

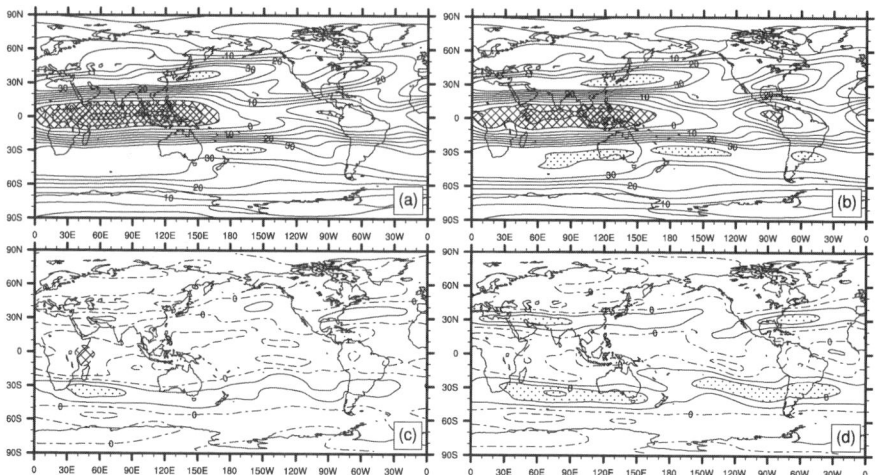

Fig. 5.12 *Panels* (**a**) and (**b**) show the annual average of the zonal wind (m s^{-1}) in the 10-year runs using the Emanuel and Arakawa-Schubert schemes, respectively. The contour intervals are 5 m s^{-1} between -20 and 30 m s^{-1} and 10 m s^{-1} between 30 and 70 m s^{-1}. Values smaller than -5 and larger than 40 m s^{-1} are *hatched* and *dotted*, respectively. *Panels* (**c**) and (**d**) show the model bias with the Emanuel and Arakawa-Schubert schemes from ERA-15, respectively. The contour interval is 4 m s^{-1} and negative and naught contours are *dotted* and *dash-dotted*, respectively. Values smaller than -8 and larger than 8 m s^{-1} are *hatched* and *dotted*, respectively

5.3.4 Other Modifications

Land-surface scheme MATSIRO (Minimal Advanced Treatment of Surface Interaction and Run-Off) was introduced to AFES in order to improve land-surface representation in long-term runs, in particular, as the atmospheric component of our coupled model. In the original land-surface scheme, the land is represented by the surface with its properties namely the albedo, roughness, field capacity vegetation type, and three layers in the ground. MATSIRO introduces a single layer canopy with the effects of photosynthesis and a multilayer snow model. The distribution of snow, near-surface air temperature over snow covered surface and precipitation over the continent have been found to be improved (not shown).

The grid-condensation is calculated with a statistical scheme based upon Le Treut and Li (1991). The subgrid scale moisture is assumed to vary with the uniform distribution. In AFES 2 an experimental modified version is introduced. In the modified scheme the uniform distribution is replaced with the Gaussian distribution and the moisture variance is calculated from the turbulence scheme. The modified scheme is able to enhance locality to clouds. However the calculated variance is found to be too small to produce reasonable cloud distribution and requires fine vertical invervals. The grid-condensation scheme is being modified further to include the variance of the temperature was also taken into account as in the original suggestion by Sommeria and Deardroff (1977) and Mellor (1977).

5.4 Concluding Remarks

In this article, AFES 2 with modifications from the original version (Shingu et al. 2003; Ohfuchi et al. 2004) are documented. Major changes, introduction of an alternative method to calculate the associated Legendre functions and the Emanuel convective parametrization, are described in detail. The former enables accurate Legendre transforms at higher resolutions. The latter is found to function at high resolutions (O(10–100 km)) and to be useful at low resolutions since it improves the representation of large-scale flows.

There are still modifications to be done to use computational resource efficiently and improve physical performance. Introduction of the semi-Lagrangian advection has been considered and a few implementations for tracer have been tested (Peng et al. 2005; Y.O. Takahashi, *pers. comm.*). The semi-Lagrangian advection for momentum has also been being tested with a prototypical advection code. Among the physics schemes, stratiform and convective clouds and its interaction with the radiation and turbulence need to be modified among others.

It is not our priority to incorporate the nonhydrostatic dynamical framework and cloud microphysics in order to convert AFES to a global nonhydrostatic model at this time. In order to increase the resolution an order of magnitude, say to 1-km horizontal and O(1 000) levels, it requires $O(10^4)$ times more computer power that that required for T1279L96. An alternative approach is to make use of a hybrid "sync" system of heterogeneous models such as hydrostatic global model and nonhydrostatic regional models. Hybrid systems include simple nesting but also other approaches may be pursued. Our experiences show that most of hydrostatic phenomena is represented quite well with AFES using an O(10–100 km) horizontal resolution. Nonhydrostatic effects can be best reproduced by a regional nonhydrostatic model such as CReSS (Cloud Resolving Storm Simulator; Tsuboki 2007). In such a system, nonhydrostatic effects in tropical convective regions and/or in a region of interest are calculated with the regional model and the large-scale flows are calculated by the global model accurately with minimal computational cost. Each model maintains its identity and may pursue improvements at the scales of interest.

Acknowledgments The new version of the AFES is the product of numerous contributions from the AFES team. Mr. H. Fuchigami and Mr. M. Yamada helped us with coding and optimization. Prof. T. Nakajima and Dr. Emori allowed us to use mstrnX and MATSIRO, respectively. Dr. Yamane introduced mstrnX to AFES. Comments from the anonymous reviewer and editor Prof. Kevin Hamilton were very useful in improving the manuscript. The NCAR Command Language was used to visualize the output.

References

Arakawa, A. and Schubert, W.H. (1974) Interaction of cumulus cloud ensemble with the large-scale environment. Part I. J. Atmos. Sci., 31, 671–701.
Arakawa, A. and Suarez, M.J. (1983) Vertical differencing of the primitive equations in sigma coordinates. Mon. Wea. Rev., 111, 34–45.

Belousov, S.L. (1962) Tables of normalized associated Legendre polynomials. Brown D.E. (Trans.), Pergamon Press, Oxford.

Bony, S. and Emanuel, K.A. (2001) A parameterization of the cloudiness associated with cumulus convection; evaluation using TOGA COARE data. J. Atmos. Sci., 58, 3158–3183.

Emanuel, K.A. (1991) A scheme for representing cumulus convection in large-scale models. J. Atmos. Sci., 48, 2313–2335.

Emanuel, K.A. and Živković-Rothman M. (1999) Development and evaluation of a convection scheme for use in climate models. J. Atmos. Sci., 56, 1766–1782.

Enomoto, T., Fuchigami, H., Shingu, S. (2004) Accurate and robust Legendre transforms at large truncation wavenumbers with the Fourier method., Proc. the 2004 workshop on the solution of partial differential equations on the sphere, 20–23 July 2004, Yokohama, Japan, 17–19.

ERBE Science Team (1986) First Data From the Earth Radiation Budget Experiment (ERBE). Bull. of the Amer. Meteor. Soc., 67, 818–824.

Gibson, J.K., Kållberg, P., Uppala, S., Hernandez, A., Nomura, A., Serrano, E. (1999) ERA-15 description (version 2 – January 1999), ECMWF Re-Analysis Project Report Series, European Centre for Mid-Range Weather Forecasts, Reading, United Kingdom.

Hobson, E.W. (1931) The theory of spherical and ellipsoidal harmonics. Cambridge Univesity Press, Cambridge.

Hoskins, B.J. and Simmons, A.J. (1975) A multi-layer spectral model and the semi-implicit method. Quart. J. Roy. Meteor. Soc., 101, 637–655.

Kistler, R., Kalnay, E., Collins, W., Saha, S., White, G. Woollen, J., Chelliah, M., Ebisuzaki, W., Kanamitsu, M., Kousky, V. van den Dool, H., Jenne, R. and Fiorino, M. (1999) The NCEP/NCAR 50-year reanalysis. Bull. Amer. Meteor. Soc., 82, 247–267.

Komori, N., Kuwano-Yoshida, A., Enomoto, T., Sasaki, H. and Ohfuchi, W. (2007) High-resolution simulation of the global coupled atmosphere-ocean system: description and preliminary outcomes of CFES (CGCM for the Earth Simulator). In High Resolution Numerical Modelling of the Atmosphere and Ocean. W. Ohfuchi and K. Hamilton (Eds.), Springer, New York.

Le Treut, H. and Li, Z.-X. (1991) Sensitivity of an atmospheric general circulation model to prescribed SST changes: feedback effects associated with the simulation of cloud optical properties. Clim. Dynam., 5, 175–187.

Mellor. G.L. (1977) The Gaussian cloud model relations. J. Atmos. Sci., 34, 356–358.

Nakajima, T., Tsukamoto, M., Tsushima, Y., Numaguti, A., and Kimura, T., 2000: Modeling of the radiative process in an atmospheric general circulation model. Appl. Opt., 39, 4869–4878.

Numaguti, A., Takahashi, M., Nakajima, T., and Sumi, A. (1997) Description of CCSR/NIES/Atmospheric General Circulation Model. CGER's Supercomputer Monograph Report, 3, National Institute of Environmental Sciences, Tsukuba, Japan, 1–48.

Ohfuchi, W., Nakamura, H., Yoshioka, M.K., Enomoto, T., Takaya, K., Peng, X., Yamane, S., Nishimura, T., Kurihara, Y., and Ninomiya, K. (2004) 10-km mesh meso-scale resolving simulations of the global atmosphere on the Earth Simulator: preliminary outcomes of AFES (AGCM for the Earth Simulator). J. Earth Simulator, 1, 8–34.

Peng, M.S., Ridout, J.A. and Hogan, T.F. (2004) Recent modification of the Emanuel convective scheme in the Navy operational global atmospheric prediction system. Mon. Wea. Rev., 132, 1254–1268.

Peng, X., Xiao, F., Ohfuchi, W., and Fuchigami, H. (2005) Conservative semi-Lagrangian transport on a sphere and the impact on vapor advection in an atmospheric general circulation model. Mon. Wea. Rev. 133, 504–520.

Reynolds, R.W., Rayner, N.A., Smith, T.M., Stokes, D.C. and Wang, W., 2002: An improved in situ and satellite SST analysis for climate. J. Climate, 15, 1609–1625.

Sekiguchi, M. (2004) A study on evaluation of the radiative flux and its computational optimization in the gaseous absorbing atmosphere. Science Doctoral Dissertation, Tokyo University, 121 pp (in Japanese).

Sekiguchi, M., Nakajima, T., Suzuki, K., Kawano, K., Higurashi, A., Rosenfeld, D., Sano, I. and Mukai, S. (2003) A study of the direct and indirect effects of aerosols using global satellite data sets of aerosol and cloud parameters. J. Geophys. Res., 108, (10.1029), 2002JD003359.

Shingu, S., Takahara, H., Fuchigami, H., Yamada, M., Tsuda, Y., Ohfuchi, W., Sasaki, Y., Kobayashi, K., Hagiwara, T., Habata, S., Yokokawa, M., Itoh, H., and Otsuka, K. (2002) A 26.58 Tflops

References

global atmospheric simulation with the spectral transform method on the Earth Simulator. Proceedings of Supercomputing 2002, http://www.sc-2002.org/paperpdfs/pap.pap331.pdf.

Shingu, S., Fuchigami, H., Yamada, M. (2003) Vector parallel programming and performance of a spectral atmospheric model on the Earth Simulator in Realizing Teracomputing: Proceedings of the Tenth ECMWF Workshop on the Use of High Performance Computing in Meteorology, W. Zwieflhofer and N. Kreitz (Eds.), World Scientific, 29–46.

Sommeria, G. and Deardroff, J.W. (1977) Subgrid-scale condensation in models of nonprecipitation clouds. J. Atmos. Sci., 34, 344–355.

Swartztrauber, P.N. (1993) The vector harmonic transform method for solving partial differential equations in spherical geometry. Mon. Wea. Rev., 121, 3415–3437.

Swartztrauber, P.N. (2002) Computing the points and weights for Gauss-Legendre quadrature. SIAM Journal on Scientific Computing, 24, 945–954.

Takahashi, Y.O., Hamilton, K., and Ohfuchi, W. (2006) Explicit global simulation of the mesoscale spectrum of atmospheric motions, Geophys. Res. Lett., 33, L12812, doi:10.1029/2006GL026429.

Takata, K., Emori, S., Watanabe, T. (2003) Development of the minimal advanced treatments of interaction and runoff. Global Planet. Change, 38, 209–222.

Teixeira. J. and Hogan, T.F. (2002) Boundary layer clouds in a global atmospheric model: simple cloud cover parameterizations. J. Climate, 15, 1261–1276.

Temperton, C. (1991) On scalar and vector transform methods for global spectral models. Mon. Wea. Rev., 119, 1303–1307.

Tsuboki, K. (2007) High-Resolution Simulations of High-Impact Weather Systems Using the Cloud-Resolving Model on the Earth Simulator. In High Resolution Numerical Modelling of the Atmosphere and Ocean. W. Ohfuchi and K. Hamilton (Eds.), Springer, New York. Chapter 9.

Uppala, S.M., Kållberg, P.W., Simmons, A.J., Andrae, U., da Costa Bechtold, V., Fiorino, M., Gibson, J.K., Haseler, J., Hernandez, A., Kelly, G.A., Li, X., Onogi, K., Saarinen, S., Sokka, N., Allan, R.P., Andersson, E., Arpe, K., Balmaseda, M.A., Beljaars, A.C.M., van de Berg, L., Bidlot, J., Bormann, N., Caires, S., Chevallier, F., Dethof, A., Dragosavac, M., Fisher, M., Fuentes, M., Hagemann, S., Hólm, E., Hoskins, B.J., Isaksen, L., Janssen, P.A.E.M., Jenne, R., McNally, A.P., Mahfouf, J.-F., Morcrette, J.-J., Rayner, N.A., Saunders, R.W., Simon, P., Sterl, A., Trenberth, K.E., Untch, A., Vasiljevic, D., Viterbo, P., and Woollen, J. 2005: The ERA-40 re-analysis. Quart. J. Roy. Meteor. Soc., 131, 2961-3012.doi:10.1256/qj.04.176

Xie, P.-P. and Arkin, P.A. (1997) Global Precipitation: a 17-year monthly analysis based on gauge observations, satellite estimates, and numerical model outputs. Bull. Amer. Meteor. Soc., 78, 2539–2558.

Chapter 6
Precipitation Statistics Comparison Between Global Cloud Resolving Simulation with NICAM and TRMM PR Data

M. Satoh, T. Nasuno, H. Miura, H. Tomita, S. Iga, and Y. Takayabu

Summary A "global cloud resolving simulation" with horizontal grid interval of 3.5 km is conducted using a nonhydrostatic icosahedral atmospheric model (NICAM). NICAM is a cloud resolving model in the sense that updraft cores of deep cumulus that have a few km in horizontal size are marginally represented using explicit microphysical schemes. Results from the aqua-planet experiment of NICAM are compared with the TRMM PR (Tropical Rainfall Measurement Mission, Precipitation Radar) data that have a horizontal resolution close to the grid interval of NICAM.

Precipitation statistics of NICAM are compared to those of the TRMM PR data over oceans. Probability distribution functions of rainfall rate show that relative occurrence of rainfall rate of NICAM is similar to that of the TRMM PR data for strong rains. Spectral representation of rainfall rate shows that the result of NICAM has generally higher rain-top height than that of TRMM PR. NICAM produces stronger rainfall especially for deep convective rains. Both NICAM and TRMM PR have two peaks of rainfall rate in shallow and high rain-top heights for stratiform rains.

One of the advantages of using the global cloud resolving model is that the model results provide plenty of information concerning physical quantities, which are not easily retrieved from the observation, such as vertical velocity and ice phase concentrations. In particular, since the precipitation intensity of the TRMM PR data does not take account of updraft velocities of the air, the rainfall speed might be upward relative to the ground in very strong cloud cores of deep cumulus, even when the positive rainfall rate is analyzed by the TRMM PR algorithm.

6.1 Introduction

A first result from the global cloud resolving simulation was reported by Tomita et al. (2005) using the Nonhydrostatic ICosahedral Atmospheric Model (NICAM) with horizontal mesh interval of 3.5 km. The numerical experiment was based on the aqua-planet where the global surface was assumed to be ocean everywhere

(Neale and Hoskins 2000). The result shows a multiscale structure of tropical convection: meso-scale circulation in 10 km-order horizontal scale, cloud clusters in 100 km-order scale, super cloud clusters in 1,000 km-oder scale, and convectively coupled Kelvin waves in planetary scale. Similar to the observational results (Nakawaza 1988), eastward propagation of super cloud clusters embedded in westward motions of cloud clusters is reproduced in the model results. The numerical simulation was performed using the Earth Simulator, and the computational performance of NICAM on the Earth Simulator is reported in Satoh et al. (2005). Miura et al. (2005) further argued the climate sensitivity of NICAM and found that the response of cloud amount to the increase in surface temperature is opposite to the existing atmospheric general circulation model. Although global cloud resolving models have a potential for more reliable prediction of future climate change, many ambiguity remains in global cloud resolving simulations: results will depend on physical schemes such as microphysics, and numerical convergence is not obtained at 3.5 km-mesh interval (Tomita et al., 2005), which implies requirement for subgrid schemes. Numerical results must be investigated by comparing with observational data in various perspectives.

The "cloud resolving model" of this study means a model with grid interval of a few km, and is sometimes called a cloud-system resolving model. In such a model, explicit cloud physics schemes are used and meso-scale circulation are reasonably represented by marginally permitting updraft cores of deep convection. In regional scale, much higher resolution can be used to express finer scales of clouds. Since effective resolution of numerical models is generally larger than the grid interval (Bryan et al. 2003; Skamarock 2004), one may require finer grid intervals to achieve numerical convergence of moist models. For the global domain, however, resolution with a few km grid interval is only achievable using the existing computer facility. A global model with grid interval of a few km will shed new light since it is essentially different from heuristic general circulation models in the sense that it does not require cumulus parameterization based on statistical assumptions (Randall et al. 2003). Since it is quite a new approach, its realism must be tested by various perspectives. As one of these directions, simulation results of a global cloud resolving model should be examined by comparing wide area, high-resolution, and uniformly observed data given by satellites.

In this study, precipitation properties of NICAM are compared with the TRMM PR data (Tropical Rainfall Measurement Mission, Precipitation Radar). The TRMM PR data provides high-resolution and uniform precipitation data in three-dimensional domain. Although many studies on precipitation have been conducted by comparing cloud resolving numerical results to high-resolution satellite data in regional domains, this study is unique in arguing statistics of precipitation of very wide domain using a result from the global cloud resolving simulation. Since the NICAM simulation was conducted under the idealized aqua-planet condition, this study is restricted to a statistical comparison of general characteristics over the ocean. In reality, however, the rain properties are different between areas of the ocean: e.g., Eastern Pacific vs. Western Pacific.

The structure of this paper is as follows: NICAM and the aqua-planet experimental setup are briefly described in the next section and the precipitation pattern of the experiment is reviewed. In Sect. 6.4, probability distribution functions of rainfall rate are compared between NICAM and the TRMM PR data. Then, in Sect. 6.5, the precipitation data of NICAM is analyzed in the same method presented in Takayabu (2002) using spectral representations of rainfall rate, and statistics of vertical velocity and ice phase quantities are also discussed. The final section summarizes the results and future directions.

6.2 Model and the Experimental Setup

NICAM adopts the icosahedral grid shown in Fig. 6.1 (Tomita and Satoh, 2004; Satoh et al., 2007). To gain numerical accuracy, the original grid configuration is modified by using the spring dynamics (Tomita et al. 2001, 2002). The recursive division of the

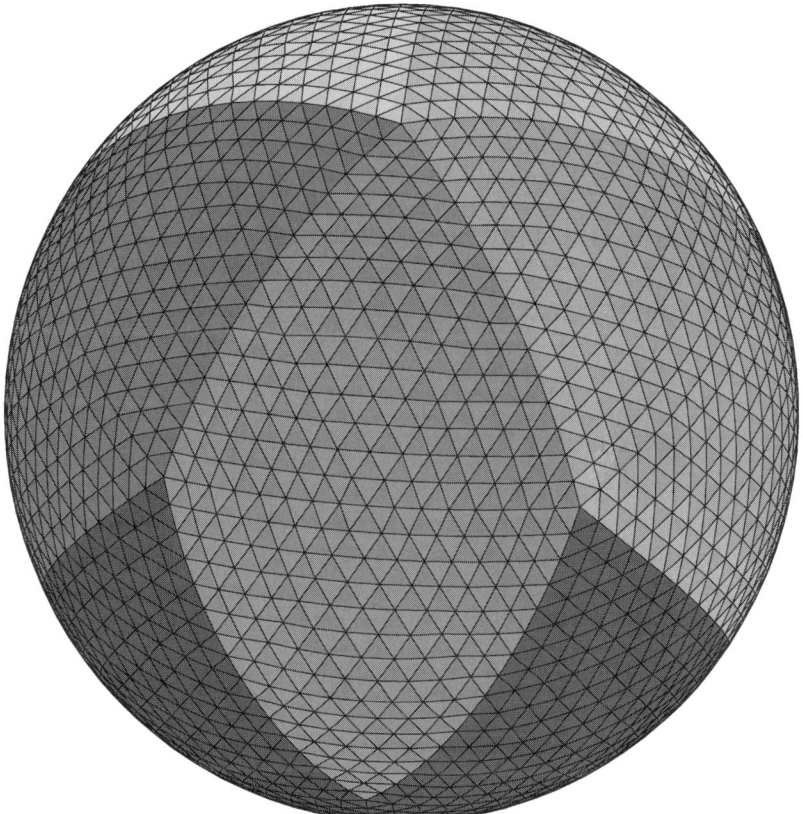

Fig. 6.1 An example of the icosahedral grid. This grid is generated from the original icosahedron by dividing triangles 4-times recursively. We obtain grid interval of 3.5 km by dividing the original icosahedron 11-times

original triangles and the spring relaxation give quasiuniform gird system. Figure 6.1 is the grid structure obtained after dividing the original icosahedron four times. By dividing nine, ten, and eleven times, one obtains quasiuniform grids with grid interval of about 14, 7, and 3.5 km, respectively.

We conduct the aqua-planet experiment using NICAM with global cloud resolving calculation (Tomita et al. 2005). The aqua-planet experiment is an idealized case where the earth is assumed to be covered by ocean everywhere. Neale and Hoskins (2000) proposed as a standard experiment for intercomparison of global atmospheric models with moist processes. We used the control distribution of the sea-surface temperature, the ozone distribution, and the equinox solar radiation. The cloud microphysical scheme is based on Grabowski (1998), in which simplified ice phases are introduced. Only two categories of prognostic variables are used in Grabowski (1998): airborne- and sedimentation-hydrometers. Airborne-hydrometer consists of cloud water and cloud ice, while sedimentation-hydrometer represents rain and snow. Liquid and ice phase partition is diagnostically determined and depends only on temperature. Radiation scheme of Nakajima et al. (2000), boundary turbulence of the mixed layer scheme of Mellor-Yamada level 2, and the surface scheme of Louis (1979) are used. The 3.5 km-mesh experiment is conducted for 10 days after spin up integrations for 60 days with 14 km-mesh interval run and 20 days with 7 km-mesh interval run. The number of vertical levels is 54 where vertical intervals increase with height.

6.3 Precipitation Distribution

Figure 6.2 shows global distribution of 90 min-averaged precipitation at day 5 of the 3.5 km-mesh experiment of NICAM. Near the equator, more precipitation is seen in the eastern hemisphere (centered at longitude 90°) than in the western hemisphere (centered at longitude −90°), and intertropical convergence zone (ITCZ) is formed around the equatorial area. Since the boundary condition is zonally symmetric in this aqua-planet experiment, there should be no systematic difference between the precipitation rates in the two hemispheres. The precipitation pattern is propagating along the equator and has a planetary scale structure with zonal wave number one similar to the observation (Takayabu et al. 1999). As described in Tomita et al. (2005), the planetary scale structure is propagating eastward with phase speed of about $15\,\mathrm{m\,s^{-1}}$ and looks like a convectively coupled Kelvin wave. Within the hemispheric scale precipitating region in the eastern hemisphere, there exist super cloud clusters with horizontal scale of a few thousand kilometers; centers of the super cloud clusters are located at longitudes about 50°, 100°, and −60° in Fig. 6.2. Within the super cloud clusters, cloud clusters are moving westward having horizontal scale of a few hundred kilometers. Within each cloud clusters, 10 km-scale convective systems with mesoscale circulations are represented (Nasuno et al. 2007). This hierarchical structure of tropical convection is very similar to the observed convective system (Nakazawa 1988).

6.4 Precipitation Frequency

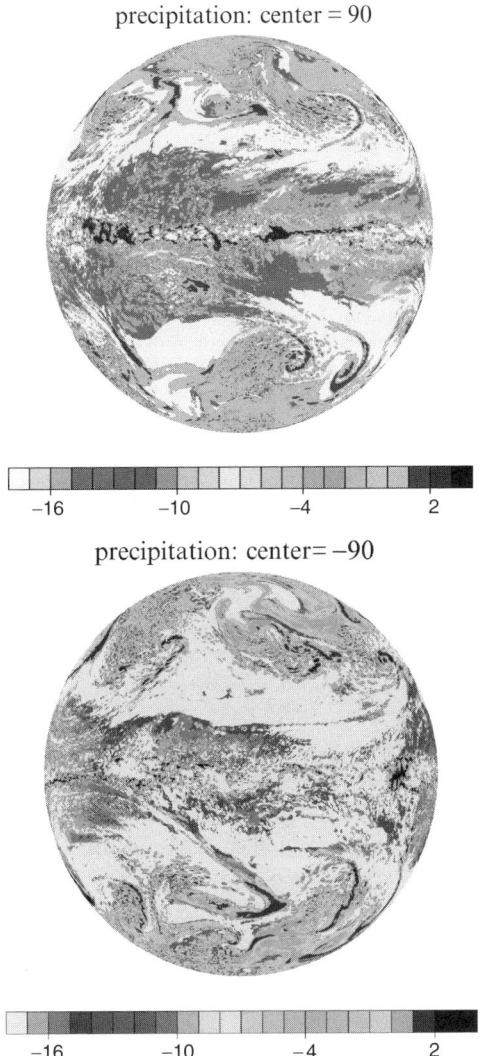

Fig. 6.2 Precipitation for the 3.5 km-mesh experiment. Ninety minutes average between 0:00–1:30 at 85 days. Value is $\log_{10}$ (precipitation [mm h^{-1}]). Longitude of center line is located at 90° (*top*) and −90° (*bottom*). Each corresponds to eastern and western hemispheres of the aqua-planet model, respectively

6.4 Precipitation Frequency

For quantitative evaluation of simulated tropical convection, we investigate statistics of the precipitation by comparing the NICAM data with the TRMM PR data. We use a snapshot data in the region within latitudinal belt of 30°N–30°S at day 5 of

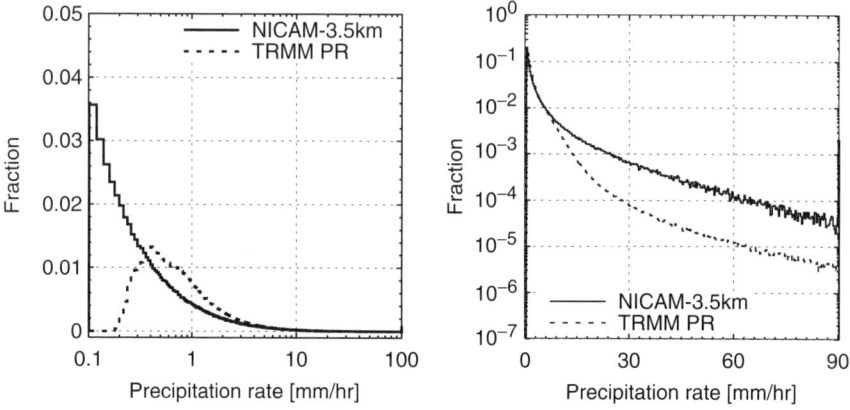

Fig. 6.3 Comparison of the probability distribution functions of rainfall rate between NICAM (*solid*) and TRMM PR data (*dash*) with 0.02 mm h^{-1}-bin in log-scale (*left*) and with 0.3 mm h^{-1}-bin in linear scale (*right*)

the NICAM simulation. The numerical data are interpolated at the latitude–longitude grid points with equidistant interval of 0.039° (about 4 km), rather than the original icosahedral grid points. There are 10,240 points along the equator for the interpolated data. For the TRMM PR data, we analyze the ocean region in the domain of longitudes 150°E–120°W and latitudes 10°S–10°N for six year period from 1998 to 2003. Although we separately analyze El Nino/La Nina/Normal year periods, overall profiles are similar; we only show the result of the normal years (June 2000 – May 2002 and Jan. 2003 – Dec. 2003). The nadir resolution of the TRMM PR data is about 4 km, which is close to the grid interval of the NICAM simulation.

Figure 6.3 compares the precipitation frequency of NICAM and that of the TRMM PR data. Bin size of (a) is 0.02 mm h^{-1}, while that of (b) is 0.3 mm h^{-1}. The vertical axis shows the ratio of the number of columns of each precipitation bin to the number of total rain columns. As shown in (a), the frequency of TRMM PR becomes less as the rain rates become lower; this comes from the fact that rainfall rates less than 0.3 mm h^{-1} are not reliable for the TRMM PR data. Note that the relative fraction, or the slope of the curves, is informative in Fig. 6.3, rather than the absolute value of the fraction. The number of total rain columns is different between the TRMM PR and the NICAM data, since weak rain less than 0.3 mm h^{-1} is hardly detectable in the TRMM PR data, while NICAM counts much weaker rain less than 0.3 mm h^{-1}. The comparison between NICAM and the TRMM PR data in Fig. 6.3 (b) shows that the slopes of two curves are very similar in the heavy precipitation region (>20 mm h^{-1}), while those in the weak precipitation region (<20 mm h^{-1}) is different. That is, the TRMM PR data contains weaker rain columns than the NICAM data, at least in the detectable regime >0.3 mm h^{-1}. Although the NICAM simulation gives heavier rainfall compared to the TRMM PR data, the relative fraction larger than 20 mm h^{-1} has a similar profile.

6.5 Spectral Representations of Rain-Top Height

The NICAM precipitation data is analyzed by the same spectral method given by Takayabu (2002). Spectral representations of rainfall rate are made by sorting precipitation profiles according to rain-top heights. For the detection of rain tops, we use the threshold of $0.3 \, \text{mm} \, \text{h}^{-1}$, which is the same value used in Takayabu (2002). Figure 6.4 shows spectral representations of rain profiles for NICAM and the TRMM PR data. The TRMM PR data is analyzed for the ocean regions.

In Fig. 6.4, spectral representations are produced for convective and stratiform rains. It should be remarked, however, that classification of convective/stratiform rains is not definite for NICAM. For the TRMM PR data, the classification of rain types is based on the PR2a23 algorithm in Takayabu (2002). Note that the result of the TRMM PR in Fig. 6.4 is different from the original figure shown as Fig. 1 in Takayabu (2002) because, in this study, the isolated shallow rain pixels are reclassified from the stratiform to the convective rain adapting the suggestion of Schumacher and Houze (2003); more convective rains are counted particularly for shallow clouds, which would be categorized as stratiform rains by the PR2a23 algorithm. As a result,

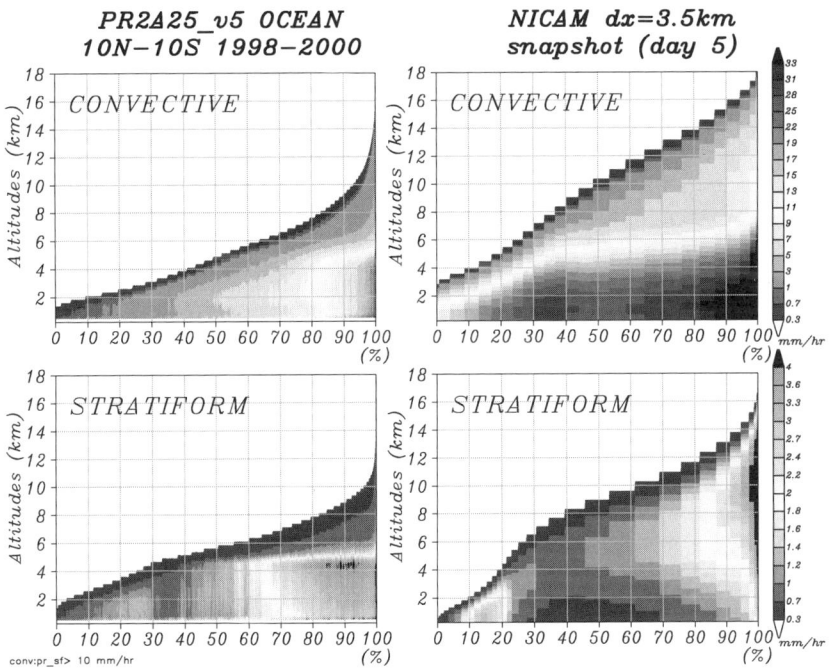

Fig. 6.4 Comparison of spectral representations of rainfall rate between TRMM PR and NICAM. *Top left*: TRMM PR, convective rains over ocean; *bottom left*: TRMM PR, stratiform rains over ocean; *top right*: NICAM, convective rains; *bottom right*: NICAM, stratiform rains. Rainfall rate of NICAM is defined by the sum of rain and snow fall rate relative to the air.

more shallow convective rain appears in Fig. 6.4 compared to Fig. 1 of Takayabu (2002). For example, the ratio of shallow convective rains with rain-top height less than 5.2 km (corresponding to the bright band level) is 36 % in Fig. 6.1 of Takayabu (2002), while it is increased to 54% in Fig. 6.4. Since the stratiform rains are more populated, decrease in the ratio of shallow stratiform rains is small: 43 % vs 38 %, because of the larger total number of pixels.

As for the NICAM data, classification of convective/stratiform rains is rather arbitrary. The same algorithm as the TRMM PR data could be applied to the simulated data (TRMM Precipitation Radar Team, 2005); it requires detection of the bright band and radar echo intensity. For numerical results, radar echo is indirectly obtained from the concentration of rain drop using an empirical formula (Jorgensen et al. 1997; Montmerle et al. 2000). Other heuristic methods for classification can be used; Lang et al. (2003) investigated six kinds of classification method and found that the resulting properties are qualitatively similar. We have tested different classification methods according to diagnostically calculated radar echo, rainfall intensity at the ground, and maximum vertical velocity: convective rains are defined (1) when the calculated radar echo is $Z_{max} > 39$ dBZ, which is the same threshold used in the TRMM PR algorithm, (2) when the rainfall rate at the ground exceeds 10 mm h^{-1} (Grabowski 2003), and (3) when the maximum vertical velocity exceeds 1 m s^{-1}. We found that the overall characteristics are similar as described in Lang et al. (2003), though minor quantitative difference appears: more convective rains are detected when the radar echo is used than when the rainfall rate is used. In the following analysis, we take a simple approach following Grabowski (2003); we use the rainfall rate at the ground with a threshold of 10 mm h^{-1} to define convective rains.

In addition to the classification, we should note that the effect of vertical velocity of air is not taken into account for the estimation of rainfall rate of the TRMM PR data. From the numerically simulated data of NICAM, on the other hand, we obtain vertical velocity of air w and terminal velocity of rain relative to the air V_t. Directly comparable value of rainfall rate of the TRMM PR data is the rainfall rate due to the terminal velocity V_t, while the real rainfall rate should be calculated using the absolute rainfall velocity relative to the ground: $V_t - w$. Within strong updraft cores of convective rains, the upward velocity w might be strong enough so that rain drops have upward motion relative to the ground; that is the rainfall rate is negative. Thus, we made two kinds of spectral representations of rainfall rate: one calculated using terminal velocity V_t relative to the air, and the other calculated using absolute velocity relative to the ground $V_t - w$, given by

$$Pr' = V_t(Q_R + Q_S),$$
$$Pr = (V_t - w)(Q_R + Q_S),$$

respectively. $Q_R = \rho q_r$ and, $Q_S = \rho q_s$ are mass concentrations of rain and snow, respectively. For cloud schemes of Grabowski (1998), the sum of rain and snow $Q = Q_R + Q_S$ is a prognostic variable, and terminal velocity V_t for Q is determined by diagnostically calculated ratio of rain and snow. We use the sum of rain

6.5 Spectral Representations of Rain-Top Height   107

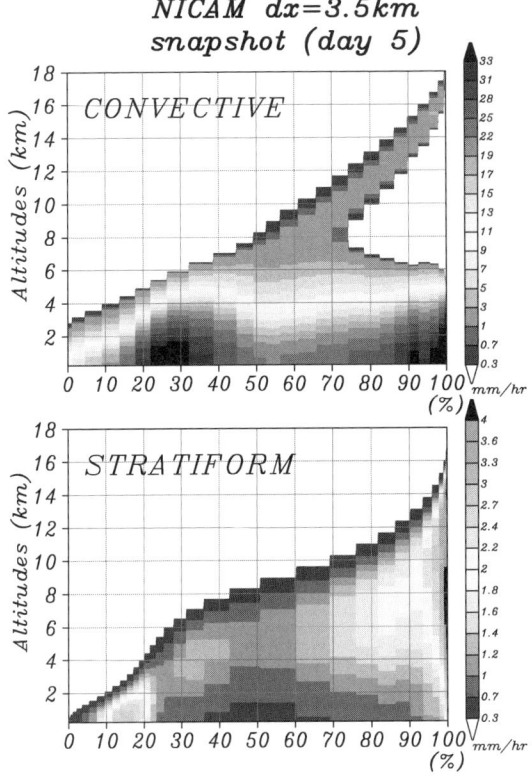

Fig. 6.5 Spectral representations of rainfall rate of NICAM. Rainfall rate is defined by the sum of rain and snow fall rate relative to the ground

and snow concentrations to define rain-top height and rainfall rate. We have calculated corresponding figures using only the rain concentration, and found that the results are completely different from the analysis of the TRMM PR data; it is because the rain and snow partition is diagnostically determined from temperature so that the rain concentration is unevenly distributed in lower layers.

Right panels of Fig. 6.4 are Pr' and two panels of Fig. 6.5 are Pr of NICAM for convective/stratiform rains. The TRMM PR data (Fig. 6.4, left) should be compared to Pr' of NICAM (Fig. 6.4, right). From this comparison, we found the following characteristics:

1. The rain-top height of NICAM is generally higher and the rainfall rate is stronger than that of TRMM. The ratios of shallow clouds with rain-top height less than the bright band level about 5.2 km are 23% for convective rains and 27% for stratiform rains. These ratios are much smaller than those of TRMM.
2. The rain profiles suddenly decreases above the bright band level for both convective and stratiform rains of TRMM, while NICAM has a similar jump only for

convective rains and stratiform rains of NICAM show upward increases of rain at the bright band level for the rain-top height $z = 4$–10 km; the rainfall rate of stratiform rains increases with height near the bright band level, and it becomes smaller below 5 km. In particular, the stratiform rain profile for the rain-top height $z = 8$–10 km has a maximum near the bright band level both for TRMM and NICAM.

3. Rainfall rate in the lower level has two peaks with respect to the abscissa at around rain-top heights $z < 4$ km and $z > 10$ km (cumulative frequency about 10–20% and >80%, respectively). These peaks are commonly seen for the stratiform rain rates of both NICAM and TRMM.

Figure 6.5 shows spectral representations of rainfall rate relative to the ground. In this case, rain tops are detected when rainfall rate relative to the ground exceeds 0.3 mm h^{-1}. Upper level rainfall rate is negative for deep convective rains with rain-top height $z > 12$ km. This results from strong vertical velocity in convective cores, where rain/snow particles are blown upward. They will fall down in surrounding areas as stratiform rains. The above mentioned effect of upward velocity is not considered in the TRMM PR data. It is one of the advantages of using numerical models to gain information of quantities that cannot easily be retrieved from observation.

Figure 6.6 shows spectral representations of concentrations of rain and snow $q_R + q_S$ (upper left), vertical velocity w (lower left), total precipitation (sum of convective and stratiform rains: upper right), and total precipitation relative to the air (lower

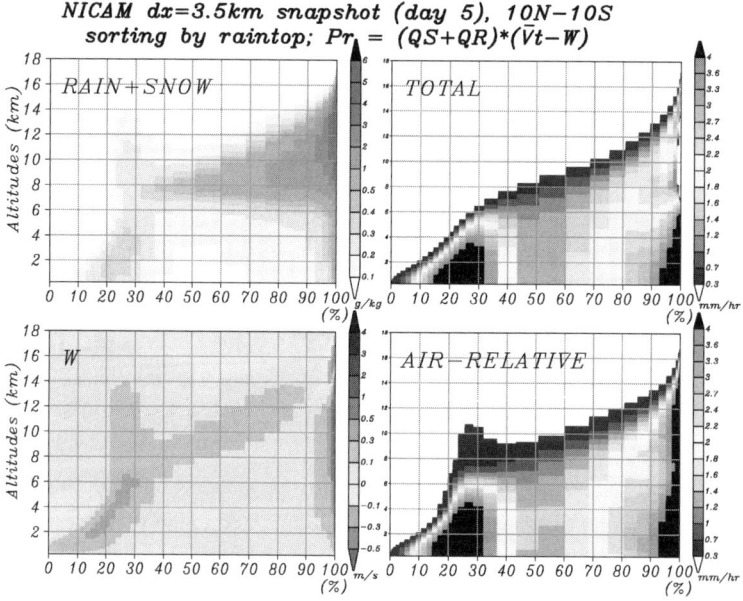

Fig. 6.6 Spectral representations of the sum of rain and snow concentrations (*top left*), vertical velocity (*bottom left*), total precipitation relative to the ground (*top right*), and total precipitation relative to the air (*bottom right*)

right), sorted against to rain-top heights using the threshold of rainfall rate relative to the ground (Fig. 6.5). For rain-top height $z = 4$–5 km, it is interesting that a strong upward motion with rain/snow particles exists above the rain-top height. It seems that this kind of rain system with the rain-top height $z = 4$–5 km is independent of the upward motion above the rain-top height; the upper system is located on the periphery of anvil clouds. In the upper layer of deep convective rains, rainfall rate could be negative because of upward motion $w - V_t > 0$. Rainfall rate is generally smaller in the upper layer than that in the lower layer, though large amounts of snow exist in the upper layer. This is because sedimentation rate of snow is small. Note that we do not consider graupel that has large sedimentation speed in cloud physical scheme.

Two peaks of rainfall rate below the bright band level are both seen in NICAM and TRMM stratiform rains: rain-top height $z > 10$ km and $z = 4$–5 km. From the NICAM results, we conjecture that these peaks represent two typical types of rain: cold rain and warm rain. In the case of cold rain ($z > 10$ km), deep convective cores with large amount of snow produce precipitation peak in the lower level. In the case of warm rain ($z = 4$–5 km) (i.e., below the bright band level 0°C), precipitation is mainly produced from liquid clouds. For the rain-top height $z = 4$–10 km, which occurs in large frequency, peak level of precipitation is in the mid-troposphere. Such distribution implies snow in these regions originates from deep convection in the vicinity. Snow area almost corresponds to the area of low OLR, and rain area to the intense surface precipitation. In general, snow area is wider than rain area, so that averaging over both areas will enhance the characteristics of snow area. One of the reasons for higher rain-top height of NICAM relative to that of TRMM is that the rain-top height of NICAM is defined by using total precipitation intensity of rain and snow. Snow particles are not detectable by the TRMM PR.

6.6 Summary

We have developed a global cloud resolving model called NICAM (Nonhydrostatic ICosahedral Atmospheric Model) at Frontier Research Center for Global Change, JAMSTEC. The model used the icosahedral grid and a nonhydrostatic dynamical core that guarantees conservation of mass and energy, and is suitable for long-term statistical experiments. The first global cloud resolving simulation with horizontal mesh interval 3.5 km was performed under an idealize aqua-planet condition. The numerical results show multiscale structure of tropical convection from meso-scale circulation to global scale convectively coupled Kelvin waves.

In this study, we studied statistical properties of precipitation and compared with the analysis of the TRMM PR data. Since the experiment is under an idealize condition on global ocean, we compared the numerical results with the TRMM PR data over ocean. The grid interval of NICAM is about 3.5 km is close to the nadir resolution of TRMM PR, 4 km. Although the effective resolution of numerical models is much larger than the grid interval (Bryan et al. 2003; Skamarock 2004), we do not introduce any corrections to the statistics of NICAM results and directly compared with the TRMM PR data.

Comparison of probability distribution functions of rainfall rate shows that relative occurrence of heavy rain (>20 mm h^{-1}) is similar between NICAM and TRMM, while weak rain (<20 mm h^{-1}) is less frequent for NICAM than for the TRMM PR data. This result implies that shallow clouds with weaker rain are not well resolved in NICAM and that NICAM tends to produce stronger rainfall rate than the observations.

Next, we investigate spectral representations of rainfall rate for convective and stratiform rains according to the rain-top heights. We used the cloud scheme proposed by Grabowski (1998), where one category is used for sedimentation-hydrometers of prognostic variables, so that precipitation represents sum of rain and snow fall. The ratio of shallow clouds of rain-top height less than the bright band level about 5 km is much smaller for NICAM than for TRMM, and NICAM generally shows higher rain-top rains. Rainfall rate of NICAM is much stronger particularly for convective rains. These results suggest that NICAM with 3.5 km-grid interval requires suitable sub-grid scheme to obtain quantitatively reasonable rainfall rate. Similar spectrum properties were calculated by Shige et al. (2004) using the Goddard Cumulus Ensemble (GCE) model, though they ascribed the stronger precipitation of their model results to regional differences. For stratiform rains, NICAM and the TRMM PR data have a similar behavior of spectral representations where rainfall rate has two peaks at shallow rain-top height and deep rain-top height, indicating successful representation of both warm and cold rain systems. As one of the advantages of using the numerical results, we analyzed corresponding spectral representations of vertical velocity and ice phase concentrations, which are difficult to be retrieved from observations. We found that rainfall rate in the upper layer of deep convective rains is negative due to strong updraft velocity, whose effect is not considered in the TRMM PR algorithm. Although model results might be sensitive to choice of classification method of convective and stratiform rains, the above-mentioned results are qualitatively unchanged (Lang et al. 2003).

This study is a first comparison between the global cloud resolving data (NICAM with 3.5 km-grid interval) and the high-resolution satellite data (TRMM PR). As a starting point, we showed statistical properties of precipitation. It is a big step for numerical models that direct comparison of the precipitation statistics between the numerical results and the satellite data became available using high-resolution wide area numerical simulations such as a global cloud resolving simulation of this study. This kind of approach will reveal new scientific aspects of precipitation systems, and lead to improvements of numerical models and satellite data analysis algorithms, though it is not straightforward because results might be sensitive to threshold values. In particular, statistical comparisons of precipitation systems will give useful information for improvements of cloud microphysical schemes, which are one of the most ambiguous ingredients of cloud resolving models.

Acknowledgments We appreciate Prof. K. Nakamura and Mr. T. Sasaki for providing the TRMM PR analysis data. Numerical experiments of NICAM were done using the Earth Simulator. This research was supported by Core Research for Evolutional Science and Technology, Japan Science and Technology Agency (CREST, JST).

References

Bryan, G. H., J. C. Wyngaard, and J. M. Fritsch, 2003: Resolution requirements for the simulation of deep moist convection. *Mon. Wea. Rev.*, **131**, 2394–2416.
Grabowski, W. W., 1998: Toward cloud resolving modeling of large-scale tropical circulation: A simple cloud microphysics parameterization. *J. Atmos. Sci.*, **55**, 3283–3298.
Grabowski, W.W., 2003: Impact of ice microphysics on multiscale organization of tropical convection in two-dimensional cloud-resolving simulations. *Q. J. R. Meteorol. Soc.*, **129**, 67–81.
Jorgensen, D.P., M.A. LeMone and S. Trier, 1997: Structure and evolution of the 22 February 1993 TOGA COARE squall line: Aircraft observations of precipitation, circulation, and surface energy fluxes. *J. Atmos. Sci.*, **54**, 1961–1985.
Lang, S., W.-K. Tao, J. Simpson, and B. Ferrier, 2003: Modeling of convective-stratiform precipitation processes: Sensitivity to partitioning methods. *J. Appl. Meteor.*, **42**, 505–527.
Louis, J. F., M. Tiedke, and J.-F. Geleyn, 1982: A short history of the PBL parameterization at ECMWF. Workshop on Planetary Boundary layer Parameterization, ECMWF, Reading, U.K. 59–80.
Miura, H., H. Tomita, T. Nasuno, S. Iga, M. Satoh, T. Matsuno, 2005: A climate sensitivity test using a global cloud resolving model under an aqua planet condition, *Geophys. Res. Lett.*, **32**, L19717, doi:10.1029/2005GL023672.
Montmerle, T., J.-p. Lafore and J.-L. Redelsperger, 2000: A tropical squall line observed during TOGA COARE: Extended comparisons between simulations and Doppler radar data and the role of midlevel wind shear. *Mon. Weath. Rev.*, **128**, 3709–3730.
Nakajima, T., M. Tsukamoto, Y. Tsushima, A. Numaguti, and T. Kimura, 2000: Modeling of the radiative process in an atmospheric general circulation model. *Appl. Opt.*, **39**, 4869–4878.
Nakazawa, T., 1988: Tropical super clusters within intraseasonal variations over the western Pacific. *J. Meteor. Soc. Japan*, **66**, 823–839.
Nasuno, T., H. Tomita, S. Iga, H. Miura, M. Satoh, Multi-scale organization of convection simulated with explicit cloud processes on an aqua planet. *J. Atmos. Sci.* **64**, 1902–1921.
Neale, R.B., and B.J. Hoskins, 2000: A standard test for AGCMs including their physical parameterizations: the proposal. *Atmos. Sci. Lett.*, **1**, doi: 10.1006/asle.2000.0019.
Randall, D. A., M. Khairoutdinov, A. Arakawa, W. Grabowski, 2003: Breaking the cloud-parameterization deadlock. *Bull. Amer. Meteor. Soc.*, **84**, 1547–1564.
Satoh, M., H. Tomita, H. Miura, S. Iga, and T. Nasuno, 2005: Development of a global cloud resolving model – A multi-scale structure of tropical convections. *J. Earth Simulator*, **3**, 11–19.
Satoh, M., T. Matsuno, H. Tomita, H. Miura, T. Nasuno, and S. Iga, 2007: Nonhydrostatic Icosahedral Atmospheric Model (NICAM) for global cloud resolving simulations. *J. Comp. Phys.*, doi: 10.1016/j.jcp.2007.02.006.
Schumacher, C. and R. A. Houze Jr., 2003: The TRMM precipitation radar's view of shallow, isolated rain. *J. Appl. Meteor.*, **42**, 1519–1524.
Shige, S., Y. Takayabu, W.-K. Tao, and D. E. Johnson, 2004: Spectral retrieval of latent heating profiles from TRMM PR Data. Part I: Development of a model-based algorithm. *J. Appl. Meteor.*, **43**, 1095–1113.
Skamarock, W. C., 2004: Evaluating mesoscale NWP models using kinetic energy spectra. *Mon. Wea. Rev.*, **132**, 3019–3032. doi: 10.1175/MWR2830.1.
Takayabu, Y. N., 2002: Spectral representation of rain profiles and diurnal variations observed with TRMM PR over the equatorial area. *Geophys. Res. Let.*, **29**, doi:10.1029/2001GL014113.

Takayabu, Y. N., T. Iguchi, M. Kachi, A. Shibata, and H. Kanzawa, 1999: Abrupt termination of the 1997–98 El Nino in response to a Madden-Julian oscillation. *Nature*, **402**, 279–282.

Tomita, H. M. Tsugawa, M. Satoh, and K. Goto, 2001: Shallow water model on a modified icosahedral geodesic grid by using spring dynamics. *J. Comput. Phys.*, **174**, 579–613.

Tomita, H. M. Satoh, and K. Goto, 2002: An optimization of the icosahedral grid modified by the spring dynamics. *J. Comput. Phys.*, **183**, 307–331.

Tomita, H. and M. Satoh, 2004: A new dynamical framework of nonhydrostatic global model using the icosahedral grid. *Fluid Dyn. Res.*, **34**, 357–400.

Tomita, H., H. Miura, S. Iga, T. Nasuno, and M. Satoh, 2005: A global cloud-resolving simulation : Preliminary results from an aqua planet experiment. *Geophys. Res. Lett.*, **32**, L08805, doi:10.1029/2005GL022459.

TRMM Precipitation Radar Team, 2005: Tropical rainfall measuring mission (TRMM) precipitation radar algorithm instruction manual for version 6, Japan Aerospace Exploration Agency (JAXA) and National Aeronautics and Space Administration (NASA), 175 pp., available at http://www.eorc.nasda.go.jp/TRMM/document/text/TRMM_V6.pdf.

Chapter 7
Global Warming Projection by an Atmospheric General Circulation Model with a 20-km Grid

Akira Noda, Shoji Kusunoki, Jun Yoshimura, Hiromasa Yoshimura, Kazuyoshi Oouchi, and Ryo Mizuta

Summary A global warming projection was conducted on the Earth Simulator by using a very high horizontal resolution atmospheric general circulation model with 20-km grid. Tropical cyclones (TCs) and the rain band (Baiu) during the East Asian summer monsoon season are selected as the main targets of this study, because these bring typical extreme events but so far the global climate models have not given reliable simulations and projections due to their insufficient resolutions. The model reproduces TCs and a Baiu rain band reasonably well under the present-day climate conditions. In a warmer climate at the end of this century, the model projects, under the IPCC SRES A1B scenario, that the annual mean occurrence number of TCs decreases by about 30% globally (but increased in the North Atlantic) and TCs with large maximum surface winds increase. The Baiu rain band activities tend to intensify and last longer until August, suggesting more damages due to heavy rainfalls in a warmer climate.

7.1 Introduction

Meteorologists and climatologists have been trying to answer the question as to how the tropical cyclones (TCs), including typhoons and hurricanes, will behave differently in a future warmer climate with increased greenhouse gases. Considering that TCs are potentially devastating to human life and society, the projection of their occurrence in future warmer climates is critical. The warming of the global sea surface temperature (SST) leads to an increase in moisture amount in the lower atmosphere, which contributes to enhance the release of latent heat, a potential as the energy source of TCs. Thus, it is generally recognized that the future climate changes are favorable for TC development.

However, comprehensive and reliable projections of TCs remain challenging. Therefore, a climate model capable of handling the complexities of various physical processes is necessary. Former projections of TCs derived from traditional models are controversial, and most have coarse horizontal resolutions limited to 100 or a few 100 km, which are inadequate for modeling TCs (Henderson-Sellers et al. 1998; Walsh and Ryan 2000). As a way to ameliorate these inadequacies, a regional nesting approach with high-resolution atmospheric models has been introduced to simulate more realistic TCs (Knutson et al. 1998; Knutson and Tuleya 2004). Nevertheless, due to the intrinsic nature of regional modeling, these tactics have a fatal defect, that is, the target area is limited. Consequently, it is obvious that comprehensive global analysis and evaluation of TCs are problematic with a regional modeling approach.

Our tactics is to break through the dead-end situation; a global warming projection was conducted with the use of a 20-km-mesh atmospheric general circulation model (AGCM) and the Earth Simulator (Habata et al. 2003, 2004). The Earth Simulator is one of the fastest supercomputers in the world. The 20-km-mesh horizontal resolution is unprecedented for use as a global climate model for projecting global warming. In contrast to the less-sophisticated models, this one reveals detailed structures of TCs, including eyes, eyewalls, and spiral rainbands (see, e.g., Fig. 1 of Oouchi et al. 2006). Therefore, the model is likely to provide more reliable projections of TCs.

Another target is the rainy season or Baiu rain band in the East Asia summer monsoon season. It is called the Mei-yu in China and the Changma in Korea. The Baiu has a great impact in East Asia, as do TCs. Many AGCM studies have attempted to reproduce the Baiu rain band by forcing the models with observed or climatological SST. However, the models have commonly underestimate the rainfall (Lau et al. 1996; Lau and Yang 1996; IPCC 2001; Liang et al. 2001; Kusunoki et al. 2001). Moreover, they fail to simulate the observed northward migration of the Baiu rain band (Kang et al. 2002). Research on the horizontal resolution dependence with respect to the reproducibility of the Baiu rain band has yielded optimistic results, namely, that a higher resolution model increases the rainfall amount of the Baiu rain band, while lower resolution models do not. Nevertheless, even the highest resolution simulation does not simulate the Baiu rain band realistically (Sperber et al. 1994; Kawatani and Takahashi 2003; Kobayashi and Sugi 2004). An observational study (Ninomiya and Akiyama 1992) stressed the fact that the disturbances of multiscales from mesoscale to synoptic and planetary scales play important roles in the formation and maintenance of the Baiu rain band and that the Baiu rain band is characterized by mutual interactions between phenomena within this multiscale hierarchical system. This suggests the difficulty in simulating the Baiu rain band by AGCM is still attributable to an insufficient horizontal resolution. Hence, we have attempted to evaluate the ability of the very high resolution 20-km-mesh AGCM to simulate the present-day climatology of the Baiu rain band and project its change in a warmer climate.

This chapter summarizes the recent results about tropical cyclones and the Baiu rain band from a time-slice global warming projection conducted using a very high horizontal resolution AGCM with 20-km grid developed and optimized on the Earth Simulator (Mizuta et al. 2006; Oouch et al. 2006; Kusunoki et al. 2006). The models and experimental design used for the experiment are described in Sec. 7.2. Major findings and results are shown in Sec. 7.3. Finally, discussion and concluding remarks are given in Sec. 7.4.

7.2 Methods

7.2.1 Models

High horizontal resolution AGCM experiments were achieved by adopting the "time-slice" method (Bengtsson et al. 1996; IPCC 2001), which is a two-tier global warming projection approach using an atmosphere-ocean general circulation model (AOGCM) and an AGCM with a horizontal resolution that is higher than that of the atmospheric part of the AOGCM. The two models used in this study are briefly described in the following.

The AOGCM used in the first step of the time-slice experiment is an MRI-CGCM2.3 (Yukimoto et al. 2006a). The atmospheric part of this model has a horizontal spectral truncation of T42 corresponding to about a 270-km horizontal grid spacing and has 30 levels with a 0.4-hPa top. The oceanic part of this model is a Bryan-Cox-type grid. The horizontal grid spacing is $2.5°$ in longitude and $2°$ in latitude. To resolve equatorial Kelvin and Rossby waves, the latitudinal grid spacing is decreased near the equator between $4°S$ and $4°N$ down to the minimum of $0.5°$ at the equator. In the sea ice model, compactness and thickness are predicted based on thermodynamics. The sea ice is advected by the surface ocean current. To keep the model climatology close to the observation, a flux adjustment technique was applied to the heat and freshwater globally and to the wind stress near the equator.

The 20-km-mesh AGCM used in the second step of the time-slice experiment is a unified model of the Meteorological Research Institute (MRI)/Japan Meteorological Agency (JMA) based on an operational numerical weather prediction model of JMA (JMA-GSM0103; JMA 2003). The Arakawa-Schubert scheme (Arakawa and Schubert 1974) with prognostic closure similar to that of Randall and Pan (1993) was employed for deep, moist convection. The scheme was further improved to include the entrainment and detrainment effects between the cloud top and cloud base in convective downdraft instead of reevaporation of convective precipitation (Nakagawa and Shimpo 2004). A level-2 turbulence closure scheme, based on Mellor and Yamada (1974), was used for vertical diffusion. The surface turbulent fluxes were calculated by the bulk formulae following the Monin–Obukhov similarity theory. The radiation and land surface schemes in JMA-GSM0103 were modified to be consistent with MRI-CGCM2.3 (Mizuta et al. 2006). The time integration was accelerated by introducing a semi-Lagrangian scheme (Yoshimura and Matsumura 2005). The model has a horizontal spectral truncation of TL959 corresponding to about a 20-km horizontal grid spacing and has 60 levels with a 0.1-hPa top. It was found that a mere increase in the horizontal resolution gives rise to large model biases in precipitation and temperature, much less organization of convection, and suppression of tropical cyclone generation. Therefore, we carefully tuned the model to improve the model's present-day climatology by changing the parameters in the evaporation process, cloud water content diagnosis, vertical transport of horizontal momentum in cumulus, and gravity wave drag (Mizuta et al. 2006). The integration of the 20-km-mesh AGCM and data processing were performed on the Earth Simulator.

7.2.2 *Experimental Design*

The time-slice experiment was conducted as follows. The 20-km-mesh AGCM was integrated for 10 years as a time slice, present-day climate simulation, with observed climatological SSTs (Reynolds and Smith 1994) averaged from 1982 to 1993. The general performance of the simulation is discussed in detail by Mizuta et al. (2006). As a time slice, future climate simulation, the AGCM was integrated for 10 years with projected SSTs and a concentration scenario for greenhouse gases and aerosols from 2080 to 2099. The projected SSTs used for the present experiment were a superposition of the observed SSTs and the differences between the present SSTs (1979–1998, 20-year mean taken from twentieth century climate simulations) and the future SSTs (2080–2099, 20 year mean) (Fig. 7.1), which were obtained from a climate change simulation (Yukimoto et al. 2006b) with the MRI-CGCM2.3 based on the Intergovernmental Panel on Climate Change (IPCC) SRES A1B emission scenario (IPCC 2000). The A1B scenario is an intermediate emission scenario characterized by a future world of very rapid economic growth, global population that peaks in the middle of the twenty-first century and declines thereafter, and by a balanced introduction of new and more efficient technologies of all energy supply (IPCC 2000, 2001). Around the period of 2080 to 2099, with the concentration of CO_2 nearly doubled, the global mean surface air temperature increases by about 2.5° for the MRI-CGCM2.3 simulation. The concentrations of greenhouse gas and aerosols were assumed as those for the year 2090 prescribed by the A1B scenario. The spatial structure of SST change by the MRI-CGCM2.3 simulation resembles an El Niño-like response in which warming is larger in the tropical central and east Pacific than in the west Pacific.

7.3 Results

7.3.1 *Change in Tropical Cyclones*

Oouchi et al. (2006) performed an objective tracking of TCs using model outputs. The criteria for selection of TCs are basically the same as those used by Sugi et al. (2002) and originally in line with those of Bengtsson et al. (1996) as follows:

(1) Across the 45°S–45°N latitudinal belt, the grid point corresponding to a TC-center candidate was defined as the one where the minimum surface pressure

7.3 Results

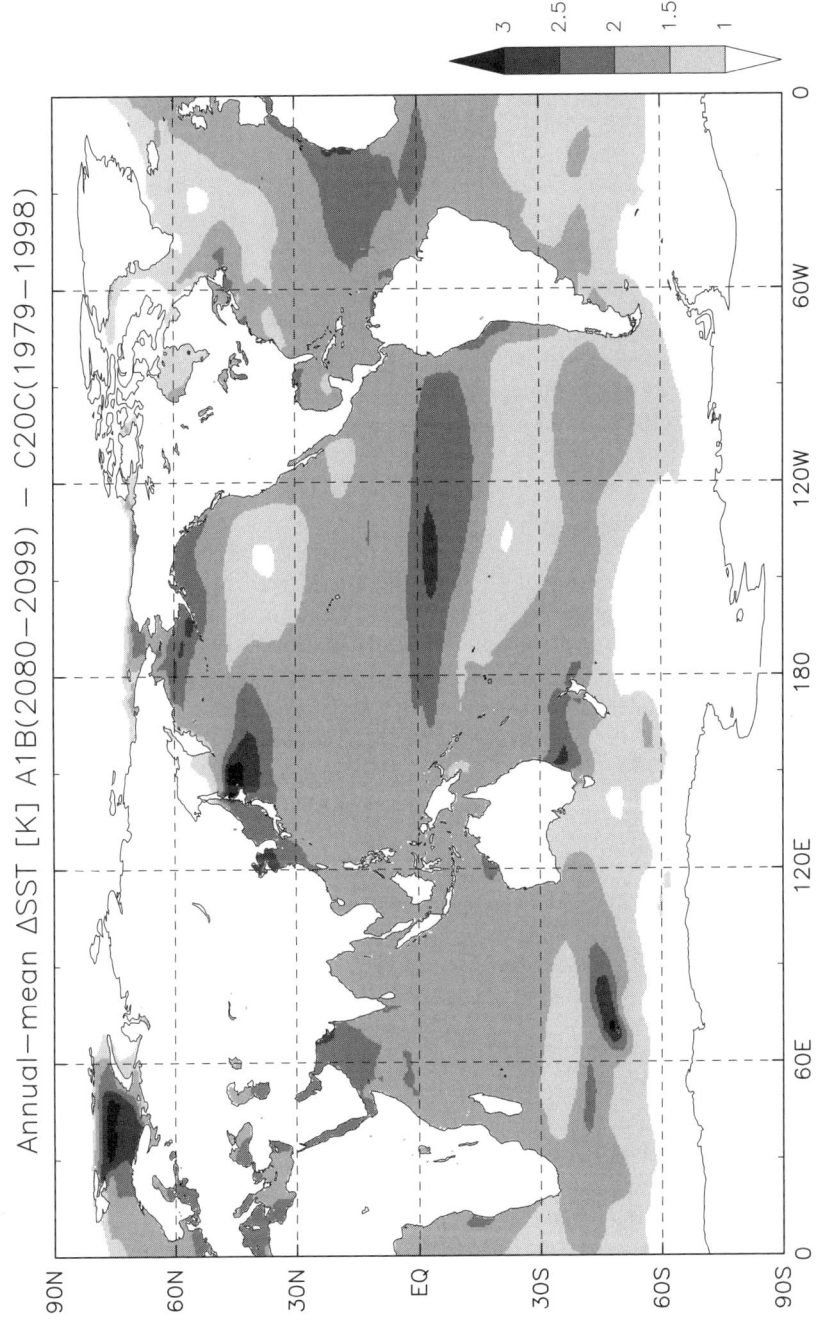

Fig. 7.1 Difference in annual mean sea surface temperature between the future and present-day experiments

is at least 2 hPa lower than the mean surface pressure over the surrounding 7° latitude × 7° longitude grid box.
(2) The magnitude of the maximum relative vorticity at 850 hPa exceeds $3.5 \times 10^{-5}/s$.
(3) The maximum wind speed at 850 hPa is larger than 15 m s^{-1}.
(4) The temperature structure aloft has a marked warm core such that the sum of the temperature deviations at 300, 500, and 700 hPa exceeds 2 K.
(5) The maximum wind speed at 850 hPa is larger than that at 300 hPa.
(6) The duration is not shorter than 36 h.

The data set used in the tracking includes six-hourly, 20-km-mesh original outputs of sea-level pressure, 850- and 300-hPa wind velocities, and temperature fields monitored over grid points between 45°S and 45°N. The initial position of each TC is restricted to oceanic grid points between 30°S and 30°N. For the evaluation of the tracked results, a comparison was made with the observational global TC data set (1979–1998) obtained from the Unisys Corporation website at http://weather.unisys.com/hurricane.

With respect to the number of the simulated TCs, the annual mean global frequency in the present-day climate simulation is almost equivalent to the observation (Global in Table 7.1). The agreement is evident on the basin-wide scale as well, though the model overestimates/underestimates the occurrence in the Southern/Northern hemisphere. In future climate simulations, the number is generally reduced, showing an approximate 30% reduction worldwide. A similar trend is observed across most of the oceanic basins, except for the North Atlantic, where an increase in the TC number is detected.

To check the simulated geographical distribution of the TCs, the simulated tracks were compared with the observation. The tracks obtained from the observation, present-day and future climate simulations are plotted in Fig. 7.2. A general comparison among the panels reveals that the geographical distribution of the simulated tracks in the present-day and future climate simulations resembles those from the observations. This indicates that the geographical distribution of TCs will not be significantly altered in a warmer climate. Figure 7.2 also illustrates that the frequency of TCs is reduced in future climate simulations, which is consistent with Table 7.1.

Another great concern is what to extent the occurrence of devastating TCs would change in a future warmer climate, because an increase in the number of stronger TCs would enhance the risk of natural disasters. To answer this concern, the maximum surface wind of each TC was selected as a measure of the TC intensity. Figure 7.3 shows the annual mean of the simulated TCs for the present-day and future climate simulations together with the observation as a function of the maximum surface wind speed. Although a very high AGCM with a 20 km horizontal resolution was used, the maximum surface winds of the present-day climate simulation are generally lower than those obtained from observations. In view of the preceding investigation suggesting that the TC intensity is sensitive to the detailed treatment of physical processes in the model (Knutson and Tuleya 2004), the underestimation of maximum

7.3 Results

Table 7.1 Frequency of tropical cyclone (TC) formation in the observational data and in the present-day and future simulations (Oouchi et al. 2006)

Regions	Latitudes	Longitudes	Observation	Present	Future
			20 years	10 years	10 years
Global	45 S–45 N	All	83.7 (10.0)	78.3 (8.4)	54.8 (8.4)[a]
Northern hemisphere	0–45 N	All	58.0 (7.1)	42.9 (7.0)	30.8 (5.8)[a]
Southern hemisphere	0–45 S	All	25.7 (5.6)	35.4 (3.8)	24.0 (6.1)[a]
North Indian Ocean	0–45 N	30E–100E	4.6 (2.4)	4.4 (2.5)	2.1 (1.7)[a]
Western North Pacific Ocean	0–45N	100E–180	26.7 (4.2)	12.4 (3.7)	7.7 (2.6)[a]
Eastern North Pacific Ocean	0–45N	180–90W	18.1 (4.8)	20.5 (3.4)	13.5 (5.7)[b]
North Atlantic Ocean	0–45N	90W–0	8.6 (3.6)	5.6 (2.7)	7.5 (1.8)[c]
South Indian Ocean	0–45S	20E–135E	15.4 (3.8)	25.8 (3.0)	18.6 (5.1)[a]
South Pacific Ocean	0–45S	135E–90W	10.4 (4.0)	9.4 (3.3)	5.4 (1.9)[a]

Annual-mean numbers are shown for the globe, the Northern and Southern hemispheres, and six ocean basins conveniently defined on the basis of the latitudes and longitudes indicated in the table. Standard deviations are shown in parentheses. The Student's t-test was applied to the difference between the values of the present-day simulations and those of future simulations

[a] Statistically significant decrease at 99% confidence level
[b] Statistically significant decrease at 95% confidence level
[c] Statistically significant increase at 95% confidence level

surface wind in this case could be attributed partly to the formulation of a cumulus parameterization scheme. Actually, among the parameter tuning processes of the present 20-km-mesh AGCM, the most sensitive process to improve the representation of tropical cyclones was the vertical transport process of horizontal momentum in the convection scheme (Mizuta et al. 2006). Moreover, the possibility cannot be excluded that this discrepancy may originate from insufficient horizontal resolution of the 20-km-mesh model.

In the future climate simulation in Fig. 7.3, TCs of relatively intense classes (>45 m s^{-1}) are found to increase, while those of weak-to-moderate classes are reduced. Similar frequency characteristics were confirmed in 850-hPa wind and surface pressure of TCs (figure not shown). As a measure of extreme events associated with intense TCs, we have monitored the largest maximum surface wind speed attained throughout the lifetime of all TCs for respective regions in respective years and then calculated statistics for 10 years. Table 7.2 compares the largest maximum surface wind speed in the present-day climate simulation with that from the future climate simulation. The most intense TCs in the future climate simulation tend to become much stronger than that in the present-day climate simulation worldwide.

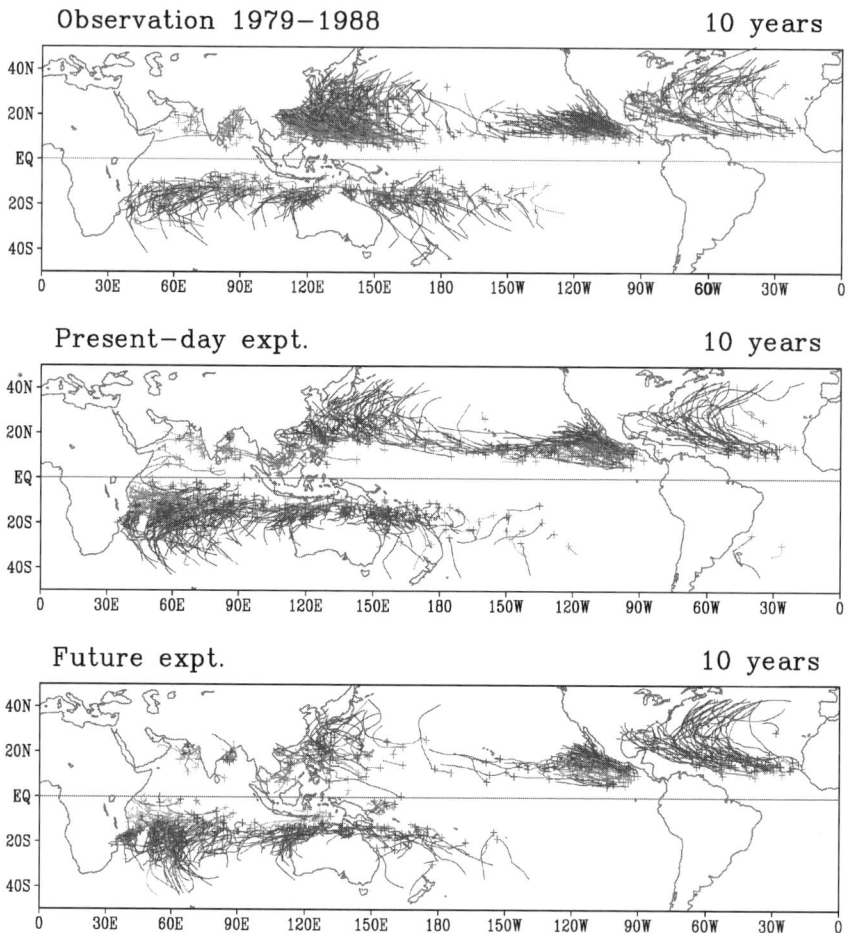

Fig. 7.2 Tropical cyclone tracks of the observational data (*top*), the present-day (*middle*), and the future climate experiments (*bottom*). The initial positions of tropical cyclones are marked with "*plus*" signs. The tracks detected at different seasons of each year is in *different colors* (*blue* for January, February, and March; *green* for April, May, and June; *red* for July, August, and September; *orange* for October, November, and December) (Oouchi et al. 2006)

7.3.2 Change in Baiu Rain Band

Kusunoki et al. (2006) and Yasunaga et al. (2006) analyzed the change in Baiu rain band in the time-slice experiment. First we examined the model's ability to reproduce the present-day climatology of precipitation. Figure 7.4a shows the seasonal marching of observed pentad precipitation climatology averaged for the longitudinal sector including Japan (Kusunoki et al. 2006). In the beginning of June, an area of

7.3 Results

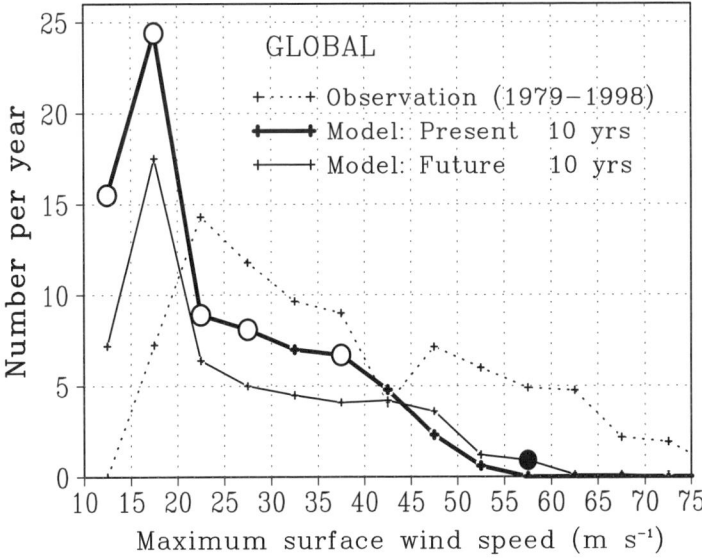

Fig. 7.3 Frequency distribution of TCs shown as a function of the maximum surface wind speed (Oouchi et al. 2006). The abscissa is the largest maximum surface wind speed attained through the lifetime of each TC. The ordinate is the annual-mean number of TCs. For the difference between the present and the future results, the plot at the 95% statistical significance level is marked with an *open* or *closed circle* (according to the two-sided Student's *t*-test). Note that the plots from the observation (*dotted lines*) include no tropical cyclones with the maximum surface wind less than 17.2 m s^{-1}

large precipitation is observed around 25–30°N. This heavy rainfall area then gradually migrates northward until the middle of July and disappears afterward. This northward-migrating rain band corresponds to the Japanese rainy season, Baiu. The present-day climate simulation (Fig. 7.4b) generally well reproduces the seasonal march of the Baiu rain band, although the model generally underestimates rainfall, especially through the end of May to the beginning of June. On the other hand, the future climate simulation (Fig. 7.4c) and the change (Fig. 7.4d) show that, in the middle of June, the position of the Baiu rain band shifts to the north compared to the present-day position. The northward migration of the Baiu rain band is not as evident as in the present-day climate, and the Baiu rain band tends to stagnate around 30°N till the beginning of August with a large precipitation increase in mid-July. This means that the termination of the Baiu period is delayed until August, later than the present-day termination in the middle of July (Fig. 7.4a).

The geographical distribution of the observed climatological precipitation and 850-hPa wind for July is shown in Fig. 7.5a. The model accurately reproduces the observed large rainfall over the western part of Japan and the Korean peninsula associated with the clockwise 850-hPa level wind over the subtropical high to the south of Japan (Fig. 7.5b). It is particularly noteworthy that the model succeeded in simulating the amount of precipitation over Western Japan and the Korean peninsula

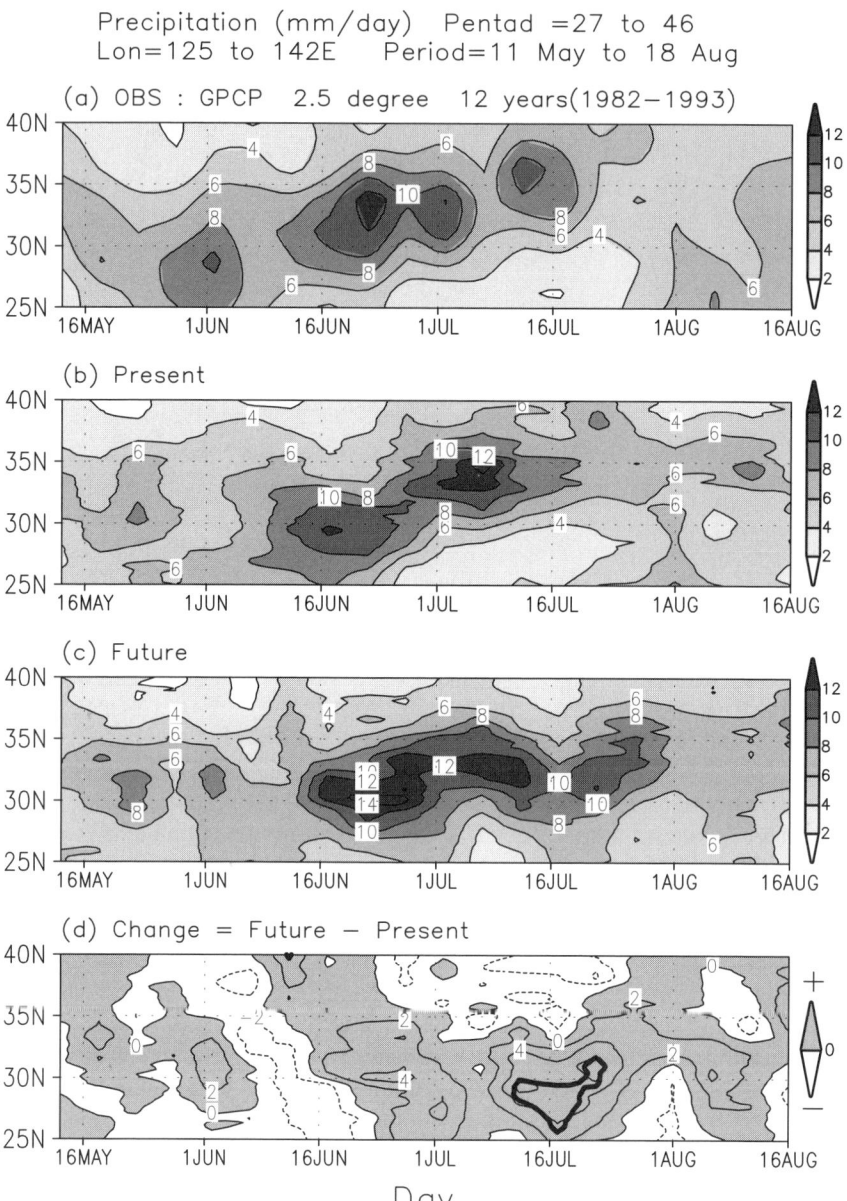

Fig. 7.4 Time-latitude cross section of climatological pentad precipitation (mm day^{-1}) along the longitudinal sector averaged for 125°E–142°E. The period is from pentad 27 (11–15 May) to 46 (14–18 Aug.). (**a**) Observation based on the global precipitation climatology project (GPCP) 2.5° data (Adler et al. 2003) for 12 years from 1982 to 1993. (**b**) Present-day climate simulation for 10 years. (**c**) Future climate simulation for 10 years. (**d**) Change as future (**c**) minus present-day climatology (**b**). Positive values are shaded. The *thick contour* shows a 90% significance level

7.3 Results

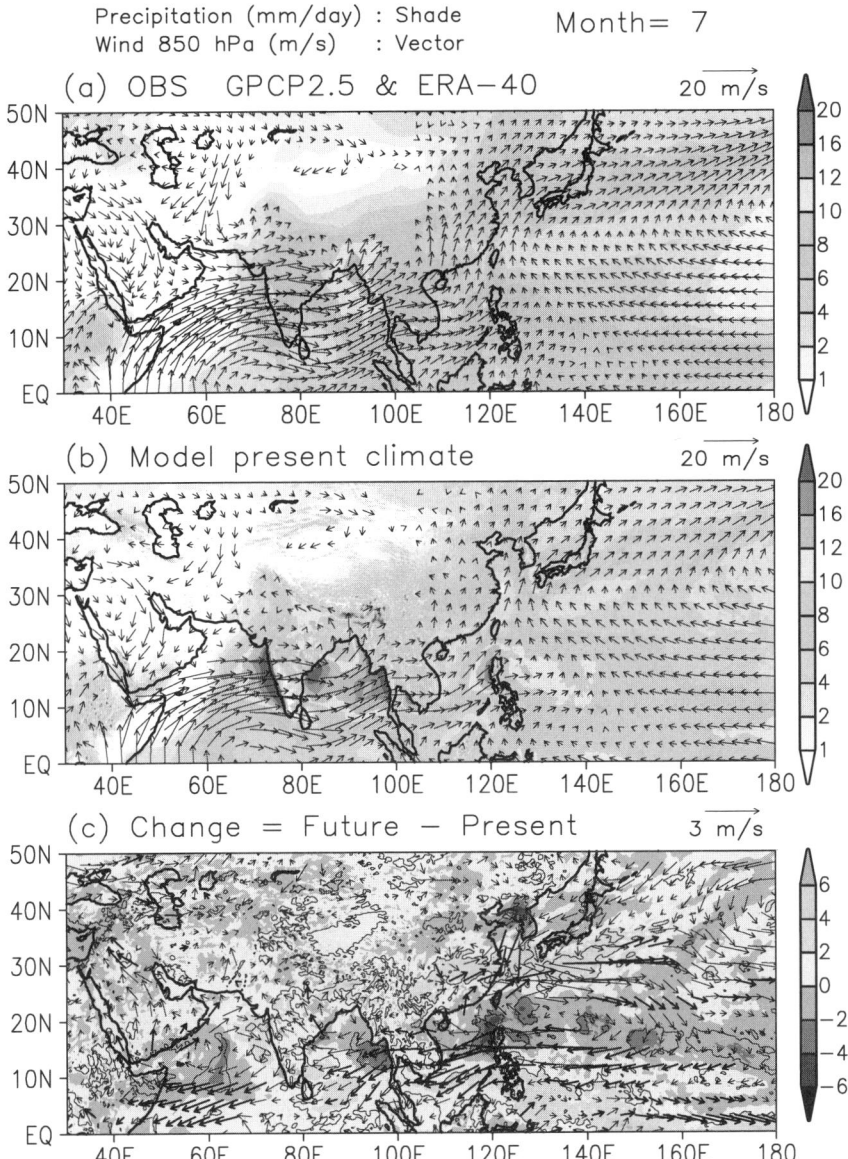

Fig. 7.5 Distribution of climatological precipitation (*color*, mm day^{-1}) and 850-hPa wind vector (*arrow*, m s^{-1}) for July. (**a**) Observed precipitation by GPCP 2.5° data (Adler et al. 2001) for 12 years from 1982 to 1993. Observed wind by ERA-40 data (Simmons and Gibson 2000) for 30 years from 1971 to 2000. (**b**) Model's present-day climate simulation. (**c**) Change as future minus present-day simulation. The *contour* and *thick arrow* show a 90% significance level of the precipitation and wind, respectively

Table 7.2 Maximum surface wind speed (m s^{-1}) of TCs in the present-day and the future simulations (Oouchi et al. 2006)

Regions	Present	Future
	10 years	10 years
Global	49.3 (2.2)	56.0 (5.3)[a]
Northern hemisphere	47.0 (5.4)	54.2 (6.4)[b]
Southern hemisphere	47.6 (2.3)	50.8 (4.5)[c]
North Indian Ocean	30.2 (9.7)	25.2 (14.6)
Western North Pacific Ocean	45.8 (5.0)	44.8 (12.0)
Eastern North Pacific Ocean	39.9 (4.6)	37.9 (7.8)
North Atlantic Ocean	43.4 (7.1)	52.2 (4.9)[b]
South Indian Ocean	46.5 (2.7)	50.3 (5.0)[c]
South Pacific Ocean	44.0 (3.7)	34.1 (9.9)[d]

For each year, the largest maximum surface wind speed attained throughout the lifetime of all TCs was selected, and the values for 10 years were then averaged. Standard deviations are shown in parentheses. The definition of the region is the same as that in Table 7.1. The two-sided Student's t-test was applied to the difference between the values of the present-day climate simulations and those of future climate simulations

[a] Statistically significant increase at 99% confidence level
[b] Statistically significant increase at 95% confidence level
[c] Statistically significant increase at 90% confidence level
[d] Statistically significant decrease at 95% confidence level

as well as the seasonal march, because neither AGCMs nor AOGCMs have ever realistically simulated such spatial structure and seasonal marching of the Baiu rain band. The change in precipitation and 850-hPa wind due to global warming is illustrated in Fig. 7.5c. Precipitation increases over the Yangtze River valley of China, the East China Sea, Western Japan and ocean to the south of Japanese archipelago. On the contrary, precipitation decreases over the Korean peninsula and Northern Japan. Strengthening of the clockwise circulation is evident over a subtropical high (20°N, 120–150°E), which means the intensification of a subtropical high. Moreover, the southward flow over the East China Sea and south-westward flow over Japan suggest that the convergence of the flow becomes stronger over the East China Sea, which leads to an increase in precipitation over there. In the future climate simulation, water vapor in the lower troposphere is expected to increase by the Clausis–Clapeyron relation because of the temperature increase due to global warming. Indeed, an analysis of vertically integrated moisture transport has revealed that the moisture convergence and divergence distribution converted into the unit of mm/day quantitatively explains the precipitation change pattern shown in Fig. 7.5c. It should also be noted from Fig. 7.5c that the Indian summer monsoon circulation weakens in a warmer climate although the water vapor transport is intensified (Kitoh et al. 1997), while the circulation associated with the Pacific high pressure system is further intensified due to the local subtropical high pressure anomaly (20°N, 120–150°E). As a result, the latter effect will more dominate the water vapor convergence than the former in the Baiu rain band in a warmer climate.

7.4 Discussion and Concluding Remarks

Previous studies using global models with a relatively high horizontal resolution at that time (e.g., 100 km) suggested that the frequency of TC would be significantly reduced worldwide in a warmer climate (Bengtsson et al. 1996; Sugi et al. 2002; Yoshimura et al. 2006). However, such a resolution is still insufficient for representing hierarchical convective structures of the observed tropical cyclone. It is, therefore, generally recognized that the future change of the TC frequency is uncertain (IPCC 2001). In the present study, the same tendency was obtained by using the 20-km-mesh AGCM which realistically reproduces the present-day climatology, and the small-scale structure of a TC. However, the present study confirmed the tendency in a more reliable and credible way. Though the mechanisms responsible for the frequency reduction in a warmer future climate have not been thoroughly analyzed in the present experiment, we assume that the stabilization of the atmosphere caused by global warming should be a primary reason (Sugi et al. 2002; Yoshimura et al. 2006) because the large-scale environmental changes are similar among the AGCMs of different resolutions and different warm SSTs. In addition to the stabilization effect, Yoshimura and Sugi (2005) have shown that the reduction of infrared radiative cooling due to the increase in CO_2 in the troposphere also contributes to reduce the frequency. As for the regional increase of the frequency in the North Atlantic Ocean, Oouchi et al. (2006) discussed a relevance to the positive region of SST spreads over the southeastern part of the North Atlantic Ocean (Fig. 7.1). The region approximately corresponds to the area of generation for most of the hurricanes that approach and threaten the eastern coast of the United States. Of the several signatures of such warm tongues over the globe, the one over the North Atlantic is conspicuous in size, and, more importantly, coincides with the tropical cyclone source.

We have also seen that the number of intense TCs will increase in a warmer climate. These results agree with those from previous studies using regional models (Knutson et al. 1998; Walsh and Ryan 2000; Knutson and Tuleya 2004) as well as with theoretical estimations based on the thermodynamic factors controlling TC intensity (Emanuel 1987; Holland 1997; Henderson-Sellers et al. 1998). The present time-slice experiment reinforces the preceding findings in a more comprehensive manner, with a long-term climate integration of a 20-km-mesh global model. Although the change in the dynamical effects of the environmental flow fields surrounding a TC, including the effect of vertical shear, is essential in assessing the influences of global warming upon TCs (Walsh and Ryan 2000), such dynamical effects have not been considered in most of the previous studies. This means that our approach including all possible dynamics that might affect the structure of TC in a globally high-resolution model is physically more consistent than studies that have ignored dynamic changes in environmental flow due to the limitation from the regional modeling.

In relation to the Baiu, an observational study by Hu et al. (2003) has demonstrated the conspicuous summertime precipitation trend for the period from 1951 to 2000, which is characterized by the negative (drying) trend to the north of the

Yangtze River valley and the positive (wetting) trend over the Yangtze River valley, the "North-drying south-wetting pattern." It is noteworthy that the precipitation change pattern in a warmer climate (Fig. 7.5c) qualitatively resembles the North-drying south-wetting pattern, suggesting that this observed pattern will be enhanced in future warmer climates.

It is well known by long-range forecasters of JMA that, under the El Niño condition, total precipitation during the Baiu period tends to increase in association with the delay of the termination of the Baiu period. This tendency is consistent with the change of the Baiu in the future warmer climate shown in the present study. In view of that, the SST change predicted by the parent AOGCM shows the El Niño-like pattern, which has a large temperature increase in the tropical eastern Pacific. Therefore, the future change in a warmer climate in East Asia can be interpreted as a response of the atmosphere to the El Niño condition of the ocean.

Recently Tanaka et al. (2005) and Yamaguchi and Noda (2006) have pointed out that in low latitudes the stabilization and the general weakening of the main circulations, including the Indian monsoon circulation and Walker circulation, are rather common features among many scenario experiments with different AOGCMs. Therefore, although we have discussed the projection based on the results from a specific AOGCM and a specific scenario experiment, the present projection about the tropical cyclones and the Baiu rain band is expected to be robust to some extent.

Finally, to examine the effect of interannual variability of SST, we have performed a time-slice experiment by prescribing interannually varying, monthly mean SSTs for 20 years for the present-day climate and a future warmer climate. Preliminary results suggest that the main conclusions about changes in tropical cyclone and Baiu are little affected by introducing the interannual variability. Details of the dependence of SSTs prescribed as the boundary condition will be discussed elsewhere.

Acknowledgments This study was conducted under the framework of the Research Revolution 2002 (RR2002) project "Development of Super-High-Resolution Global and Regional Climate Models" funded by Ministry of Education, Culture, Sports, Science and Technology (MEXT). This project is called "Kyosei-4 project." We thank Drs. T. Matsuo and T. Aoki and members of the Kyosei-4 global modeling subgroup (in alphabetical order): M. Hosaka, H. Ishizaki, A. Kitoh, H. Tsujino, T. Shibata, T. Uchiyama, T. Yasuda, and S. Yukimoto (Meteorological Research Institute, MRI), M. Hirai, T. Hosomi, T. Kadowaki, K. Katayama, H. Kawai, M. Kazumori, H. Kitagawa, C. Kobayashi, M. Kyoda, S. Maeda, T. Matsumura, S. Murai, M. Nakagawa, A. Narui, T. Ose, R. Sakai, T. Sakashita, Y. Takeuchi, K. Yamada, and M. Yamaguchi (Japan Meteorological Agency, JMA), K. Fukuda, K. Horiuchi, K. Miyamoto, and H. Murakami (Advanced Earth Science and Technology Organization, AESTO).

References

Adler, R. F., G. J. Huffman, A. Chang, R. Ferraro, P. Xie, J. Janowiak, B. Rudolf, U. Schneider, S. Curtis, D. Bolvin, A. Gruber, J. Susskind, P. Arkin, and E. Nelkin. The version-2 global precipitation climatology project (GPCP) monthly precipitation analysis (1979-present). J. Hydrometeor., 4, 1147–1167 (2003).

References

Arakawa, A. and W. H. Schubert. Interaction of cumulus cloud ensemble with the large-scale environment. Part I. J. Atmos. Sci., 31, 674–701 (1974).

Bengtsson, L., M. Botzet and M. Esch. Will greenhouse gas-induced warming over the next 50 years lead to higher frequency and greater intensity of hurricanes? Tellus, 48A, 57–73 (1996).

Emanuel, K. A. The dependence of hurricane intensity on climate. Nature, 326, 483–485 (1987).

Habata, S., M. Yokokawa and S. Kitawaki. The development of the Earth Simulator. IEICE TRANSACTIONS on Information and systems. E86-D, 1947–1954 (2003).

Habata, S., K. Umezawa, M. Yokokawa and S. Kitawaki. Hardware system of the Earth Simulator. Parallel Comput., 30, 1287–1313 (2004).

Henderson-Sellers, A. et al. Tropical cyclones and global climate change: A post-IPCC assessment. Bull. Am. Meteorol. Soc., 79, 19–38 (1998).

Holland, G. J. The maximum potential intensity of tropical cyclones. J. Atmos. Sci., 54, 2519–2541 (1997).

Hu, Z.-Z., S. Yang and R. Wu. Long-term climate variations in China and global warming signals. J. Geophys. Res., 108(D19), 4614 (2003). doi:10.1029/2003JD003651.

IPCC (Intergovernmental Panel on Climate Change). Special Report on Emissions Scenarios. A Special Report of Working Group III of the Intergovernmental Panel on Climate Change. Cambridge University Press, Cambridge, UK (2000).

IPCC (Intergovernmental Panel on Climate Change). Climate Change 2001: The Scientific Basis. Contribution of Working Group I to the Third Assessment Report of the Intergovernmental Panel on Climate Change. Cambridge University Press, Cambridge, United Kingdom and New York, NY, USA, 881pp (2001).

JMA. Outline of the operational numerical weather prediction at the Japan Meteorological Agency (Appendix to WMO numerical weather prediction progress report). Japan Meteorological Agency, 157pp. (2003) (available online: http://www.jma.go.jp/jma/jma-eng/jma-center/nwp/outline-nwp/pdf/ol4_2.pdf)

Kang, I. S., K. Jin, B. Wang, K.-M. Lau, J. Shukla, V. Krishnamurthy, S. Schubert, D. Wailser, W. Stern, A. Kitoh, G. Meehl, M. Kanamitsu, V. Galin, V. Satyan, C.-K. Park, and Y. Liu. Intercomparison of the climatological variations of Asian summer monsoon precipitation simulated by 10 GCMs. Clim. Dyn., 19, 383–395 (2002).

Kawatani, Y. and M. Takahashi. Simulation of the Baiu front in a high resolution AGCM. J. Meteorol. Soc. Jpn., 81, 113–126 (2003).

Kitoh, A., S. Yukimoto, A. Noda and T. Motoi. Simulated changes in the Asian summer monsoon at times of increased atmospheric CO_2. J. Meteorol. Soc. Jpn., 75, 1019–1031 (1997).

Knutson, T. R. and R. E. Tuleya. Impact of CO_2-induced warming on simulated hurricane intensity and precipitation: Sensitivity to the choice of climate model and convective parameterization. J. Clim., 17, 3477–3495 (2004).

Knutson, T. R., R. E. Tuleya and Y. Kurihara. Simulated increase of hurricane intensities in a CO_2-warmed climate. Science, 279, 1018–1020 (1998).

Kobayashi, C. and M. Sugi. Impact of horizontal resolution on the simulation of the Asian summer monsoon and tropical cyclones in the JMA global model. Clim. Dyn., 93, 165–176 (2004).

Kusunoki, S., M. Sugi, A. Kitoh, C. Kobayashi, and K. Takano. Atmospheric seasonal predictability experiments by the JMA AGCM. J. Meteorol. Soc. Jpn., 79, 1183–1206 (2001).

Kusunoki, S., J. Yoshimura, H. Yoshimura, A. Noda, K. Oouchi and R. Mizuta. Change of Baiu rain band in global warming projection by an atmospheric general circulation model with a 20-km grid size. J. Meteorol. Soc. Jpn., 84, 581–611 (2006).

Lau, K.-M. and S. Yang. Seasonal variation, abrupt transition, and intraseasonal variability associated with the Asian summer monsoon in the GLA GCM. J. Clim., 9, 965–985 (1996).

Lau, K.-M., J. H. Kim and Y. Sud. Intercomparison of hydrologic processes in AMIP GCMs. Bull. Am. Meteorol. Soc. 77, 2209–2227 (1996).

Liang, X. Z., W. C. Wang and A. N. Samel. Biases in AMIP model simulations of the east China monsoon system. Clim. Dyn., 17, 291–304 (2001).

Mellor, G. L. and T. Yamada. A hierarchy of turbulence closure models for planetary boundary layers. J. Atmos. Sci., 31, 1791–1806 (1974).

Mizuta, R., K. Oouchi, H. Yoshimura, A. Noda, K. Katayama, S. Yukimoto, M. Hosaka, S. Kusunoki, H. Kawai, and M. Nakagawa. 20-km-mesh global climate simulations using JMA-GSM model – Mean climate states. J. Meteorol. Soc. Jpn., 84, 165–185 (2006).

Nakagawa, M and A. Shimpo. Development of a cumulus parameterization scheme for the operational global model at JMA. RSMC Tokyo-Typhoon Center Technical Review, No. 7, 10–15 (2004).

Ninomiya, K. and T. Akiyama. Multi-scale features of Baiu, the summer monsoon over Japan and the East Asia. J. Meteorol. Soc. Jpn., 70, 467–495 (1992).

Oouchi, K., J. Yoshimura, H. Yoshimura, R. Mizuta, S. Kusunoki, and A. Noda. Tropical cyclone climatology in a global-warming climate as simulated in a 20-km-mesh global atmospheric model: Frequency and wind intensity analyses. J. Meteorol. Soc. Jpn., 84, 259–276 (2006).

Randall, D. and D.-M. Pan. Implementation of the Arakawa-Schubert cumulus parameterization with a prognostic closure. Meteorological Monograph/The representation of cumulus convection in numerical models 46, 145–150 (1993).

Reynolds, R. W. and T. M. Smith. Improved global sea surface temperature analyses using optimum interpolation. J. Clim., 7, 929–948 (1994).

Simmons, A. J. and J. K. Gibson. The ERA-40 project plan. ERA-40 Project report series Vol. 1. European Centre for Medium-Range Weather Forecasts, Reading, U.K., 62 pp. (2000).

Sperber, K. R., S. Hameed, G. L. Potter and J. S. Boyle. Simulation of the northern summer Monsoon in the ECMWF model: Sensitivity to horizontal resolution. Mon. Weather Rev., 122, 2461–2481 (1994).

Sugi, M, A. Noda and N. Sato. Influence of global warming on tropical cyclone climatology: An experiment with the JMA Global Model. J. Meteorol. Soc. Jpn., 80, 249–272 (2002).

Tanaka, H. L., N. Ishizaki and D. Nohara. Intercomparison of the intensities and trends of Hadley, Walker and monsoon circulations in the global warming projections. SOLA, 1, 77–80 (2005). doi:10.2151/sola.2005-021.

Walsh, K. J. E. and B. F. Ryan. Tropical cyclone intensity increase near Australia as a result of climate change. J. Clim., 13, 3029–3036 (2000).

Yamaguchi, K., and A. Noda. Global warming patterns over the North Pacific: ENSO versus AO. J. Meteorol. Soc. Jpn., 84, 221–241 (2006).

Yasunaga, K., M. Yoshizaki, Y. Wakazuki, C. Muroi, K. Kurihara, A. Hashimoto, S. Kanada, T. Kato, S. Kusunoki, K. Oouchi, H.Yoshimura, R. Mizuta and A. Noda. Changes in the Baiu frontal activity in the future climate simulated by super-high-resolution global and cloud-resolving regional climate models. J. Meteorol. Soc. Jpn., 84, 199–220 (2006).

Yoshimura, H. and T. Matsumura. A two-time-level vertically-conservative semi-Lagrangian semi-implicit double Fourier series AGCM. CAS/JSC WGNE. Res. Act. Atmos. Ocean Model., 35, 3.27–3.28 (2005).

Yoshimura, J. and M. Sugi. Tropical cyclone climatology in a high-resolution AGCM – Impacts of SST warming and CO_2 increase. SOLA, 1, 133–136 (2005). doi: 10.2151/sola.2005-035.

Yoshimura, J., M. Sugi and A. Noda. Influence of greenhouse warming on tropical cyclone frequency. J. Meteorol. Soc. Jpn., 84, 405–428 (2006).

Yukimoto, S., A. Noda, A. Kitoh, M. Hosaka, H. Yoshimura, T. Uchiyama, K. Shibata, O. Arakawa and S. Kusunoki. The Meteorological Research Institute coupled GCM, version 2.3 (MRI-CGCM2.3) – Control climate and climate sensitivity. J. Meteorol. Soc. Jpn., 84, 333–363 (2006a).

Yukimoto, S., A. Noda, T. Uchiyama and S. Kusunoki. Climate change of the late nineteenth through twenty-first centuries simulated by the MRI-CGCM2.3. Meteorol. Geophys., 56, 9–24 (2006b).

Chapter 8
Simulations of Forecast and Climate Modes Using Non-Hydrostatic Regional Models

Masanori Yoshizaki, Chiashi Muroi, Hisaki Eito, Sachie Kanada, Yasutaka Wakazuki and Akihiro Hashimoto

Summary Two applications with a cloud-resolving model are shown utilizing the Earth Simulator. The first application is a case in the winter cold-air outbreak situation observed over the Sea of Japan as a forecast mode. Detailed structures of the convergence zone (JPCZ) and formation of mechanism of transverse convective clouds (T-modes) are discussed. A wide domain in the horizontal (2000 × 2000) was used with a horizontal resolution of 1 km, and could reproduce detailed structures of the JPCZ as well as the cloud streets in the right positions. It is also found that the cloud streets of T-modes are parallel to the vertical wind shears and, thus, similar to the ordinary formation mechanism as longitudinal convective ones. The second application is changes in the Baiu frontal activity in the future warming climate from the present one as a climate mode. At the future warming climate, the Baiu front is more active over southern Japan, and the precipitation amounts increase there. On the other hand, the frequency of occurrence of heavy rainfall greater than 30 mm h^{-1} increases over the Japan Islands.

8.1 Introduction

The Earth Simulator (ES) has 640 nodes (1 node has 8 CPU) and its total peak performance is about 40 TFLOPS. This was the fastest supercomputer in the world during 2002–2004. When the ES can be applied to meteorological and oceanic phenomena, higher resolutions of the models/wider calculation ranges and longer simulations can be available.

In this study, two meteorological applications of the ES are shown by utilizing cloud-resolving nonhydrostatic models (NHMs). The first application is to problems of a cloud band associated with the Japan Sea polar air mass convergence zone (JPCZ) and the formation mechanism of cloud streets transverse to the northwesterly prevailing winds (T-modes) in winter as a forecast mode. The second application is to changes in the Baiu frontal activity in the future warming climate from the present one as a climate mode. This is a continuation of a study on the atmospheric global circulation model (AGCM) presented by Noda et al. (2006). In the forecast mode, the initial conditions, such as short-range forecasts, are important. On the other hand, in the climate mode, the boundary conditions, such as the sea surface temperature (SST), are essential.

8.1.1 Forecast Mode: JPCZ and the Formation Mechanism of T-Modes

Under the northwesterly prevailing winds in winter, the dry and cold continental air mass drastically moistens and warms as it moves over the Sea of Japan. As a result, we frequently observe a cloud band with convergence zones and various types of cloud streets, such as longitudinal cloud streets parallel to the prevailing winds (L-modes) and T-modes. These detailed patterns are seen in Figs. 8.1a and 8.4b. Many studies have been conducted to study the structures of the JPCZ and formation mechanisms of these cloud streets. By utilizing observation data, Hozumi and Magono (1980) and Research group on mesoscale meteorology of Marine Department, Japan Meteorological Agency (JMA) (1988) tried to make conceptual models to represent characteristic features of the JPCZ. Nagata et al. (1986) and Nagata (1991) reproduced the JPCZ using a hydrostatic model with a horizontal resolution of about 40 km and showed that three lower-boundary conditions (the orographic effect on the northern part of the Korean Peninsula, horizontal distribution of SST, and land-ocean contrast) are essential to form the JPCZ. However, observations and numerical simulations are not sufficient, and details of the structures of the JPCZ and formation mechanisms of T-modes remain to be determined.

During the period of 12–17 January 2001, cold-air outbreaks were remarkable over the Sea of Japan. In the present study, the January 14 case was selected as a typical one and studied in detail. Figure 8.1 shows the horizontal patterns of clouds observed by meteorological satellite and sea surface winds estimated by the QuikSCAT satellite. From the satellite image, the cloud band was defined as the V-shaped cloud area, which extends from the neck of the Korean Peninsula to a southeast direction (Fig. 8.1a). Meanwhile, an intense convergent zone is observed (Fig. 8.1b) and this is called the JPCZ. The JPCZ roughly corresponds to the cloud band. Large horizontal shear is also found around the southwestern side of the JPCZ. When we see the periphery of the JPCZ, L-modes are found on its both sides. T-modes are also seen in the JPCZ or on its northeastern side, when the figure is enlarged (for example, see Fig. 8.4b).

On the same day, an instrumented aircraft traversed the JPCZ at the height of 5.6 km (Fig. 8.2). From the air-borne down-looking cloud radar observation, shallow, deep, and shallow clouds are found on the northeastern, middle, and southwestern sides of the JPCZ, respectively. From photographs taken from the aircraft, shallow, deep, and shallow clouds were seen, similarly to cloud radar observation, and stratiform clouds are found over the JPCZ.

8.1 Introduction

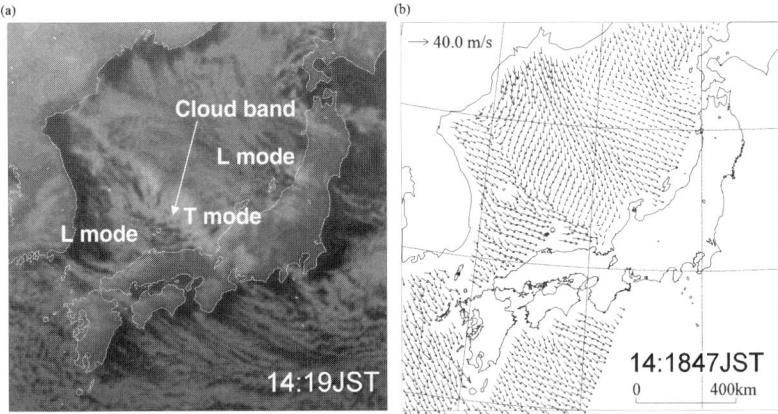

Fig. 8.1 Horizontal patterns of (**a**) clouds observed by meteorological satellite at 19 JST (Japan Standard Time; JST = UTC + 9h) on 14 January 2001 and (**b**) sea surface winds estimated by QuikSCAT satellite at 1847 JST. (From Yoshizaki et al. 2004.)

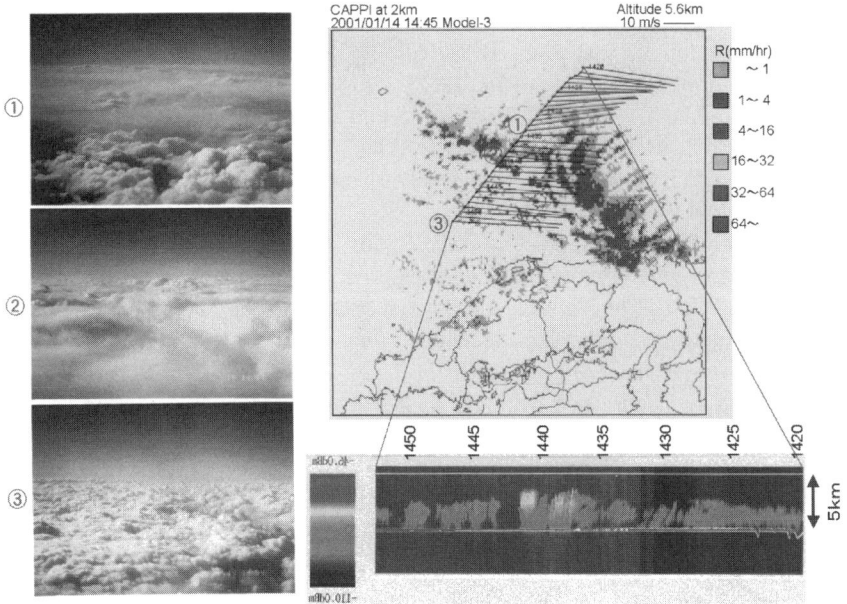

Fig. 8.2 (Upper right) Flight course and observed wind vectors at the height of 5.6 km, and radar pattern at 2 km obtained by JMA radar. Intense radar echo areas indicate the JPCZ. (Lower right) Time-height section of the air-borne down-looking cloud radar. A middle line indicates the sea surface. (Left) Photographs taken from the aircraft, looking to the southwest. (From Murakami et al. 2002.)

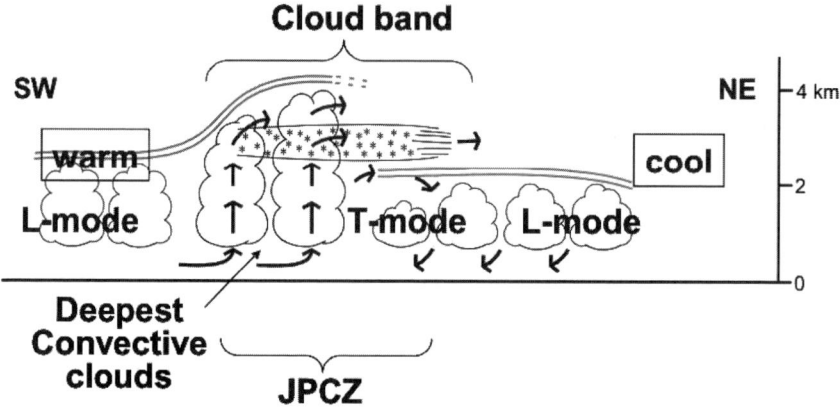

Fig. 8.3 Schematic figure of the JPCZ and its environment. (From Murakami et al. 2002.)

Figure 8.3 is a schematic figure of the JPCZ and its environment obtained by aircraft and satellite observations. This is a vertical cross section in the southwest-northeast direction, normal to the JPCZ. The JPCZ is generated where the cold air mass on the northeastern side of the JPCZ collides with the warm air mass on its southwestern side. The deepest clouds are found near the southwestern edge of the JPCZ. According to the aircraft observation, the clouds over the T-modes appeared to be blown off from the deepest convective clouds. On the other hand, convective clouds were seen in T-modes when visible images obtained by satellite were examined. Similar features to convective clouds were also found by Shimizu and Tsuboki (2005), who observed the T-modes on 26 December 2000 by utilizing the dual Doppler radars. They indicated that the direction of T-modes is similar to that of the vertical wind shear obtained near an upper-air sounding site, and new cells inside them form in the upshear direction.

To confirm such observed features, an NHM with a higher resolution of 1 km (1 km-NHM) is applied to this case. The domain is 2,000 × 2,000 and has 38 vertical layers (Fig. 8.4). Compared with the observed satellite image (Fig. 8.1a), the cloud band accompanied by the JPCZ and various cloud streets are well simulated. To study the cloud patterns in detail, a domain enclosed by white lines is enlarged (Fig. 8.4b). The detailed structures of the JPCZ and convective cells in L-modes are reproduced.

Along the line in Fig. 8.4b, the vertical sections of the mixing ratio of snow, potential temperature, and horizontal winds are shown (Fig. 8.5). From Fig. 8.5a, the deepest cumulus-type cloud is found around the southwestern edge of the JPCZ. On both southwestern and northeastern sides of the JPCZ, shallow convective clouds, which are grouped in L-modes, are simulated, but the cloud heights on its southwestern side are taller. From Fig. 8.5b, the air mass on the southwestern side is warmer than that on the northeastern one. Because of this distribution, the equivalent potential temperature is higher on the southwestern side of the JPCZ (not shown).

8.1 Introduction

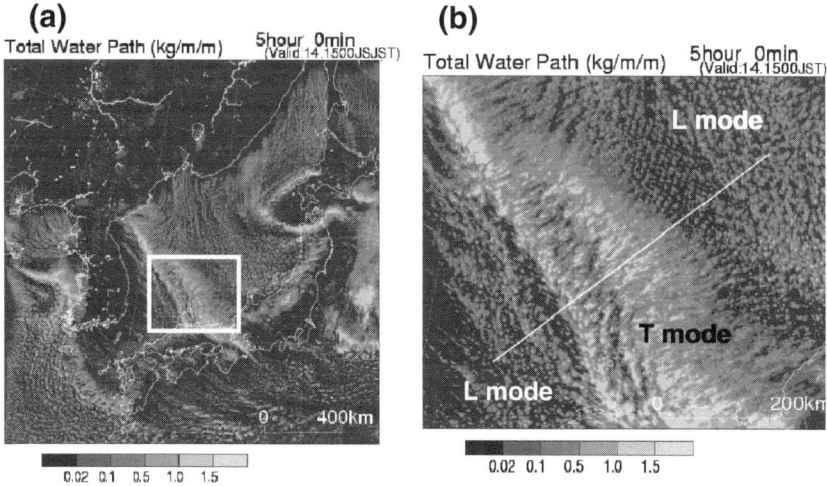

Fig. 8.4 Horizontal patterns of vertically integrated condensed water obtained by 1 km-NHM. (**a**) Nearly whole domain and (**b**) enlarged domain of the white box in (**a**). (From Eito et al. 2004.)

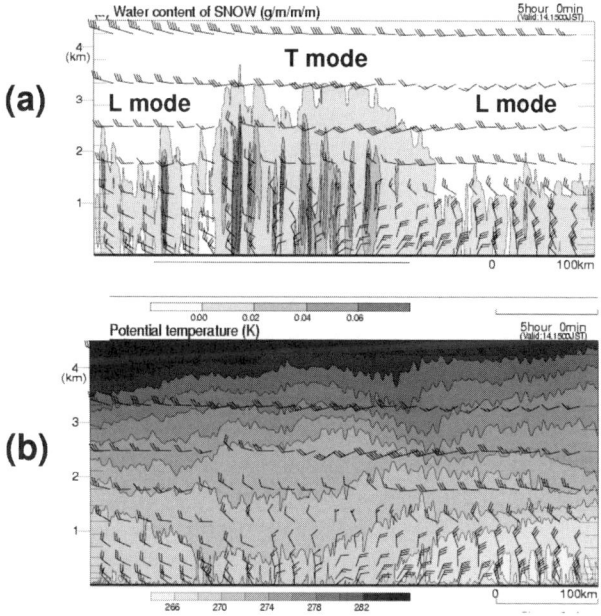

Fig. 8.5 Vertical cross sections of (**a**) mixing ratio of snow and (**b**) potential temperature along the white line in Fig. 8.4b. Horizontal winds are drawn in both figures. Snow areas are drawn by hatched regions in (**a**). (From Eito et al. 2004.)

Here, the formation mechanisms of T-modes are discussed. Over the T-mode areas, simulated winds are northerly near the surface and westerly above the height of 2 km. Then, the vertical wind shear is southwesterly, i.e., parallel to the direction of cloud lines of T-modes. Therefore, the simulated results suggest that T-modes form similarly to longitudinal convective cloud streets. In this manner, the highly resolved NHM could attain the detailed structures of the JPCZ and find the formation mechanisms of T-modes, which had been long disputed.

8.1.2 Climate Mode: Changes in the Baiu Frontal Activity in the Future Warming Climate from the Present Climate

Here the ES applies to the climate problem. The primary concern in this section is to study changes in Baiu frontal activity in a future warming climate due to the increase of greenhouse gases, such as CO_2. In order to attack this problem, the following steps were tried. First, we assumed that the CO_2 amount increases following the emission scenario A1B. The scenario A1B is characterized by a future world of very rapid economic growth, global population that peaks in middle of the twenty-first century and declines thereafter, and by a balanced introduction of new and more efficient technologies of all energy supply (IPCC, 2001). The CO_2 amount increases about twice in concentration around 2080–2099. By assuming such environmental changes, a coupled atmosphere-ocean general circulation model (AOGCM) was run with a coarse horizontal resolution, for example, about 300 km. Second, the AGCM with a horizontal resolution of 20 km was run to simulate atmospheric parts in detail. The SST information is given as a lower-boundary condition. Third, a regional NHM was run by obtaining the lateral data from the AGCM results. A project of Kyousei – 4 modeling in JMA/MRI (Meteorological Research Institute) is underway (2002–2006) to obtain the results of AGCM and NHM. Because the AGCM results appear in a study by Noda et al. (2006), only NHM results are shown in the present study.

The NHM was used to simulate June and July cases in the future warming and present climates. The horizontal resolution of the NHM is set to be 5 km. The model domain is 800 × 600 horizontal grids (4,000 km × 3,000 km) and has 48 vertical layers (top height: 22 km). The 40-day simulation, which is a unit of calculation, requires about 64 h using 30 nodes of the ES.

In the cloud physics, the mixing ratios and number concentrations of cloud water, rain, cloud ice, snow, and graupel are predicted. No cumulus-cloud parameterization is included in the NHM. Thus, the treatments of cumulus clouds are considerably different from those of the AGCM among physical processes. Such NHMs with small horizontal grids are called cloud-resolving ones.

The initial and lateral boundary conditions of the NHM were obtained from the outputs of the AGCM in a one-way nesting manner. In order to combine the AGCM and NHM smoothly, it is necessary to reduce the horizontal phase differences of propagating large-scale disturbances between both models. A spectral boundary coupling

8.1 Introduction

(SBC) method was adopted. In this method, large-scale components from the AGCM are combined with small-scale ones from the NHM using the wave-number decomposition (separation wave number, K_b) (Kida et al. 1991; Yasunaga et al. 2005). The SBC method was applied to the fields of horizontal winds and potential temperature above a height of 5 km for every 20 min, and the inverse K_b was selected to be about 1,000 km.

Figure 8.6 shows the horizontal distributions of the monthly-mean daily accumulated precipitation in the present climate, and the differences between those in the future warming climate and those in the present one. Since the NHM obtains the lateral boundary conditions from the outputs of the AGCM, the general features simulated by the NHM were similar to those simulated by the AGCM (Fig. 8.6a). In the future warming climate (Fig. 8.6b), positive anomalies of precipitation from the present climate are found, especially over southern Japan where the anomalies reach 6–7 mm day^{-1}. On the other hand, negative anomalies are seen over northern Japan and the Korean Peninsula, reaching -4 mm day^{-1}.

In East Asia, the onset and end of the Baiu season are important information as well as indicators of remarkable seasonal changes. Figure 8.7 shows the time – latitude sections of precipitation averaged in 130 E – 135 E in the present and future warming climates. Here, the Baiu front is defined by large temperature gradients and

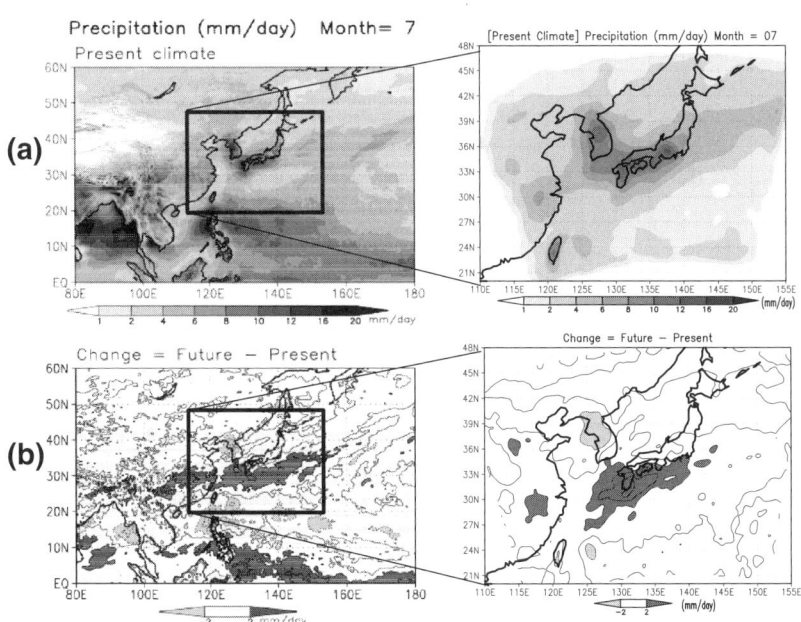

Fig. 8.6 Horizontal distributions of (**a**) the monthly-mean daily accumulated precipitation in July in the present climate and (**b**) the differences between the future warming climate and the present one. Left (right) panels show horizontal distributions obtained by AGCM (NHM). Units are mm day^{-1}. (From Yoshizaki et al. 2005.)

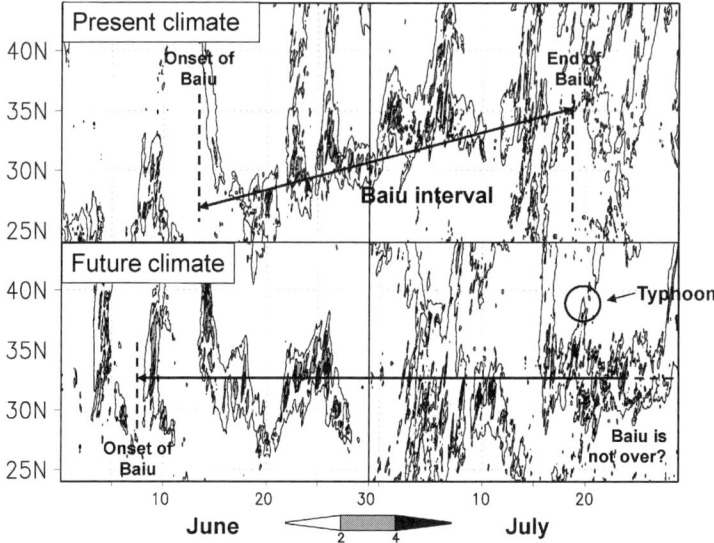

Fig. 8.7 Time–latitude sections of precipitation (mm h^{-1}) averaged in 130 E–135 E in the present climate (upper figure) and in the future warming climate (lower figure). Some years are selected arbitrarily. (From Yoshizaki et al. 2005.)

wind shears near the surface, and wind jets at 850 hPa (not shown). In the present climate, the Baiu front moves northward gradually and disappears in the middle of July. On the other hand, in the future warming climate, the northward shift of the Baiu front is not seen and the end of the Baiu season is not found even until the end of July.

The NHM precipitation in the present and future warming climates is compared over the Japan Islands (all) and five separate regions; SW (Southwest), KS (Kyushu), CJ (Central Japan), EJ (Eastern Japan), and NJ (Northern Japan) (Fig. 8.8c). Figure 8.8a shows the mean precipitation amounts in June and July. It is noteworthy that the precipitation in the future warming climate increases by 10% on average. The rate of increase is 30% larger in the KS region but 10% smaller in the NJ region. These features are consistent with the previous results in Fig. 8.6.

The frequency of heavy rainfalls is shown in Fig. 8.8b. Here, heavy rainfalls are defined as those with precipitation intensity greater than 30 mm h^{-1}. It is found that the frequency of heavy rainfall increases over all regions in the future warming climate.

Next, the NHM precipitation in the present climate is compared with the observations. For that purpose, radar-AMeDAS-analyzed rainfall, estimated by meteorological radars and calibrated using AMeDAS precipitation data, was utilized as observation data. AMeDAS is the Automated Meteorological Data Acquisition System of JMA. It is found that the NHM precipitation in each area is generally smaller but acceptable on the northern part of the Japan Islands (not shown) compared with the

8.1 Introduction

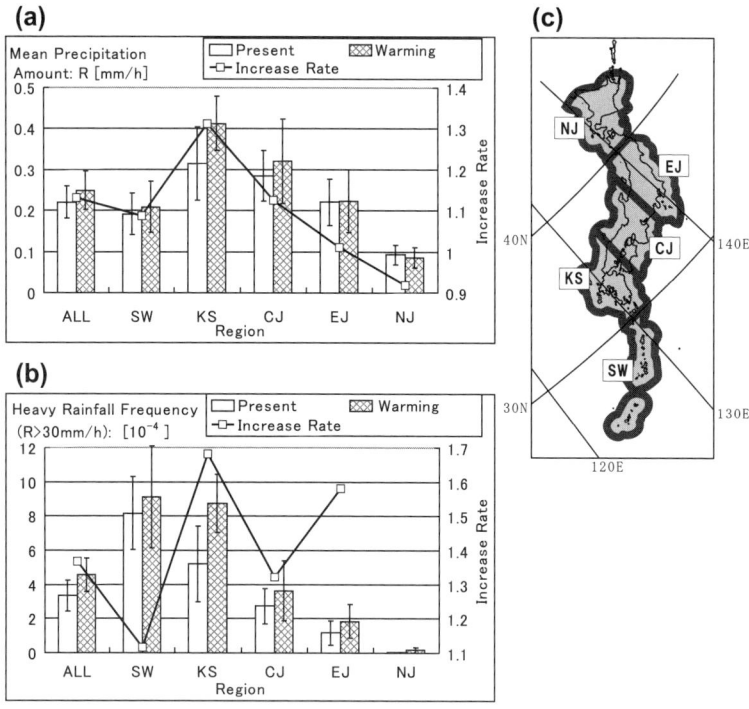

Fig. 8.8 (a) Mean precipitation amounts (the left ordinate; mm h^{-1}) and (b) frequency (the left ordinate; non-dimensional) of heavy rainfall over the Japan Islands (all) and five regions in June and July. (From Yoshizaki et al. 2005.) Five regions denoted by SW, KS, CJ, EJ, and NJ are shown in (c). In (a) and (b), left (right) columns denote the NHM results in the present (future warming) climates. Root mean square errors (bars) and rates of increase of the future warming climate from the present one (solid lines; the right ordinates) are also shown. The frequency shown in (b) is computed from hourly output of accumulated rainfall with area weight as (the time when rainfall exceeds 30 mm h^{-4}) divided by (the total time of numerical integration for the analysis). It is a very small number so that 10^4 is multiplied

observations. However, the NHM precipitation in Kyushu (KS) and southwestern (SW) regions is considerably small, and, thus, precipitation averaged in the Japan Islands is about 70%.

In order to determine the reason for differences, the dependence of the rainfall frequency on rainfall intensity is examined (Fig. 8.9). It is seen that most differences come from the underestimation of the precipitation intensity between 1 and 30 mm h^{-1}. Since such weak-to-moderate precipitation may be produced by small-scale convective cells, the present 5 km-NHM, which include no cumulus-cloud parameterization, could not resolve them. As next steps, therefore, we will adopt higher horizontal resolutions (for example, 1 km) in order to obtain further quantitative precipitation.

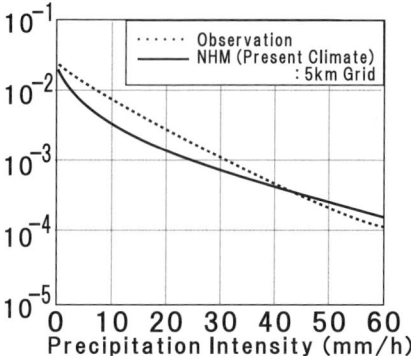

Fig. 8.9 Dependence of the rainfall frequency on its precipitation rates over all regions in June and July. Solid lines denote precipitation simulated by the 5 km-NHM, and dotted ones analyzed by radar-AMeDAS-analyzed rainfall data. (From Yoshizaki et al. 2005.)

8.2 Summary

In this study, we showed two meteorological applications of the ES with a cloud-resolving NHM as forecast and climate modes. The first topic is the cases in the winter cold-air outbreak situation observed over the Sea of Japan. The detailed structures of the JPCZ and formation mechanism of T-modes have long been disputed. By utilizing the ES, a wide domain with a horizontal resolution of 1 km could be applied, such as $2{,}000 \times 2{,}000$. It could reproduce the detailed structures of the JPCZ as well as the cloud streets in the right positions. It is also found that the cloud streets of T-modes are parallel to the vertical wind shears and, thus, similar to the L-modes in essence.

The second topic is changes in the Baiu frontal activity in the future warming climate from the present one. The Baiu front is more active over southern Japan, and the precipitation amounts increase there. On the other hand, the frequency of occurrence of heavy rainfall greater than 30 mm h^{-1} increases over the Japan Islands.

The appearance of the ES is epoch-making for the scientific community and contributed to the clarification of some unexplained phenomena, as shown in the present article. However, higher resolutions and further improvement of the models are still required to achieve a fundamental understanding; this is because meteorological and oceanic phenomena have hierarchical structures of horizontal scales, where governing physics are different if horizontal scales are different. In our community, faster and more efficient computers than the ES would always be desirable.

Acknowledgments We are indebted to an anonymous referee for giving useful comments and improving our article. Thanks are extended to Earth Simulator Center, where all calculations were done.

References

Eito, H., C. Muroi, S. Hayashi, T. Kato, and M. Yoshizaki, 2004: A high-resolution wide-range numerical simulation of cloud bands associated with the Japan Sea Polar-air mass Convergence Zone in winter using a non-hydrostatic model on the Earth Simulator. *CAS/JSC WGNE Research Activities in Atmospheric and Oceanic Modelling Report*, No. **34**, WMO/TD-No.1220, 5, 7–8.

Hozumi, K., and C. Magono, 1984: The cold structure of convergent cloud ands over the Japan Sea in winter monsoon period. *J. Meteor. Soc. Japan*, **62**, 522–533.

IPCC (Intergovernmental Panel on Climate Change), 2001: Climate Change 2001. Cambridge University Press, Cambridge, 881 pp.

Kida, H., T. Koide, H. Sasaki, and M. Chiba, 1991: A new approach for coupling a limited area model to a GCM for regional climate simulations. *J. Meteor. Soc. Japan*, **69**, 723–728.

Murakami, M., M. Hoshimoto, N. Orikasa, H. Horie, H. Okamoto, H. Kuroiwa, H. Minda, and K. Nakamura, 2002: Inner structures of snow bands associated with the Japan Sea polar-airmass convergence zone based on aircraft observations. *Proc. Int. Conf. Mesoscale Convective Systems and Heavy Rainfall/Snowfall in East Asia*, Tokyo, Japan, 522–527.

Nagata M., M. Ikawa, S. Yoshizumi, and T. Yoshida, 1986: On the formation of a convergent cloud band over the Japan Sea in winter; numerical experiments. *J. Meteor. Soc. Japan*, **64**, 841–855.

Nagata, M., 1987: On the structure of a convergent cloud band over the Japan Sea in winter; a prediction experiment. *J. Meteor. Soc. Japan*, **65**, 871–883.

Noda, A., S. Kusunoki, J. Yoshimura, H. Yoshimura, K. Oouchi, and R. Mizuta, 2006: Global warming projection by an atmospheric general circulation model with 20-km grid size. In "*High Resolution Numerical Modelling of the Atmosphere and Ocean*," edited by Wataru Ohfuchi and Kevin Hamilton, Springer, New York.

Research group on mesoscale meteorology of Marine Department, JMA, 1998: On the mesoscale structure of the cloud band system over Japan Sea in winter monsoon period – A mesoscale observation on board R/V Keifu-Maru (in Japanese). *Tenki*, **35**, 237–248.

Shimizu, K. and K. Tsuboki, 2005: Formation processes of transverse-mode snowbands observed off the Hokuriku District on 26 December 2000 (in Japanese). *Meteor. Note (Mesoscale Convective Systems)*, **208**, 243–250.

Yasunaga, K., H. Sasaki, Y. Wakazuki, T. Kato, C. Muroi, A. Hashimoto, S Kanada, K. Kurihara, M. Yoshizaki, and Y. Sato., (2005): Performance of the long-term integrations of the Japan Meteorological Agency nonhydrostatic model with use of the spectral boundary coupling method. *Weather and Forecasting*, **20**, 1061–1072.

Yoshizaki, M., T. Kato, H. Eito, S. Hayashi and W.-K. Tao, 2004: An overview of the field experiment "Winter Mesoscale Convective System (MCSs) over the Japan Sea in 2001," and comparisons of the cold-air outbreak case (14 January) between analysis and a non-hydrostatic cloud-resolving model. *J. Meteor. Soc. Japan*, **82**, 1365–1387.

Yoshizaki, M., C. Muroi, S. Kanada, Y. Wakazuki, K. Yasunaga, A. Hashimoto, T. Kato, K. Kurihara, A. Noda and S. Kusunoki, 2005: Changes of Baiu (Mei-yu) frontal activity in the global warming climate simulated by a non-hydrostatic regional model. *SOLA*, **1**, 25–28.

Chapter 9
High-Resolution Simulations of High-Impact Weather Systems Using the Cloud-Resolving Model on the Earth Simulator

Kazuhisa Tsuboki

Summary High-impact weather systems occasionally cause huge disasters to human society owing to heavy rainfall and/or violent wind. They consist of cumulonimbus clouds and usually have a multiscale structure. High-resolution simulations within a large domain are necessary for quantitatively accurate prediction of the weather systems and prevention/reduction of disasters. For the simulations, we have been developing a cloud-resolving model named the Cloud Resolving Storm Simulator (CReSS). The model is designed for a parallel computer and was optimized for the Earth Simulator in the present study. The purpose of the present research is high-resolution simulations of high-impact weather systems in a large calculation domain with resolving individual cumulonimbus clouds using the CReSS model on the Earth Simulator. Characteristic high-impact weather systems in East Asia are the Baiu front, typhoons, and winter snowstorms. The present chapter describes simulations of these significant weather systems. We have chosen for the case study of the Baiu front the Niigata–Fukushima heavy rainfall event on July 13, 2004. Typhoons for simulations are T0418, which caused a huge disaster due to strong wind, and T0423, which caused severe flood over the western Japan in 2004. Snowstorms were studied by an idealized numerical experiment as well as by a simulation of cold air outbreak over the Sea of Japan. These experiments clarified both the overall structures of weather systems and individual clouds. The high-resolution simulations resolving individual clouds permit a more quantitative prediction of precipitation. They contribute to accurate prediction of wind and precipitation and to reduction of disasters caused by high-impact weather systems.

9.1 Introduction

One of the most important objectives of limited-area models is a high-resolution simulation of high-impact weather systems for detailed studies and accurate predictions of them. High-impact weather systems are most significant phenomena in the atmosphere and sometimes cause huge disasters to human society. Understanding their mechanisms and structures is necessary for prediction and prevention/reduction of disasters. Most high-impact weather systems that cause heavy rainfalls and/or violent winds consist of cumulonimbus clouds and their organized systems. They are usually embedded within a larger weather system and occasionally have a multi-scale structure. It ranges from cloud-scale to synoptic-scale systems. Characteristic weather systems in East Asia are the Baiu front, typhoons, and winter snowstorms associated with a cold air outbreak.

In order to perform simulations and numerical experiments of the high-impact weather systems, we have been developing a cloud-resolving numerical model named "the Cloud Resolving Storm Simulator" (CReSS). Since the multiscale structure of the weather systems has a wide range of horizontal scales, a large computational domain and a very high-resolution grid to resolve individual classes of the multiscale structure are necessary to simulate evolution of the weather systems. In particular, an explicit calculation of cumulonimbus clouds is important for a quantitative simulation of precipitation associated with the high-impact weather. It is also required to formulate accurately cloud physical processes as well as the fluid dynamic and thermodynamic processes. For this type of computation, a large parallel computing with a huge memory is necessary.

The purpose of the present research is explicit simulations of clouds and their organized systems in a large domain (larger than $1\,000 \times 1\,000$ km) with resolving individual clouds using a very fine grid system (less than 1 km in horizontal). This will clarify a detailed structure of the high-impact weather systems and permit a more quantitative prediction of the associated precipitation. This will contribute for accurate prediction of precipitation and for reduction of disasters caused by the high-impact weather systems.

In this research, we have improved the CReSS model and optimized it for the Earth Simulator. Objectives of the present study are detailed simulations of the cloud and precipitation systems associated with the Baiu front, typhoons and associated rainbands, and snowstorms in cold polar air streams over a sea. In the present chapter, we will describe the basic formulation and characteristics of CReSS and summarize some results of the simulation experiments of high-impact weather systems such as a localized heavy rainfall associated with the Baiu front, typhoons, and snowstorms.

9.2 Description of CReSS

The basic formulation of CReSS is based on the nonhydrostatic and compressible equation system using terrain-following coordinates. Prognostic variables are three-dimensional velocity components, perturbations of pressure and potential temperature, water vapor mixing ratio, subgrid scale turbulent kinetic energy (TKE), and cloud physical variables. A finite difference method is used for the spatial discretization. The coordinates are rectangular and dependent variables are set on a staggered grid: the Arakawa-C grid in horizontal and the Lorenz grid in vertical. For time integration, the mode-splitting technique is used. Terms related to sound waves of the

9.2 Description of CReSS

basic equation are integrated with a small time step and other terms with a large time step.

Cloud physical processes are formulated by a bulk method of cold rain, which is based on Lin et al. (1983), Cotton et al. (1986), Murakami (1990), Ikawa and Saito (1991), and Murakami et al. (1994). The bulk parameterization of cold rain considers water vapor, rain, cloud, ice, snow, and graupel. The microphysical processes implemented in the model are described in Fig. 9.1.

Parameterizations of the subgrid scale eddy motions in CReSS are one-order closure of Smagorinsky (1963) or the 1.5-order closure with TKE. In the latter parameterization, the prognostic equation of TKE is used. All numerical experiments of the present article used the three-dimensional 1.5-order closure scheme. The surface process of CReSS is formulated by a bulk method. The bulk coefficients are formulated by the scheme of Louis et al. (1981).

Several types of initial and boundary conditions are available. For a numerical experiment, a horizontally uniform initial field provided by a sounding profile will be used with an initial disturbance of a thermal bubble or random temperature perturbation. The boundary conditions are rigid wall, periodic, zero normal-gradient, and wave-radiation types.

CReSS enables to be nested within a coarse-grid model and to perform a prediction experiment. In the experiment, the initial field is provided by interpolation of grid point values and the boundary condition is provided by the coarse-grid model. For a computation within a large domain, conformal map projections are available. The projections are the Lambert conformal projection, the polar stereographic projection, and the Mercator projection.

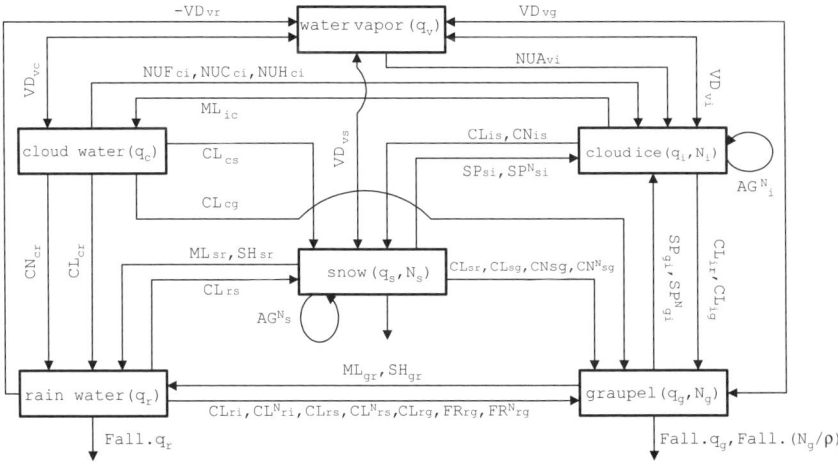

Fig. 9.1 Diagram describing of water substances and cloud microphysical processes in the bulk scheme of CReSS

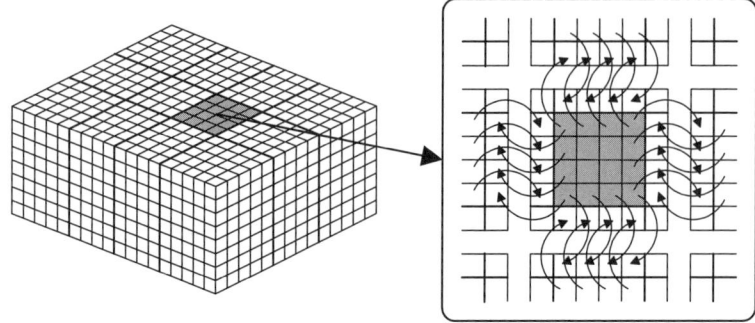

Fig. 9.2 Schematic representation of the two-dimensional domain decomposition and the communication strategy for parallel computations using MPI

For parallel computing of a large computation, CReSS provides two-dimensional domain decomposition in horizontal (Fig. 9.2). Parallel processing is performed using the Massage Passing Interface (MPI). Communications between the individual processing elements (PEs) are performed by data exchange of the outermost two grids. The OpenMP is optionally available to be used.

The readers can find the more detailed description of CReSS in Tsuboki and Sakakibara (2001, 2002).

9.3 Optimization for the Earth Simulator

The CReSS model was originally designed for parallel computers. We updated the code of CReSS from Fortran 77 to Fortran 90 and optimized it for the Earth Simulator. Communication procedures between computation nodes using MPI were also improved to be more efficient. For the intranode parallel processing, the OpenMP was introduced.

We evaluated the performance of CReSS on the Earth Simulator. The result is summarized in Table 9.1. The parallel operation ratio was measured using 128 nodes (1 024 CPUs) and 64 nodes (512 CPUs) of the Earth Simulator. The vector and parallel operation ratios are sufficiently high enough to perform a large computation on the Earth Simulator.

After the performance of CReSS was evaluated, we performed some simulation experiments of high-impact weather systems in East Asia: a localized heavy rainfall, typhoons, and winter snowstorms in cold air outbreak. The results are shown in the following sections.

Table 9.1 Evaluation of the performance of CReSS on the Earth Simulator

Vector operation ratio	99.4%
Parallel operation ratio	99.985%
Node number	128 nodes
Parallel efficiency	86.5%
Sustained efficiency	33%

9.4 Localized Heavy Rainfall

Precipitation systems associated with the Baiu front occasionally cause heavy rainfall and flood while they are also important water resources in East Asia. The Baiu front extends zonally for several thousand kilometers while a localized heavy rainfall has a horizontal scale of a few hundred kilometers. Various types of multiscale structure are indicated along the Baiu front. To clarify the processes in each class of the multiscale systems of precipitation along the Baiu front, it is necessary to perform simulation experiment with a grid fine enough to resolve cloud-scale and with a domain large enough to calculate the whole system of the Baiu front. The explicit representation of cumulonimbus clouds in the model is essentially important for accurate and quantitative simulation of the localized heavy rainfall associated with the Baiu front.

We have chosen for our case study the localized heavy rainfall occurred in Niigata and Fukushima prefectures on July 13, 2004 in Japan. Radar observation of the Japan Meteorological Agency (JMA) showed that an intense rainband extended zonally and maintained for more than 6 h. The Baiu front was located to the north of Niigata and a subsynoptic scale low (SSL) moved eastward along the Baiu front.

The experimental design of the simulation is summarized in Table 9.2. The initial and boundary conditions were provided by the JMA-RSM (the Regional Spectral Model). Initial time was 1200 UTC, July 12, 2004 and 24-h simulation was performed.

The simulation showed that the SSL moved eastward along the Baiu front. Figure 9.3 shows that the SSL reaches Japan at 0020 UTC, July 13, 2004. Moist westerly wind is intense to the south of the SSL. Large precipitation area extends to the east of the SSL. On the other hand, a very intense rainband forms to the south of the SSL. Enlarged display (Fig. 9.4) of the rainband shows that it extends from the northern part of the Noto Peninsula and reaches Niigata with intensification. The rainband forms between the southwesterly and westerly winds at the low level. The rainband is composed of intense convective cells. It maintains until the SSL moves to the Pacific Ocean. The long time maintenance of the intense rainband resulted in the severe flood in Niigata Prefecture.

Table 9.2 Experimental design of the Niigata–Fukushima heavy rainfall event

Domain	$x = 1\,792$ km, $y = 1\,536$ km, $z = 18$ km
Grid number	$x = 1\,795$, $y = 1\,539$, $z = 63$
Grid size	$H = 1\,000$ m, $V = 100$–300 m
Integration time	24 h
ES node number	128 nodes (1,024 CPUs)

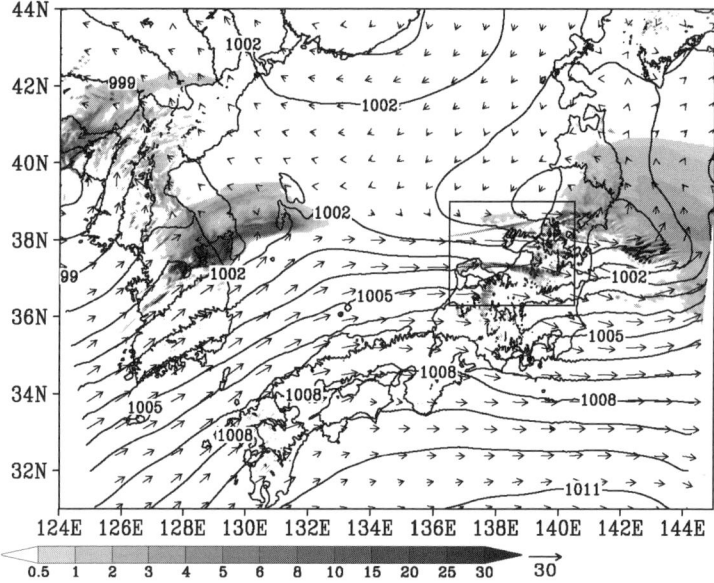

Fig. 9.3 Surface pressure (*contour lines*; hPa) and rainfall intensity (*gray levels*; mm hr^{-1}) and horizontal velocity (*arrows*) at a height of 1 610 m at 0020 UTC, July 13, 2004. The *rectangle* indicates the region of Fig. 9.4

9.5 Typhoons and the Associated Heavy Rainfall

Typhoons develop by close interaction between a large-scale disturbance and embedded intense cumulonimbus clouds. The horizontal scale of typhoons ranges from several 100 km to a few 1 000 km while that of the cumulonimbus clouds is an order of 10 km. Typhoons often bring a heavy rain and a strong wind. The heavy rain is usually localized in the eyewall and spiral rainbands that develop within typhoons. Since cumulonimbus clouds are essentially important for typhoon development, the cloud-resolving model is necessary for a detailed numerical simulation of typhoons.

Some typhoons usually attain Japan and its surroundings and cause severe disasters. In particular, ten typhoons landed over the main lands of Japan in 2004. In the present chapter, we show two simulation experiments of the typhoons. One is the typhoon T0418, which brought a very intense wind and caused huge disasters due to

9.5 Typhoons and the Associated Heavy Rainfall 147

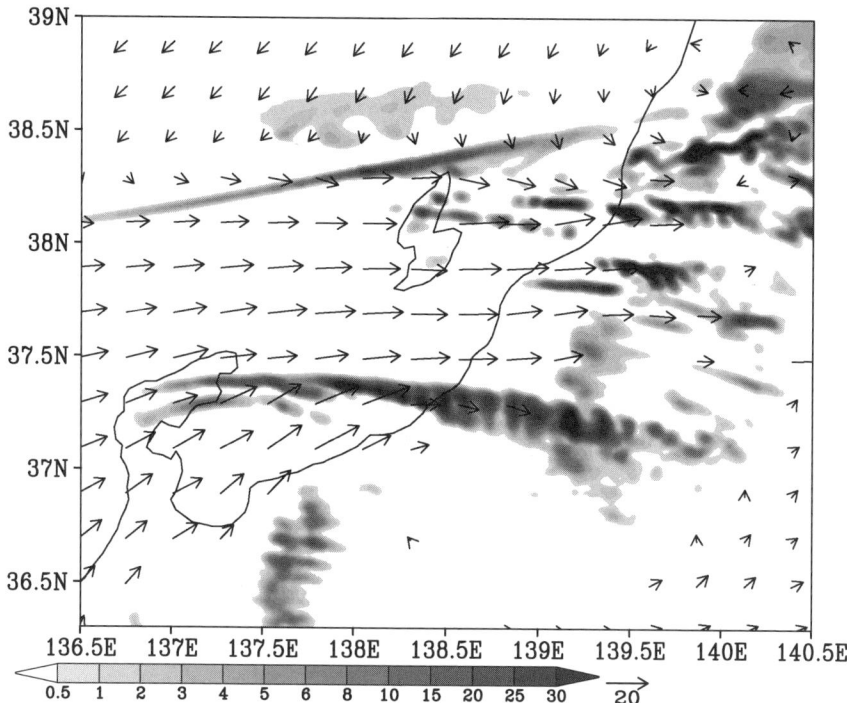

Fig. 9.4 Same as Fig. 9.3 but for the region of the rectangle in Fig. 9.3 and at a height of 436 m

the strong wind. The other is the typhoon T0423, which brought heavy rainfalls and caused severe floods.

Typhoon T0418 moved northwestward over the northwest Pacific Ocean and passed Okinawa Island on September 5, 2004. Its center passed Nago City around 0930 UTC, 5 September with the minimum sea level presser of 924.4 hPa. When T0418 pass over Okinawa Island, double eyewalls were observed. This is a distinctive feature of the typhoon. T0418 was characterized by strong winds and caused huge disasters due to the strong winds over Japan.

The main objectives of the simulation experiment of T0418 are to study the eyewall as well as spiral rainbands, and to examine structure of the strong wind associated with the typhoon around Okinawa Island. The simulation experiment of T0418 started from 0000 UTC, September 5, 2004. The experimental design of T0418 is summarized in Table 9.3.

The simulation experiment shows very detailed structure of the eye and the spiral rainbands (Fig. 9.5). Individual cumulus clouds are resolved. They are simulated within the eyewall and along the spiral rainband. A weak precipitation forms around the central part of the eye. The maximum tangential velocity is present along the

Table 9.3 Experimental design of Typhoon T0418

Domain	$x = 1\,536$ km, $y = 1\,280$ km, $z = 18$ km
Grid number	$x = 1\,539$, $y = 1\,283$, $z = 63$
Grid size	$H = 1\,000$ m, $V = 150\text{–}300$ m
Integration time	18 h
ES node number	128 nodes (1,024 CPUs)

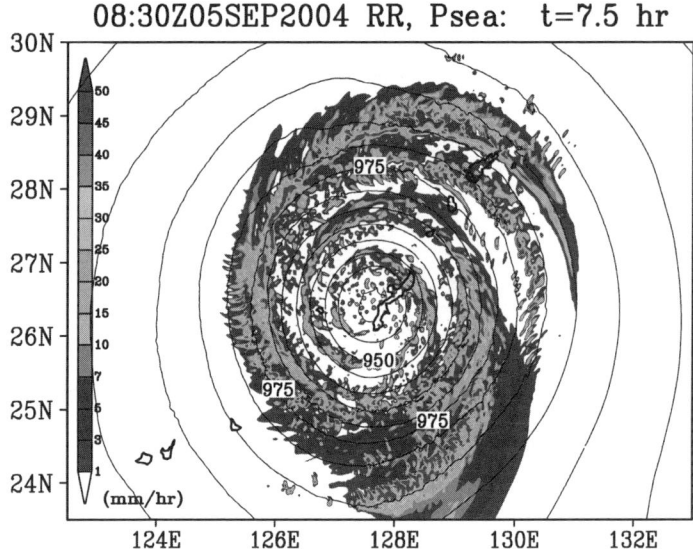

Fig. 9.5 Surface pressure (*contour lines*; hPa) and rainfall intensity (*color levels*; mm hr^{-1}) of the simulated Typhoon T0418 at 0830 UTC, September 5, 2004

eyewall and at a height of 1 km. It is larger than 70 m s^{-1}. The high-resolution experiment shows detailed structure of the cloud and precipitation systems associated with the typhoon, and simulates the overall structure of the typhoon and its movement.

Typhoon T0423 moved along the Okinawa Islands on October 19, 2004 and landed over Shikoku Island on 20 October. In contrast to T0418, T0423 is characterized by heavy rainfall over Japan. Heavy rainfalls associated with T0423 occurred in the eastern part of Kyushu, Shikoku, the east coast of the Kii Peninsula, and the Kinki District. They caused severe floods and disasters in these regions.

The purpose of the simulation experiment of T0423 is to study process of the heavy rainfall. Experimental design of T0423 is summarized in Table 9.4. At the initial time of 1200 UTC, October 19, 2004, T0423 was located to the NNE of Okinawa.

The movement of T0423 and the rainfall were successfully simulated. In the simulation, a northward moisture flux is large in the east side of the typhoon center. When the large moisture flux reaches to the Japanese Islands, heavy rainfalls occur along the Pacific Ocean side. The heavy rainfall moves eastward with the movement of the

9.5 Typhoons and the Associated Heavy Rainfall

Table 9.4 Experimental design of Typhoon T0423

Domain	$x = 1536$ km, $y = 1408$ km, $z = 18$ km
Grid number	$x = 1539$, $y = 1408$, $z = 63$
Grid size	$H = 1000$ m, $V = 200$–300 m
Integration time	30 h
ES node number	128 nodes (1,024 CPUs)

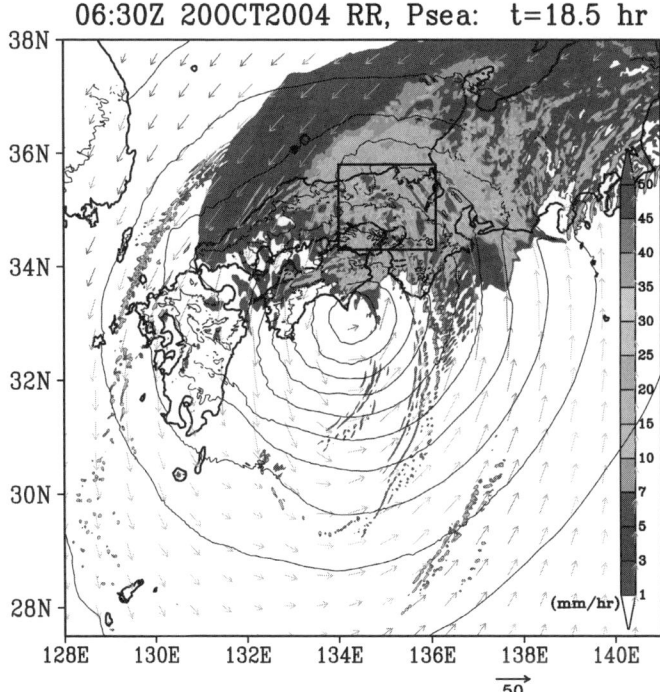

Fig. 9.6 Same as Fig. 9.5 but for the Typhoon T0423 at 0630 UTC, September 20, 2004. *Arrows* are horizontal wind velocity at a height of 974 m and *warmer-colored arrows* mean moister air. The *rectangle* indicates the region of Fig. 9.7

typhoon from Kyushu to Shikoku. When the typhoon reaches to the south of Shikoku, heavy rainfall begins in the Kinki District (the rectangle in Fig. 9.6) and intensifies at 0630 UTC, 20 October (Fig. 9.6). The distribution and intensity of precipitation correspond well to those of the radar observation.

The close view of northern Kinki shows that a large amount of solid precipitation is present around a height of 6 km (Fig. 9.7). The heavy rainfall in the region forms below the large mixing ratio of the solid precipitation. The heavy rainfall along the Pacific Ocean side moved eastward, while that in the Kinki District lasted until 12 UTC, 20 October. After the typhoon moved to the east of the Kinki District, the northeasterly was intensified significantly. Consequently, orographic rainfall formed

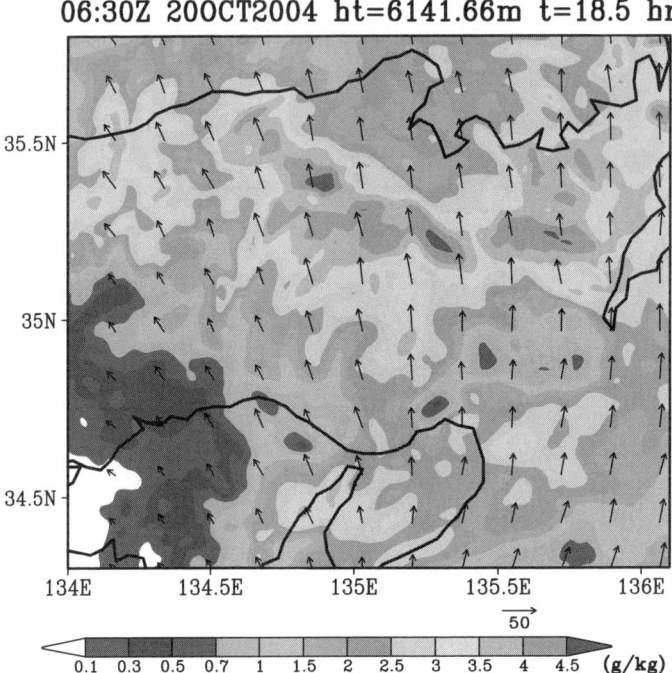

Fig. 9.7 Mixing ratio of precipitation (*color levels*; g kg^{-1}) and horizontal velocity (*arrows*) at a height of 6 142 m at 0630 UTC, September 20, 2004

in the northern Kinki District. As a result, the accumulated rainfall became a large amount and the severe flood occurred.

9.6 Snowstorms

9.6.1 Idealized Experiment of Snow Cloud Bands

When an outbreak of a cold and dry polar airmass occurs over the sea, many cloud streets or cloud bands are formed in the polar air stream. Large amounts of sensible heat and latent heat are supplied from the sea to the atmosphere. Intense modification of the airmass results in development of the mixing layer and convective clouds develop to form the cloud bands along the mean wind direction. Their length reaches an order of 1 000 km while individual convective cells have a horizontal scale ranging from a few kilometer to a few tens kilometers. In order to perform a three-dimensional simulation of cloud bands, a large computation is necessary. In order to study the detailed structure of convective cells and the formation process of the

9.6 Snowstorms

Table 9.5 Experimental design of the snow cloud bands in the cold air stream

Domain	$x = 457$ km, $y = 153$ km, $z = 11$ km
Grid number	$x = 1527$, $y = 515$, $z = 73$
Grid size	$H = 300$ m, $V = 50$–150 m
Integration time	20 h
ES node number	32 nodes (256 CPUs)

organized cloud bands, we performed three-dimensional simulation using the CReSS model on the Earth Simulator.

The experimental setting is summarized in Table 9.5. The initial condition was provided by a sounding observation on the east coast of Canada at 06 UTC, February 8, 1997.

The calculation domain in this simulation is 457 km and 153 km in x- and y-directions, respectively, with a horizontal grid spacing of 300 m. Sea ice is placed on the upstream side. Each model grid of the surface is occupied by ice or open sea according to the probability of sea ice or sea ice density. In the experiment, density of sea ice is 100% for $x = 0$–30 km and decreases linearly to 0% at $x = 130$ km and open sea of 1°C extends for $x = 130$–457 km. The sounding of a cold air outbreak is used for the initial condition.

The atmosphere over the packed ice is stably stratified and the vertical shear is large. Mixing layer develops with the distance from the edge of the packed ice. The cloud bands develop within the mixing layer. Figure 9.8 shows formation and development of cloud bands over the sea. They begin to form in the region of sea ice density of 50–70% and intensify with a distance. A large number of cloud bands form in the cold air stream. Some cloud bands merge each other and selectively develop. Consequently, the number of lines decreases with the distance along the basic flow.

Close view of the upstream region (Fig. 9.9) shows that upward and downward motions are almost uniform in the x-direction. As a result, cloud ice and precipitation extend in the x-direction uniformly. This indicates that the convections are the roll convection type in the upstream region.

In the downstream region, the roll convections change to the alignment of cellular convections (Fig. 9.10). While the upward motions are centered and downward motions are located on their both sides, cloud and precipitation show cellular pattern. In this region, the mixing layer was fully developed and the vertical shear almost vanishes in the mixing layer.

In the region of far downstream, convections change to randomly distributed cells (Fig. 9.11). Band shape of cloud almost disappears and convections become closed cell type. The morphological transformation from alignment of cells to random cells is often observed by satellite. The experiment successfully simulates the formation process of cloud bands, their extension and merging processes, and the morphological changes of convection from the roll to cellular types.

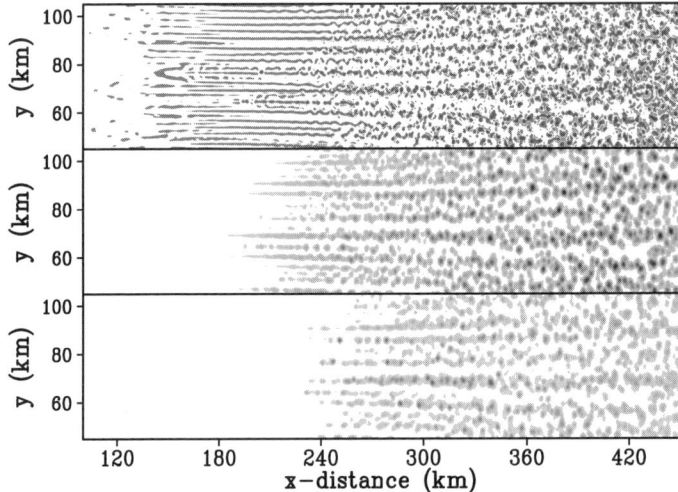

Fig. 9.8 Horizontal cross sections of vertical velocity (*upper panel*), mixing ratio of precipitation (snow, graupel, and rain) (*middle panel*) at 1 000 m in height and mixing ratio of cloud ice at 1 300 m in height (*lower panel*) for $x = 100$–450 km at 18 h from the initial time

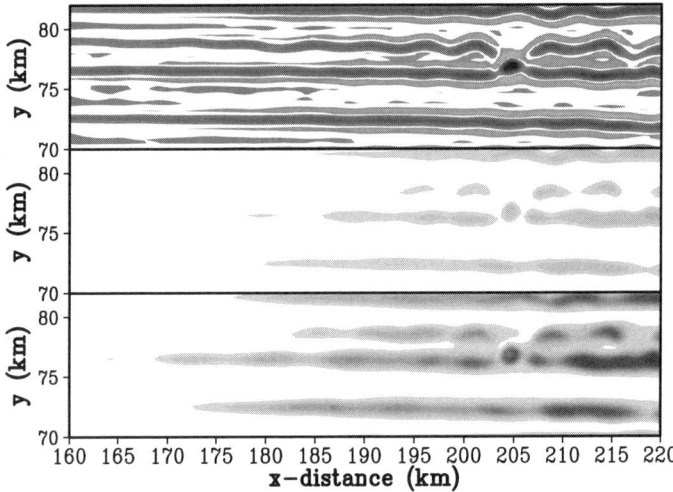

Fig. 9.9 Horizontal cross sections of vertical velocity (*upper panel*), mixing ratio of precipitation (*middle panel*) at 900 m in height and mixing ratio of cloud ice at 1 100 m in height (*lower panel*) for $x = 160$–220 km

9.6 Snowstorms

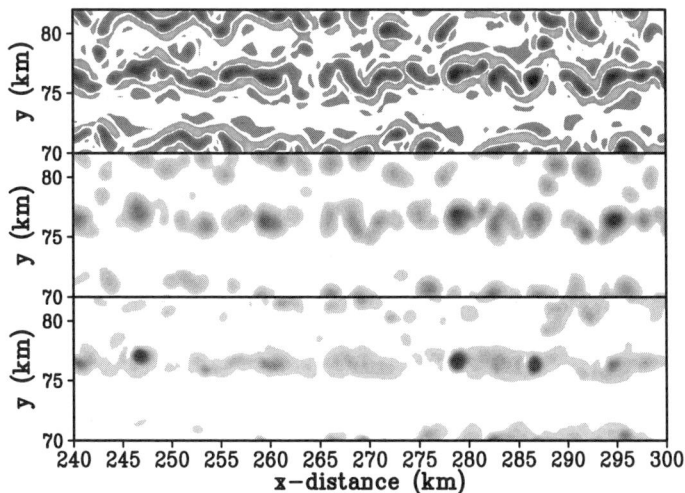

Fig. 9.10 Horizontal cross sections of vertical velocity (*upper panel*), mixing ratio of precipitation (*middle panel*) at 1 000 m in height and mixing ratio of cloud ice at 1 300 m in height (*lower panel*) for $x = 240$–300 km

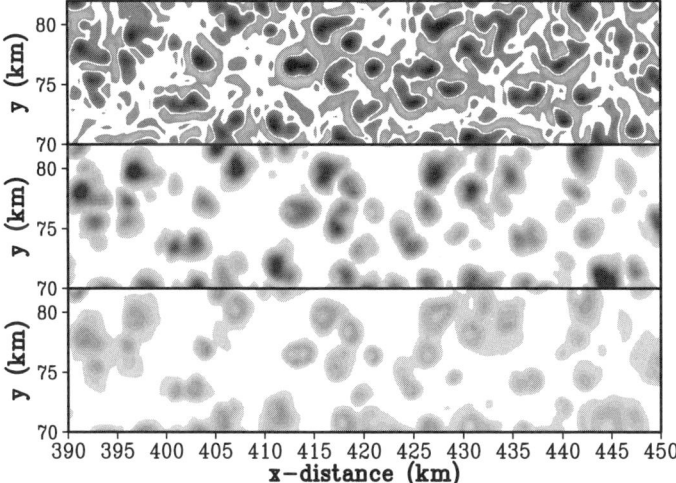

Fig. 9.11 As in Fig. 9.10, but for $x = 390$–450 km

9.6.2 Snowstorms Over the Sea of Japan

One of the major precipitation systems in East Asia is snow cloud in a cold polar air stream. In particular, various types of precipitation systems develop over the Sea of Japan: longitudinal and transversal cloud bands, convergence zone (the Japan-Sea Polar-airmass Convergence Zone; JPCZ), and vortexes. Their horizontal scale ranges from a few 100 km to 1 000 km while they are composed of convective clouds whose horizontal scale is a few kilometers.

To study development process and detailed structure of longitudinal and transversal cloud bands, we performed simulation experiment of the cold air outbreak over the Sea of Japan on January 14, 2001. The initial field at 0600 UTC, January 13, 2001 and boundary condition were provided by the JMA-RSM. The domain of the simulation covered most part of the Sea of Japan and horizontal resolution was 1 km to resolve convective clouds (Table 9.6).

Snow cloud bands over the Sea of Japan are realistically simulated (Fig. 9.12). An intense and thick cloud band composed of cumulonimbus clouds extends along the

Table 9.6 Experimental design of the snowstorm over the Sea of Japan

Domain	$x = 1350, y = 1350, z = 16$ km
Grid number	$x = 1353, y = 1353, z = 43$
Grid size	$H = 1000$ m, $V = 200$–400 m
Integration time	18 h
ES node number	36 (288 CPUs)

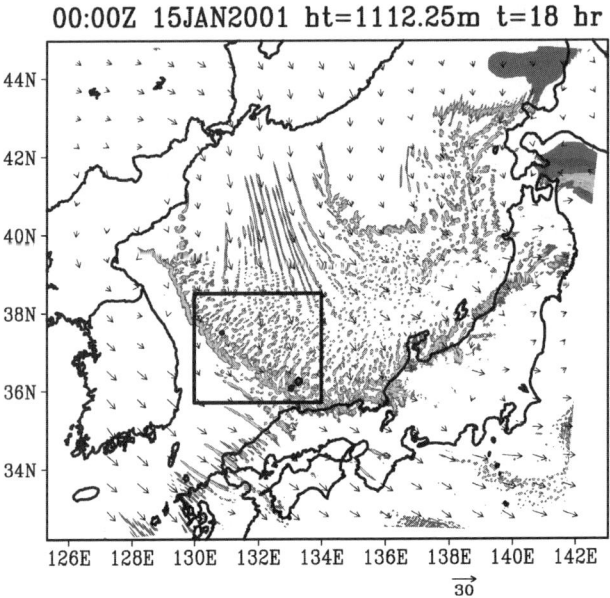

Fig. 9.12 Mixing ratio of precipitation (color levels; g kg^{-1}) and horizontal velocity (*arrows*) at a height of 1 112 m at 0000 UTC, January 15, 2001

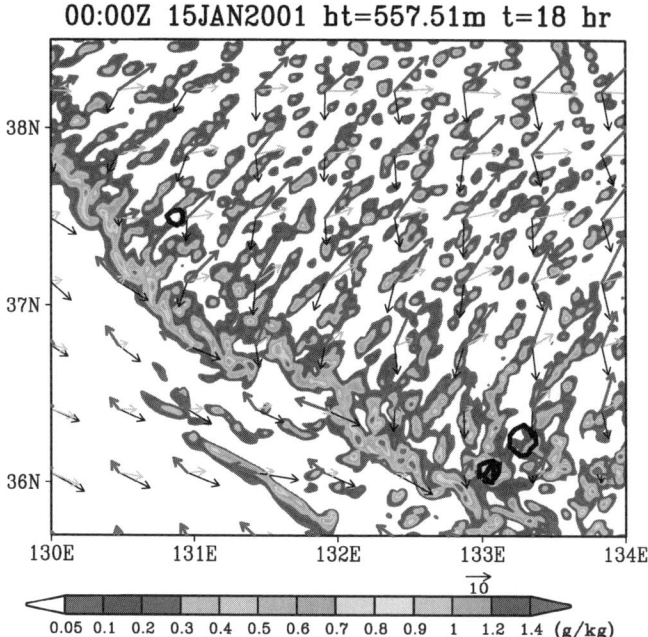

Fig. 9.13 Mixing ratio of precipitation (color levels; g kg^{-1}). *Black* and *blue arrows* are horizontal wind velocity at a height of 315 and 2 766 m, respectively. *Red arrows* are wind shear between these levels

JPCZ from the root of the Korean Peninsula to the Japanese islands. Plenty of thin cloud bands develop over the sea. Longitudinal and transversal cloud bands form to the west and east of the intense cloud band, respectively. The enlarged display of the transversal cloud bands shows that the cloud bands extend the SW–NE direction, which is almost parallel to the vertical wind shear between levels of the top of clouds and the surface (Fig. 9.13). This is consistent with the dynamic theory shown by Asai (1972).

9.7 Summary

Accurate and quantitative simulation of high-impact weather systems using a high-resolution numerical model is important for understanding mechanism and structure of conspicuous phenomena in the atmosphere. This will contribute to reliable weather prediction and to prevention/reduction of disasters due to a severe weather. Since most high-impact weather systems consist of intense cumulonimbus clouds and they have a multiscale structure, it is necessary to use a cloud-resolving model for a quantitative simulation. Each class of the multiscale components has a wide range of horizontal scales from cloud-scale to synoptic-scale. It is necessary to perform calculation within a large domain and with a very fine grid.

We have been developing a cloud-resolving numerical model named CReSS for numerical experiments and simulations of clouds and storms. Parallel computing is indispensable for these computations because most cloud systems have multiscale structures. In this study, we described the basic formulations and important characteristics of the CReSS model and showed some results of the numerical experiments of high-impact weather systems.

CReSS has been optimized for the Earth Simulator and its performance was evaluated as sufficiently high. Using CReSS on the Earth Simulator, we performed high-resolution simulations of high-impact weather systems: the localized heavy rainfall in Niigata area in 2004, typhoons of T0418 and T0423, and snowstorms in cold polar air streams. These results show that both detailed structures of individual convective clouds and overall structures of storm systems are successfully simulated using CReSS on the Earth Simulator. These experiments will contribute for accurate and quantitative prediction of high-impact weather systems and disaster prevention/reduction.

Acknowledgments The development of the CReSS model was a part of the project led by Professor A. Sumi of the University of Tokyo and was supported by the Research Organization for Information Science and Technology (RIST). The simulations and calculations of this work were performed using the Earth Simulator at the Earth Simulator Center. The author would like to thank Mr. A. Sakakibara for his indispensable contribution for the development of the CReSS model.

References

Asai, T., 1972: Thermal instability of a shear flow turning the direction with height. *J. Meteor. Soc. Japan,* **50**, 525–532.
Cotton, W. R., G. J. Tripoli, R. M. Rauber and E. A. Mulvihill, 1986: Numerical simulation of the effects of varying ice crystal nucleation rates and aggregation processes on orographic snowfall. *J. Climate Appl. Meteor.,* **25**, 1658–1680.
Ikawa, M and K. Saito, 1991: Description of a nonhydrostatic model developed at the Forecast Research Department of the MRI. *Technical Report of the MRI,* **28**, 238pp.
Lin, Y. L., R. D. Farley and H. D. Orville, 1983: Bulk parameterization of the snow field in a cloud model. *J. Climate Appl. Meteor.,* **22**, 1065–1092.
Louis, J. F., M. Tiedtke and J. F. Geleyn, 1981: A short history of the operational PBL parameterization at ECMWF. *Workshop on Planetary Boundary Layer Parameterization* 25–27 Nov. 1981, 59–79.
Murakami, M., 1990: Numerical modeling of dynamical and microphysical evolution of an isolated convective cloud – The 19 July 1981 CCOPE cloud. *J. Meteor. Soc. Japan,* **68**, 107–128.
Murakami, M., T. L. Clark and W. D. Hall 1994: Numerical simulations of convective snow clouds over the Sea of Japan; Two-dimensional simulations of mixed layer development and convective snow cloud formation. *J. Meteor. Soc. Japan,* **72**, 43–62.
Smagorinsky, J., 1963: General circulation experiments with the primitive equations. I. The basic experiment. *Mon. Wea. Rev.,* **91**, 99–164.
Tsuboki, K. and A. Sakakibara, 2001: CReSS User's Guide 2nd Edition, 210p.
Tsuboki, K. and A. Sakakibara, 2002: Large-scale parallel computing of Cloud Resolving Storm Simulator. *High Performance Computing, Springer,* H. P. Zima et al. Eds., 243–259.

Chapter 10
An Eddy-Resolving Hindcast Simulation of the Quasiglobal Ocean from 1950 to 2003 on the Earth Simulator

Hideharu Sasaki, Masami Nonaka, Yukio Masumoto, Yoshikazu Sasai, Hitoshi Uehara, and Hirofumi Sakuma

Summary An eddy-resolving hindcast experiment forced by daily mean atmospheric reanalysis data covering the second half of the twentieth century was completed successfully on the Earth Simulator. The domain covers quasiglobal from 75°S to 75°N excluding arctic regions, with horizontal resolution of 0.1° and 54° vertical levels. Encouraged by high performance of the preceding spin-up integration in capturing the time-mean and transient eddy fields of the world oceans, the hindcast run is executed to see how well the observed variations in the low- and midlatitude regions spanning from intraseasonal to decadal timescales are reproduced in the simulation. Our report presented here covers, among others, the El Niño and the Indian Ocean Dipole events, the Pacific and the Pan-Atlantic decadal oscillations, and the intraseasonal variations in the equatorial Pacific and Indian Oceans, which are represented well in the hindcast simulation, comparing with the observations. The simulated variations in not only the surface but also subsurface layers are compared with observations, for example, the decadal subsurface temperature change with narrow structures in the Kuroshio Extension region. Furthermore, we focus on the improved aspects of the hindcast simulation over the spin-up run, possibly brought about by realistic high-frequency daily mean forcing.

10.1 Introduction

Ever since the recognition of the important role played by mesoscale eddies in not only regional but also the global ocean dynamics, eddy-resolving global simulations became one of the goals for oceanic general circulation models (OGCMs). Along this line of researches, a few pilot simulations with high resolution covering either a chosen basin or the world ocean had started in the 1990s. The study of Semtner and Chervin (1992) was the first attempt to investigate the impact of so-called eddy permitting simulations on the quality of the simulated global oceanic fields. Taken together, the mean states of their simulated ocean conditions well captured the characteristics of the observational counterparts, though the simulated western boundary currents failed to separate at the expected latitudes. Smith et al. (2000) performed a simulation covering the North Atlantic with higher horizontal resolution of 0.1°, which successfully simulated the separation of the Gulf Stream and also realistic energy level of eddy activity. This result of Smith et al. (2000) was further confirmed by the works of Hurlburt and Hogan (2000) and Chassignet and Garraffo (2001), suggesting that OGCMs should have the horizontal grid spacing of the order of 0.1° or finer in order to well reproduce not only eddy-related fine structures but also realistic pathways of narrow western boundary currents including their separation points. However, all such attempts are hampered mainly by the computational performance of the contemporary supercomputers.

A couple of noteworthy eddy-resolving simulations were executed following the above-mentioned attempts. Using the Southampton Oceanography Centre's Ocean Circulation and Advanced Modelling Project (OCCAM) model (Webb, 2000), Coward et al. (2002) presented impressive preliminary results of their simulation with horizontal grid spacing of $1/12°$. A 15-year integration using the global POP with 0.1° horizontal grid spacing was done by Maltrud and McClean (2005), in which realistic global oceanic fields including eddy activity are simulated. Thus, an era of eddy-resolving global ocean simulations using sophisticated OGCMs and high-performance supercomputer systems has started with a variety of potential application, many with societal implications.

Responding to the rising tide of eddy-resolving global simulation studies, ocean simulation group at the Earth Simulator's initiative attempted first a 50-year climatological spin-up integration with horizontal grid spacing of 0.1° for a quasiglobal domain, which was successfully completed and reported by Masumoto et al. (2004) and Ohfuchi et al. (2005). Following that, a half-century long hindcast (this report) had been performed. The performance of such a high resolution model in reproducing a large variety of modes of variation in the world ocean is sure to be of great interest to both oceanic and climate communities. The main theme of the present chapter is to validate simulated results against observed ones. In what follows, we will give the overall view of our hindcast simulation focusing on dominant modes in the simulated variability, including the El Niño events, the Pacific decadal oscillation (PDO), the Pan-Atlantic decadal oscillation (PADO), and the recently discovered Indian Ocean dipole (IOD) events, as well as the intraseasonal variations in the equatorial Pacific and Indian Oceans. The structure of this chapter is as follows: The model description is given in the following Sect. 10.2. In Sect. 10.3, a few informative indexes of the model performance on the trend of time integration, on the simulated global upper-layer ocean circulations, and on the eddy activities are introduced to show the important global features of our simulation. Section 10.4 reports on the simulated variability with intraseasonal-to-decadal timescales. The final Sect. 10.5 is devoted to the summary and discussion.

10.2 Model Description

The OGCM used in this study is based on the Modular Ocean Model version 3 (MOM3) (Pacanowski and Griffies, 1999), developed at Geophysical Fluid Dynamics Laboratory/National Oceanic and Atmospheric Administration (GFDL/NOAA). It is highly optimized for the Earth Simulator (ES) and is called the OGCM for the ES (OFES). The model utilizes z-level coordinate in vertical and solves three-dimensional primitive equations in spherical coordinates under the Boussinesq and hydrostatic approximations.

The model domain covers a quasiglobal region from 75°S to 75°N excluding the Arctic Ocean, with horizontal grid spacing of 0.1°. The vertical level spacing varies from 5 m at the surface to 330 m at the maximum depth of 6,065 m, and the number of vertical levels is 54. To represent upper ocean circulation realistically, 20 levels are confined in a layer between the sea surface and the 200 m depth. The model topography is generated using 1/30° bathymetry data set provided from the OCCAM project at the Southampton Oceanography Centre (which we obtained through GFDL/NOAA). The partial cell method (Pacanowski and Gnanadesikan, 1998) implemented in MOM3, which allows bottom ocean cells to be of different thickness, is employed to represent the topography realistically.

A scale-selective damping of biharmonic operator is utilized for horizontal mixing of momentum and tracers to suppress computational noise. The viscosity and diffusivity coefficients are calculated as in Smith et al. (2000) with the value of -2.7×10^{10} m^4 s^{-1} for momentum and -9×10^9 m^4 s^{-1} for tracers, respectively, at the equator. They vary proportionally to the cube of the zonal grid spacing. Vertical viscosity and diffusivity are calculated using the K-profile parameterization (KPP) (Large et al., 1994).

The model ocean is driven by the wind stresses and surface tracer fluxes for the period from 1950 to 2003, from the last condition of the 50-year climatological spin-up integration (Masumoto et al., 2004) as the initial state. The daily mean values obtained from the NCEP/NCAR reanalysis products (Kalnay et al., 1996) are utilized for the wind stresses and the atmospheric variables that need to be specified for the tracer flux calculation. The bulk formula of Rosati and Miyakoda (1988) is adopted for the surface heat flux calculation, while the salinity flux is obtained from precipitation rate of the reanalysis data and evaporation rate derived using the bulk formula. In addition to this salinity flux, the surface salinity is restored to the climatological monthly mean value of the World Ocean Atlas 1998 (WOA98; Boyer et al., 1998a,b,c), with the restoring timescale of 6 days, so that the effects of the river run-off are included implicitly.

In the regions within 3° latitudinal distances from the northern and southern artificial boundaries at 75°N/S, temperature and salinity are restored to the monthly mean climatological values of the WOA98 (Antonov et al., 1998a,b,c; Boyer et al., 1998a,b,c) at all levels, with the restoring timescale that increases linearly from 1 day at the boundaries to 720 days at 3° from the boundaries. This buffer layers suppress the unrealistic wave propagation along the artificial boundaries and can incorporate the effects of sea ice processes. To avoid from being frozen, the surface heat flux is set to be zero in OFES when the flux is upward and the sea surface temperature (SST) becomes the value lower than $-1.8°C$, a typical freezing temperature of sea water.

To attain an efficient computational performance on the ES, OFES was highly tuned with parallelization and vectorization procedures. As the result of these optimizations, 1-year model integration using 500 processors (63 nodes) on the ES is completed in 15 h CPU time. The ES and OFES together provide us an interesting opportunity to perform several-decade integrations of the quasiglobal eddy-resolving simulations in a practical time.

We have stored snapshot data every 3 days and monthly mean fields for the entire model domain for the period of 1950–2003. The size of the snapshot data for a three-dimensional variable is about 1.09 GB (1 GB = $1,024^3$ byte), and the total amount of the stored data exceeds 37 TB (1 TB = $1,024^4$ byte). To save the storage space, therefore, we keep only the data on ocean grid points. This efficiently reduces the data size to about 57% of the total amount. In addition, a new file handling system has been developed for extracting an appropriate subdata set for further analyses, with functions to restore the compressed data and to subsample from original resolutions if required.

10.3 Overview of the Simulated Fields

The OFES hindcast integration from 1950 to 2003 forced by the daily mean atmospheric reanalysis data provide us a unique data set to investigate surface and subsurface variations at intraseasonal to decadal timescales in the tropical and midlatitude regions. Before describing some specific phenomena in detail, an overview of the simulated fields is given in this section.

10.3.1 Variations of Global Mean Values

Figure 10.1 shows time series of the global (75°S–75°N) mean potential temperature, salinity, and kinetic energy, which are dominated by the interannual and seasonal variations. In the time series of the potential temperature and salinity, relatively large adjustment to the initial biases, due to the mismatch between the initial conditions and the forcing fields, seems to appear mostly in the early 1950s. The simulated global mean potential temperature (Fig. 10.1a) and salinity (Fig. 10.1b) are about 3.63°C, and 34.727 psu, respectively. These are slightly warmer and fresher than the values of 3.58°C and 34.734 psu in the global POP simulation (Maltrud and McClean, 2005) and the values of 3.49°C and 34.749 psu in the SODA_1.4.2 (Simple Ocean Data Assimilation) reanalysis data (Carton et al., 2000a,b). The mean values are also warmer and slightly saltier than the values of 3.52°C and 34.717 psu in the observed climatology of the WOA01 (Stephens et al., 2002; Boyer et al., 2002).

10.3 Overview of the Simulated Fields

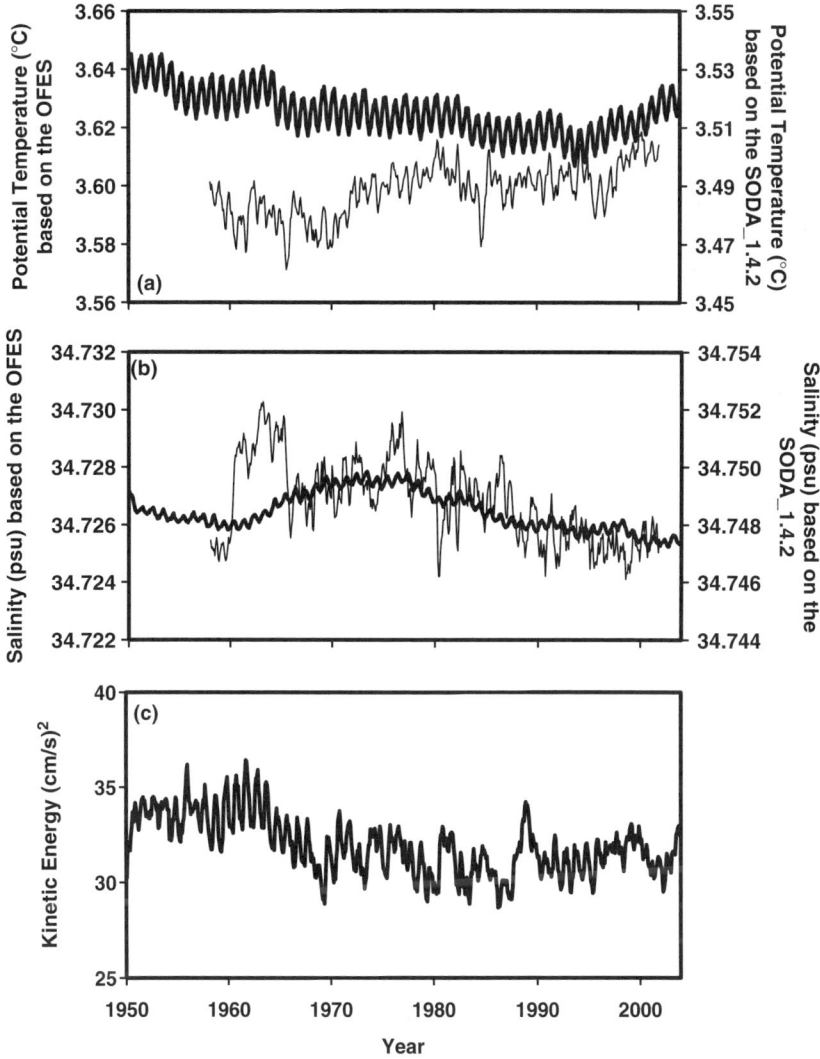

Fig. 10.1 Time series of global (75°S–75°N) mean (**a**) potential temperature (°C), (**b**) salinity (psu), and (**c**) kinetic energy (cm^2 s^{-2}) from 1950 to 2003 based on the OFES hindcast simulation. *Thin curves* in (**a**) and (**b**) show the values based on the SODA_1.4.2 reanalysis data (Carton et al., 2000a,b)

Comparison of the time series of the global mean potential temperature and salinity between the OFES result and the SODA reanalysis shows that longer timescale variations are rather similar after the mid-1980s and mid-1960s, respectively. While a warming trend is observed after 1970 in SODA, the OFES hindcast demonstrates a slight cooling trend until the mid-1990s. The difference between SODA and OFES may be mostly attributed to long-term adjustment in OFES. For about 90 years in the OFES spin-up and hindcast integrations, the global mean potential temperature decreases from 3.68 to 3.62°C. It is also apparent that seasonal-to-interannual variability in the global mean salinity is much smaller in OFES than that in SODA. One possible candidate for the differences between the two products is the omission of the high latitude processes such as the sea ice formation and the deep convection, but the issue is an open question.

The simulated global mean kinetic energy varies within a range between 30 and 35 cm^2 s^{-2}, with the mean value of 31.6 cm^2 s^{-2} for the period between 1960 and 2003. This mean value is larger than that for the climatological run, 30.3 cm^2 s^{-2}, which is the averaged over the last 10 years of the 50-year spin-up integration. The increase seems to be caused by the intraseasonal and interannual variability in the forcing fields, which induce the variability in the global mean kinetic energy at those timescales.

10.3.2 Global Upper-Layer Ocean Circulations

The overall aspects of mean global ocean circulations in the OFES hindcast simulation are almost the same as the counterparts in the 50-year climatological spin-up integration reported in Masumoto et al. (2004). A snapshot of the simulated current speed at 100 m depth (Fig. 10.2) shows that global wind-driven circulations are well represented and that the major current systems are composed of several narrow

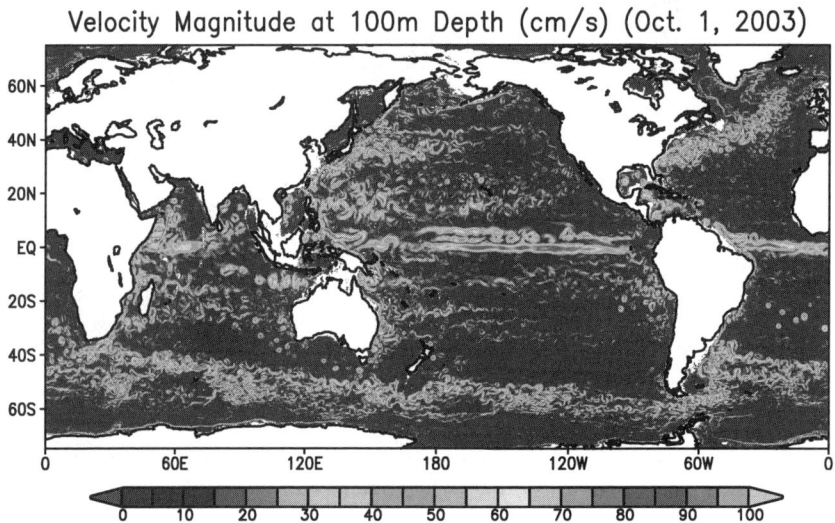

Fig. 10.2 Snapshot of simulated current speed (cm s^{-1}) at 100 m depth on October 1, 2003

jet-like structures. Horizontal distribution of root-mean-square variability of the sea surface height anomaly (SSHA) is in good agreement with the TOPEX/Poseidon-ERS altimetry observation (Ducet et al., 2000) (Fig. 10.3), supporting the idea that there is a critical horizontal resolution of about 0.1°, for the model to adequately capture the mesoscale eddy activity. The large SSHA variability, associated with strong mesoscale eddy activities, are found in the regions along swift currents, including the western boundary currents of the subtropical gyres in the major basins, the Antarctic circumpolar current (ACC), and the equatorial currents. The mesoscale eddies are also apparent even in the interior regions of the subtropical gyres. Well-known coherent rings such as the loop current rings in the Gulf of Mexico (Fig. 10.4) and the Agulhas Rings in the South Atlantic are also well represented.

The pathways of the simulated western boundary currents agree well with the observed ones. The Kuroshio and the Gulf Stream realistically separate from the western boundary at almost the right latitudes. The Equatorial Undercurrents (EUCs) are also well captured along the equator in the Pacific and the Atlantic, with realistic current structures and magnitude.

10.3.3 Improvements over the Spin-Up Integration

Distribution and magnitude of the SSHA variability in the hindcast run are basically similar to those in the OFES 50-year spin-up integration as well as the satellite observation (Fig. 10.3). However, there are some improvements over the spin-up integration in the hindcast simulation. One example can be seen in the Gulf of Mexico. The large SSHA variability associated with the loop current rings is found only in the hindcast simulation. In addition, monthly mean SSH fields for the each run (Fig. 10.4) show that the loop current rings pinch off realistically only in the hindcast simulation. However, the rings in the spin-up integration are located stationary to the north of Yucatan Channel, with another eddy to the south of it. This suggests the importance of the atmospheric forcing to the local eddy activity in the Gulf of Mexico, but detailed mechanisms for such link should be investigated in future study.

Another example of the improvement can be seen in the tropical regions and the South Pacific Ocean. The SSHA variability in the hindcast run is significantly larger than that in the spin-up run and comparable to the observed variability. The higher variability in the tropical Pacific and the equatorial Indian Ocean is attributable to the El Niño/Southern Oscillation (ENSO) phenomena and the IOD mode events (Saji et al., 1999), respectively, whose characteristics in OFES are described in detail in Sect. 10.4. The improvement in the South Pacific region, on the other hand, seems to be related to the high frequency variability in the daily mean atmospheric fields, since the similar increase in the SSHA variability is also found in the result, in which the interannual variations of the SSHA are filtered out by subtracting 1-year running mean from the original variations.

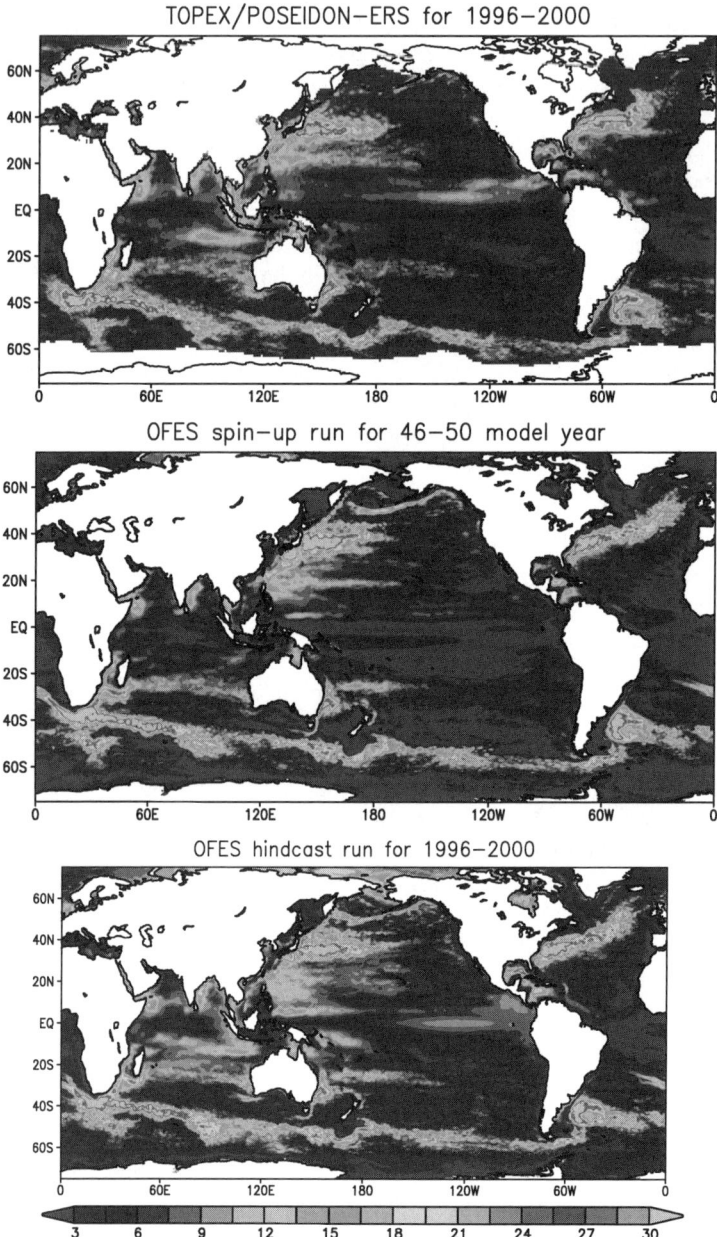

Fig. 10.3 Distributions of root-mean-square variances of sea surface height anomalies (cm) based on (**a**) the satellite observation for 1996–2000, (**b**) the spin-up integration for year 46–50, and (**c**) the hindcast simulation for 1996–2000. The observed variances are estimated from TOPEX/Poseidon-ERS merged data (Ducet et al., 2000)

10.3 Overview of the Simulated Fields

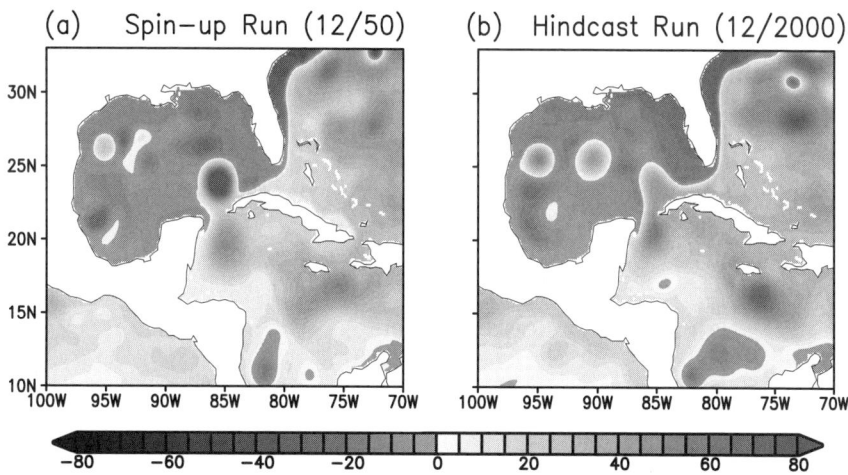

Fig. 10.4 Monthly mean distributions of sea surface height anomaly (cm) in the Gulf of Mexico in (**a**) the spin-up integration in December in year 50 and (**b**) the hindcast simulation in December 2000

10.3.4 Remaining Problems

While the OFES hindcast simulation is successful, there are still some unrealistic structures, which should be addressed in future work. One example is the pathway of the Kuroshio south of Japan. In ordinary situation, the Kuroshio flows to the northeastward along the coasts of Kyushu, Shikoku, and Honshu, west of Kii peninsula. However, the simulated Kuroshio curiously detaches from the southern coast of Kyushu and attaches to Kii peninsula, generating a cyclonic gyre to the north of the Kuroshio, before separating from the east coast of Honshu at the right latitude (Fig. 10.2). This peculiar behavior of the Kuroshio path is also found in the OFES spin-up integration, as reported in Masumoto et al. (2004).

The distribution of the SSHA variability along the Gulf Stream in the North Atlantic Ocean is not well reproduced in OFES (Fig. 10.3). In the satellite observation (Fig. 10.3a), a band of the high variability along the Gulf Stream separates into two branches, one branch turns to the north at around 40°W, extending northward along the North Atlantic Current, and another relatively weak and narrow branch extends eastward at around 35°N along the Azores Current. The OFES simulations, however, display a northeastward extension of the large variability of the SSHA along the simulated North Atlantic Current. In addition, the southern branch of the relatively high SSHA along the Azores Current does not appear in the simulated results. A possible reason for this discrepancy could be an inadequate representation of the water exchange between the Atlantic Ocean and the Mediterranean Sea through the Gibraltar Strait due to insufficient horizontal resolution even in the OFES (cf. Jia, 2000).

Another difference between the simulated results and the observed data in the SSHA variability appears in the Agulhas Ring region. The simulated rings are much stronger than those observed, and the main pathway of the rings in the South Atlantic Ocean extends to the northwest in OFES rather than to the west as observed in the satellite data. Interestingly, the above problems on the North Atlantic Current, the Azores Current, and the Agulhas Rings are also found in the global POP simulation (Maltrud and McClean, 2005).

10.4 Variability at Various Timescales

The half-century hindcast integration driven by the daily mean atmospheric fields provides us an ideal data set for investigating variability at various timescales. In this section, preliminary results, focusing on the intraseasonal, interannual, and decadal variations, are reported in some detail.

10.4.1 Intraseasonal Variability

The OFES hindcast results forced by the daily mean wind stresses are expected to have shorter timescale variability such as the intraseasonal variability. Since the spatially coherent oceanic intraseasonal variability with significant amplitude should be observed near the equatorial wave guide, as an ideal performance test, the intraseasonal variability in the equatorial Pacific and Indian Oceans are described and compared with those in observations in this subsection. In extratropics, the intraseasonal variability is largely represented as the mesoscale eddy variability and the instability of the current systems, which were described in Sect. 10.3.

10.4.1.1 Thermocline Movements in the Equatorial Pacific

Westerly wind bursts associated with the atmospheric intraseasonal variations in the tropics such as the Madden–Julian oscillation (MJO) (Madden and Julian, 1971) have been known to play a key role in triggering and terminating El Niño events by wind-induced oceanic wave processes (McPhaden, 1999; Takayabu et al., 1999). To see whether oceanic responses to these atmospheric short-term variations are represented realistically in OFES, time-longitude diagrams of 20°C isotherm depth (D20) along the equator, averaged between 2°S and 2°N, from the OFES hindcast simulation are compared with the TAO/TRITON buoy data for the 1997/1998 El Niño event in Fig. 10.5. Although there is a systematic bias in the simulated D20, which is shallower than that of the observed data by about 20 m, the OFES results capture many aspects of the intraseasonal variability as well as the interannual variations. In the

10.4 Variability at Various Timescales

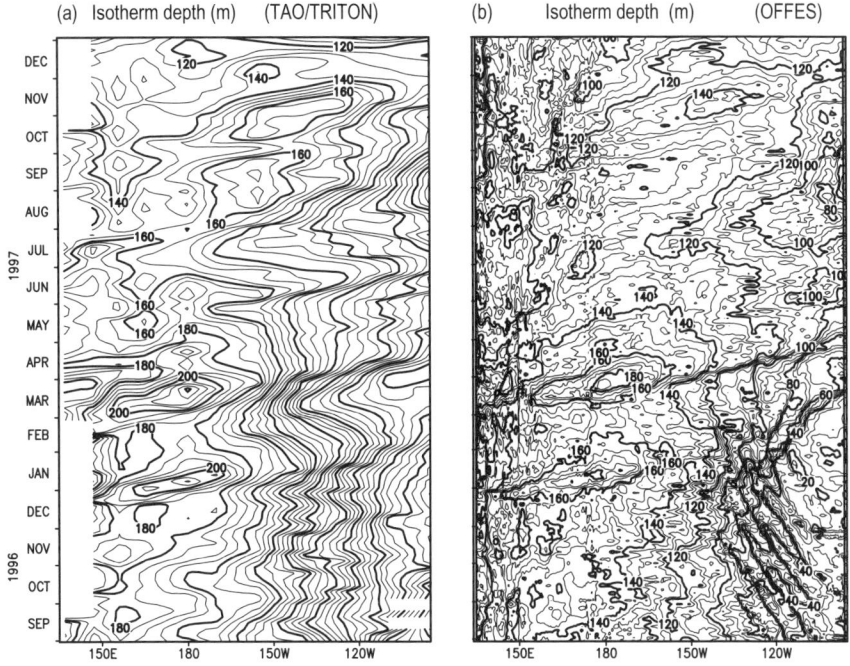

Fig. 10.5 Time-longitude diagrams of the 20°C isotherm depth (m) averaged between 2°S and 2°N based on (**a**) 5-day mean values of the TAO/TRITON array data and (**b**) 3-day snapshots of the hindcast simulation from September 1996 to December 1997. *Thick* (*thin*) *contour* intervals are 20 m (5 m)

western Pacific, a series of the Kelvin waves propagating eastward along the equator, represented by the signals with the deep isotherm, are found in November 1996, December 1996 to January 1997, and March 1997 in both the observed and simulated results. The model also captures the two intraseasonal disturbances with the deep isotherm signal, propagating in the central equatorial Pacific during the mature phase of the El Niño event in autumn 1997. The successful representation of these forced oceanic variations is mainly attributed to the high quality wind stress data used to drive the ocean model. However, the OFES results, in turn, can be considered as a valuable data set to investigate the mutual interaction between the intraseasonal and interannual variations in this particular region.

In the eastern equatorial Pacific, the westward propagating shallow isotherm signals are represented in the OFES simulation several times from September 1996 to January 1997 (Fig. 10.5b). The phase speed of the westward propagation is about 45 cm s^{-1}, with clear association with the SST anomaly, and the signals are considered as the Tropical Instability Waves, or the Legeckis Waves (Legeckis, 1977). On the other hand, the Legeckis Waves do not appear during the 1997/1998 El Niño event, which is consistent with satellite observations (Legeckis, 1977). The diagram based on the observation from TAO/TRITON array (Fig. 10.5a) is incapable of capturing the propagation of the Legeckis Waves due to the sparseness of buoys in longitudinal direction.

10.4.1.2 Upper Layer Current Variability in the Eastern Equatorial Indian Ocean

Masumoto et al. (2005) demonstrate the energetic intraseasonal variability in the upper layer currents in the eastern equatorial Indian Ocean by use of the data obtained by an upward-looking ADCP mooring located at 90°E on the equator. Dominant periods of the variability in the observed zonal and meridional currents in the layer shallower than the thermocline are 30–50 and 10–20 days, respectively. Figure 10.6 shows time-depth diagrams of zonal and meridional currents at the same location in OFES. The dominant periods of variability in the simulated zonal and meridional currents are almost the same as those reported in Masumoto et al. (2005). The vertical structures of the variability are also well reproduced in OFES. For example, the strong intraseasonal variability appears in the upper layer, both in zonal and meridional currents, with the longer timescale variability in the layer deeper than the thermocline. The eastward current in the thermocline level with the surface westward flow, which is similar to the vertical structure of the typical EUC in the Pacific and Atlantic Oceans, can be seen during February–March 2001, as observed in the ADCP measurements. Studies on mechanisms of these variability are now underway.

10.4.2 Interannual Variations

Figure 10.7 shows a time-longitude diagram of the monthly SSTA along the equator simulated in the hindcast run. In the central to the eastern Pacific Ocean, the warm and cold anomalies associated with El Niño and La Niña events are clearly found. Relatively large amplitude interannual variations are also observed in the Indian and Atlantic Oceans, with a warming trend from 1960s to 1990s in the both basins. Some preliminary comparisons of the simulated interannual variability with the observed counterparts are reported in this subsection.

10.4.2.1 El Niño/La Niña Events

To compare the ENSO variability simulated in OFES with that in the observation, we calculate the NINO3 index, the SSTA averaged in the region between 90–150°W and 5°S–5°N (Fig. 10.8). The amplitude and phase of the variability in the simulated NINO3 index (thick curve) remarkably agree with those in the observation (thin curve), which is obtained from NOAA's Climate Prediction Center. The standard deviation of the NINO3 index in OFES is 0.913°C, which is comparable to

10.4 Variability at Various Timescales

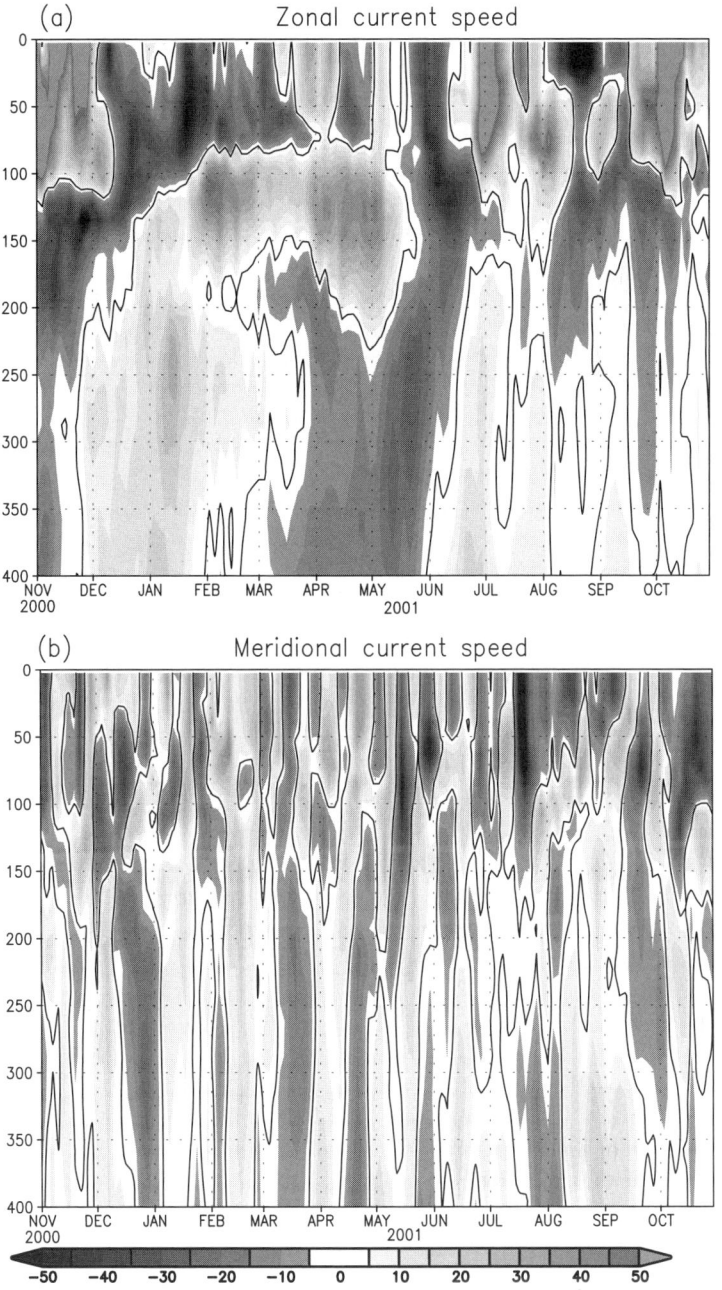

Fig. 10.6 Time-depth diagrams of the simulated (**a**) eastward current and (**b**) northward current (cm s^{-1}) at 90°E on the equator in the eastern Indian Ocean from November 2000 to October 2001. The value with the contours is zero

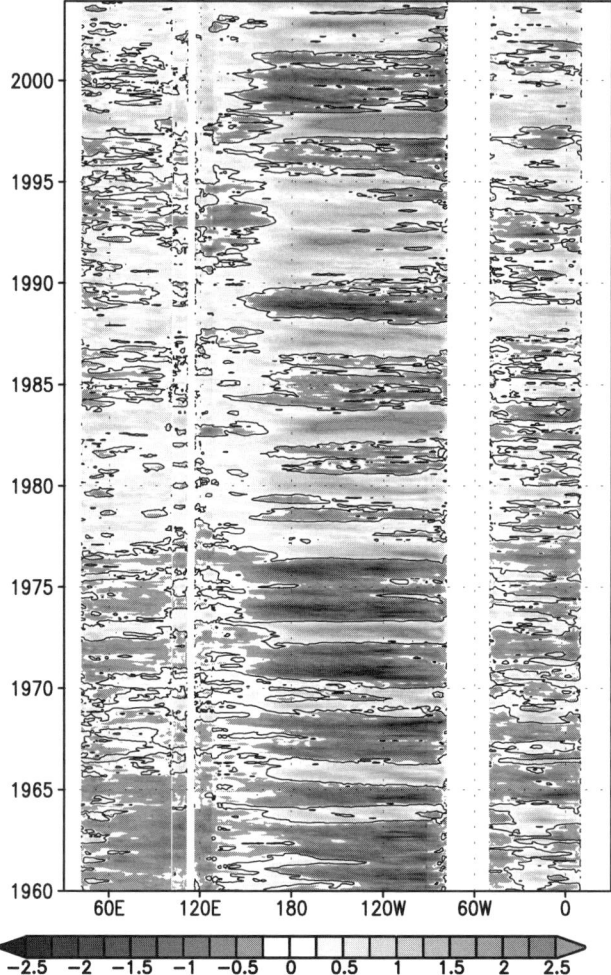

Fig. 10.7 Time-longitude diagrams of simulated sea surface temperature anomalies (°C) along the equator from 1960 to 2003. The value with the contours is zero

the value of 0.920°C in the observation. The correlation coefficient between the two time series is 0.86, suggesting the usefulness of the OFES data in analyzing oceanic responses during the evolution, mature, and termination stages of the El Niño events. Note that the simulated SST is not directly restored to the observed SST.

As an example of the subsurface variability, we focus here on the oceanic responses in the upper layer during the recent 2002/2003 El Niño event. Figure 10.9 shows meridional sections of simulated temperature, salinity, and potential density, along the dateline in autumn in 2001 and 2002, whose distributions are very similar to the observed ones taken by the TAO array (McPhaden, 2004, Fig. 10.7). In November 2002, at the peak of the El Niño event, the simulated salinity drops about 1 psu in

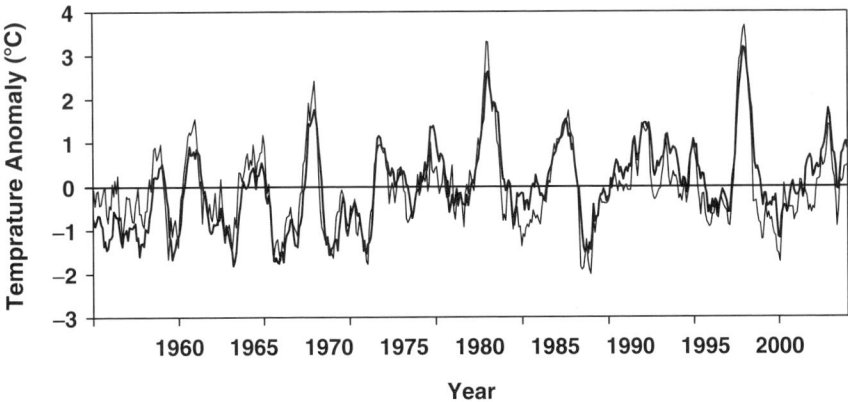

Fig. 10.8 Time series of sea surface temperature anomalies (°C) averaged in NINO3 region between 150–90°W and 5°S–5°N based on the hindcast simulation (*thick curve*) and the observation (*thin curve*) from 1960 to 2003. The observed anomalies are obtained from NOAA's Climate Prediction Center. The standard deviations in the simulation and the observation are 0.913° and 0.920°C, respectively

the upper layer between 4°S and 6°N, compared to the condition in the previous year. Such low salinity anomaly in the upper layer may result in the shallow mixed layer, which contributes to trapping of the heat within the mixed layer and to the sea surface rise due to the density change, during the evolution stage of the El Niño event, as suggested by McPhaden (2004). The simulated surface mixed layer depth in November 2002, however, is not as shallow as that in the observation, probably due to the inclusion of the restoring term for the surface salinity to the climatological field.

10.4.2.2 Indian Ocean Dipole Mode Events

The dipole mode index (DMI), the SST difference between the eastern (100–110°E, 10°S–0°) and western (60–80°E, 10°S–10°N) poles of the Indian Ocean dipole mode first introduced by Saji et al. (1999), is compared between the OFES hindcast result and the observation. The observed values are calculated using the global sea-ice and sea surface temperature data set (GISST; Rayhner et al., 1996). Figure 10.10 shows time series of 5-month running mean of the normalized DMI, excluding intraseasonal variability such as the MJO. The simulated (thick line) and observed (thin line) DMIs agree quite well, with the standard deviations of 0.257° and 0.269°C, respectively. The correlation coefficient between the two time series is 0.76.

Figure 10.11 shows the time-longitude diagrams of simulated heat content anomalies in the layer between the sea surface and 300 m depth along the equator and 5°S for the period of 1995–2000. The cold anomalies of the heat content in the eastern

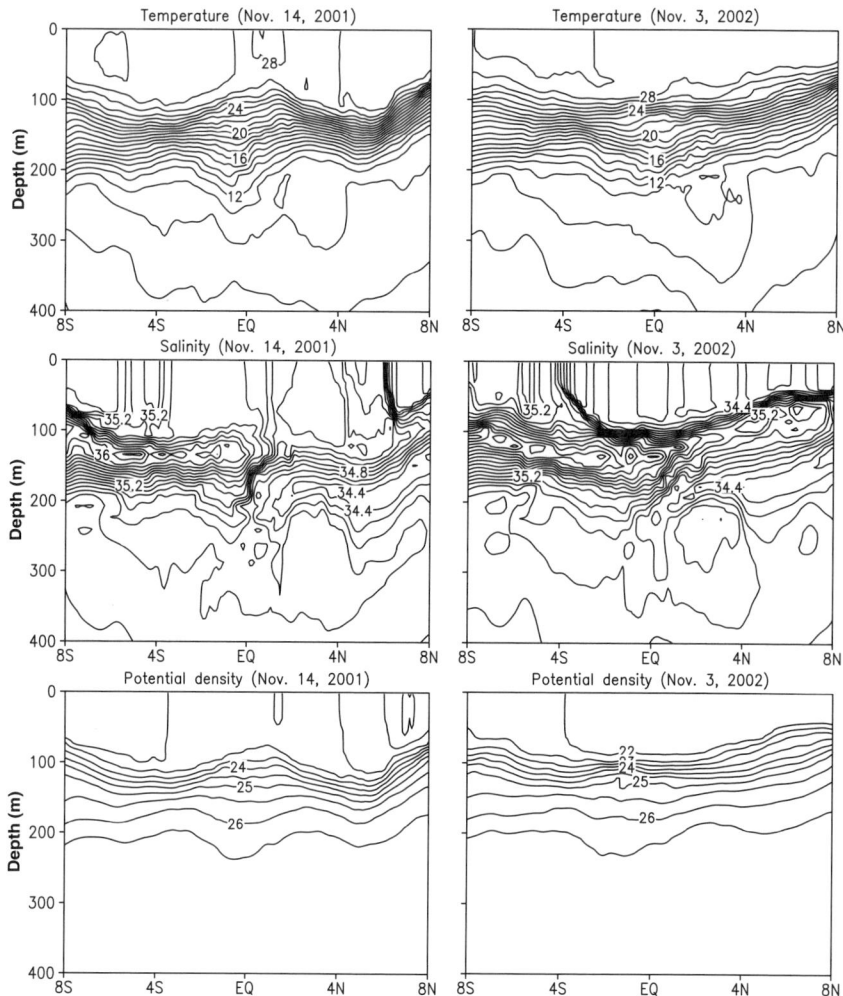

Fig. 10.9 Snapshots of latitude-depth sections of simulated (*top*) temperature (°C), (*middle*) salinity (psu), and (*bottom*) potential density (σ_θ) along the dateline in the equatorial Pacific Ocean on (*left*) November 14, 2001 and (*right*) November 3, 2002. Contour intervals are 1°C for temperature, 0.1 psu for salinity, and 0.5σ_θ for potential density, respectively

Indian Ocean are found along the both latitudes in autumn 1997. Slight phase lag of the positive heat content anomaly in the western Indian Ocean to the eastern cold anomaly is also well reproduced in OFES (cf. Saji et al., 1999). Along 5°S, the cold anomaly propagates westward with the phase speed of about 12 cm s^{-1}, which corresponds to the off-equatorial Rossby wave at this latitude, and it seems to arrive at the western Indian Ocean in about 1 year. During the previous and next years of the IOD event, i.e., in 1996 and 1998, the large positive heat content anomalies appear

10.4 Variability at Various Timescales

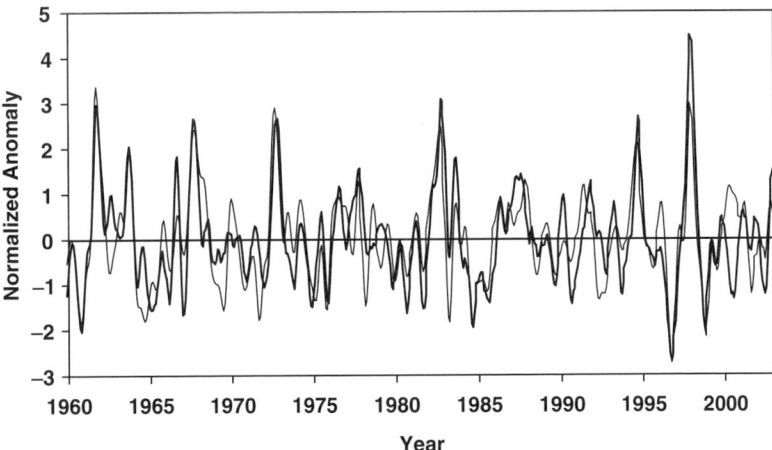

Fig. 10.10 The normalized dipole mode indexes (DMIs) based on the hindcast simulation (*thick curve*) and the observation (*thin curve*) from GISST data set (Rayhner et al., 1996) from 1960 to 2003. The indexes are normalized by their respective standard deviations and smoothed by 5-month running mean. The standard deviations of the DMIs in the simulation and the observation are 0.257° and 0.269°C, respectively

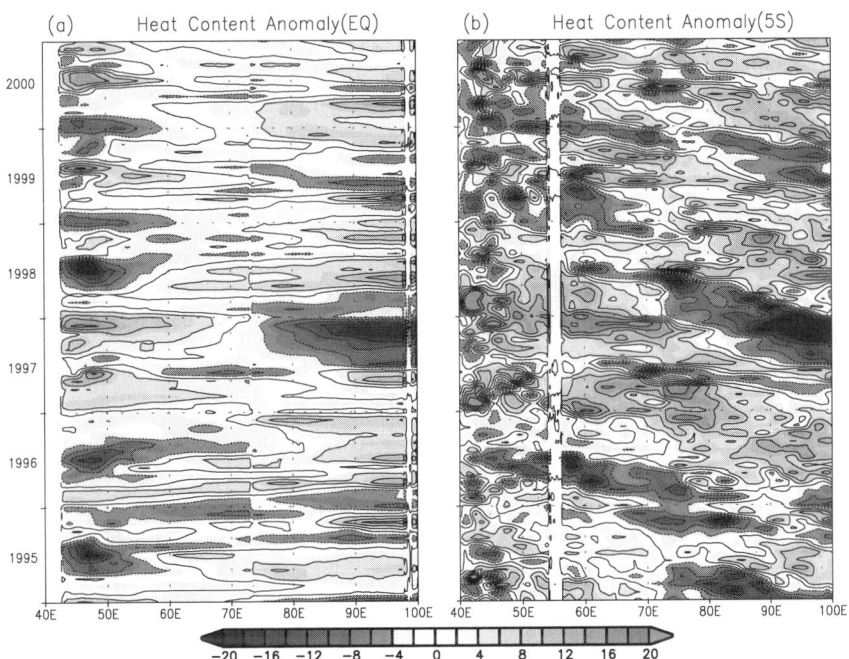

Fig. 10.11 Time-longitude diagrams of simulated heat content anomalies ($\times 10^8$ J m^{-2}) from sea surface to 300 m depth for the period of 1995–2000 along (**a**) the equator and (**b**) 5°S. Contour intervals are 4×10^8 J m^{-2}

in the eastern basin, and they propagate westward as the Rossby waves along 5°S. Importance of the oceanic wave propagation for the quasibiennial signature of the IOD event is discussed by Rao et al. (2002) and Feng and Meyers (2003), and the OFES results represent such waves in the realistic manner.

10.4.2.3 The Antarctic Circumpolar Current

Since the ACC is one of the strongest currents in the world oceans and connects the three major oceans without being blocked by landmasses, it is worth evaluating the performance of the ACC variability in the high-resolution OFES results. As shown in Fig. 10.2, the simulated ACC consists of many meandering narrow currents and is far from one simple broad current. In addition, quite complicated flow structures due to the complex bottom topography near Kerguelen Plateau in the southern Indian Ocean are represented in OFES. Such characteristics of the ACC are consistent with those in the simulation of the fine resolution Antarctic model (FRAM) (Saunders and Thompson, 1993; Lutjeharms and Webb, 1995), in the satellite observations (e.g., Gille, 1994; Hughes and Ash, 2001), and in the hydrographic observations (e.g., Orsi et al., 1995). The ACC volume transport through the Drake Passage averaged from 1960 to 2003 in the OFES hindcast simulation is 141.7 Sv (1 Sv = 10^6 m^3 s^{-1}), which is smaller than the value of 153.5 Sv in the spin-up integration, but it is slightly larger than the value of 134.0 Sv in the observation (Whitworth and Peterson, 1985).

The OFES hindcast simulation provides data of several decades to investigate mean features of the ACC variability. The period of the hindcast is longer than previous studies based on the multiyear observations (Wearn and Baker, 1980; Whitworth and Peterson, 1985; Peterson, 1988; Gille et al., 2001). The mean seasonal variation in the volume transport through the Drake Passage calculated from the results of the hindcast simulation indicates a clear semiannual signal with the maxima in March and October (not shown), which is also observed in zonal mean zonal wind stress averaged between 45°S and 65°S, suggesting the strong association between the two variables. This result is consistent with that in the spin-up integration (Masumoto et al., 2004, their Fig. 19) and with those in the direct observations (Whitworth and Peterson, 1985; Gille et al., 2001).

Figure 10.12 shows time series of monthly (thin curves) and 13-month running mean (thick curves) ACC volume transport in the hindcast simulation (Fig. 10.12a) and zonal mean zonal wind stress averaged between 45°S and 65°S (Fig. 10.12b) for the period of 1960–2003. There is a weak positive trend after 1965 in both the simulated ACC transport and the zonal wind stress fields. Superposed on these positive trends are significant interannual variations, indicated by the 13-month running mean values. The significant correlation coefficient of 0.65 between the two time series for the period from 1960 to 2003 suggests that the simulated interannual variations in the ACC transport are also related to those in the zonal wind stress, consistent with previous studies based on multiyear observations (Wearn and Baker, 1980; Peterson,

10.4 Variability at Various Timescales 175

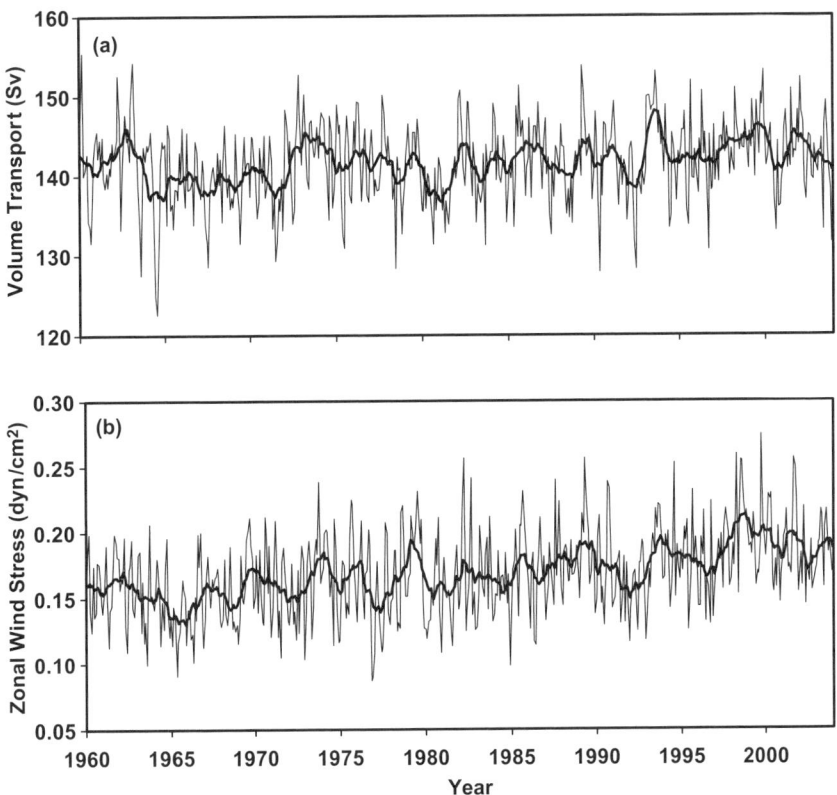

Fig. 10.12 Time series of (**a**) the simulated monthly mean ACC volume transport (Sv = 10^6 m^3 s^{-1}) through the Drake Passage and (**b**) zonal mean zonal wind stress (dyn cm^{-2}) averaged between 45° and 65°S. *Thick* (*thin*) *curves* show 13-month running mean (monthly mean) values

1988; Gille et al., 2001) and OGCMs (Gille et al., 2001). In addition to the interannual variation, intraseasonal variability is also significant in the simulated ACC transport, and the standard deviations of monthly and 13-month running mean time series are 4.96 and 2.21 Sv, respectively. Detailed investigations such as spatial or lag analyses conducted by Gille et al. (2001) are necessary to understand mechanisms for the interannual variations in the ACC, but it is beyond the scope of this chapter.

10.4.3 Decadal Variability

The 54-year hindcast of OFES makes it possible to investigate longer timescale variability in the late twentieth century. In the following subsections, simulated decadal variations in the Pacific and Atlantic Oceans are compared with those in the observations.

10.4.3.1 The Pacific Decadal Oscillation

Decadal variability in the Pacific Ocean is often described by the PDO index, a normalized index proposed by Mantua et al. (1997) that is the time coefficient of the leading principal component of the monthly SST variability in the region poleward of 20°N in the North Pacific. Figure 10.13 compares the index calculated using the OFES result (thick curve) and the one based on the observation (thin curve) obtained from the Joint Institute for the Study of the Atmosphere and Ocean (JISAO). A good agreement between the two lines indicates that the OFES simulation reproduces well the signal of the PDO, even for the rapid increase observed during the late 1970s.

To check the horizontal structure of the simulated PDO, Fig. 10.14 shows horizontal distributions of the SST differences between the positive (1977–1987) and negative (1965–1975) PDO phases from the OFES output and the Frontier Research System-Comprehensive Ocean and Atmosphere Data Set (FRS-COADS) (Tanimoto and Xie, 2002) as observed evidence. The overall structure of the simulated decadal SST difference is similar to that in the observation, although the observed distribution is much smoother due to its coarse horizontal resolution of 2°. The one advantage of the high-resolution OFES simulation can be seen in the Kuroshio Extension region, where two zonal bands of negative SST differences are found from 150°E to 170°W around 40°N in the simulation. These SST differences are associated with meridional shift of the narrow frontal structures in the Kuroshio Extension region (Nonaka et al., 2006), which can have an essential impact to the air–sea interaction in this area (Nakamura et al., 1997). Figure 10.14 demonstrates the capability of the OFES simulation in capturing such narrow structures in the SST field.

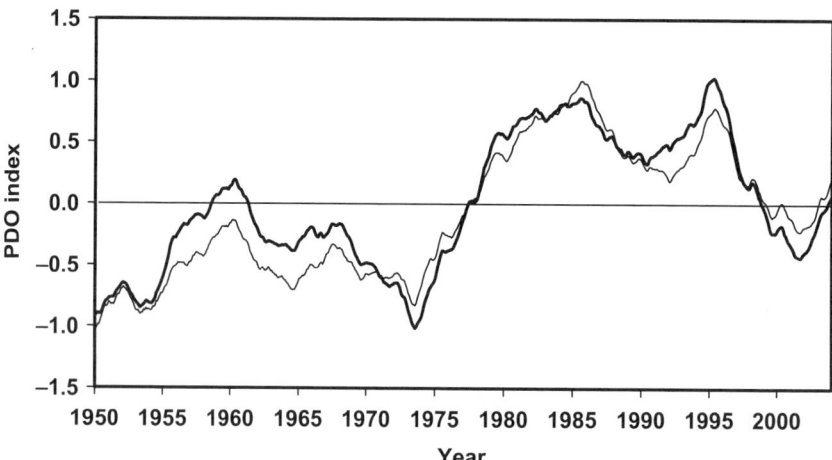

Fig. 10.13 The PDO indexes based on the hindcast simulation (*thick curve*) and the observation (*thin curve*) obtained from the JISAO. The indexes are normalized by their respective standard deviations and smoothed by 6-year running mean

10.4 Variability at Various Timescales

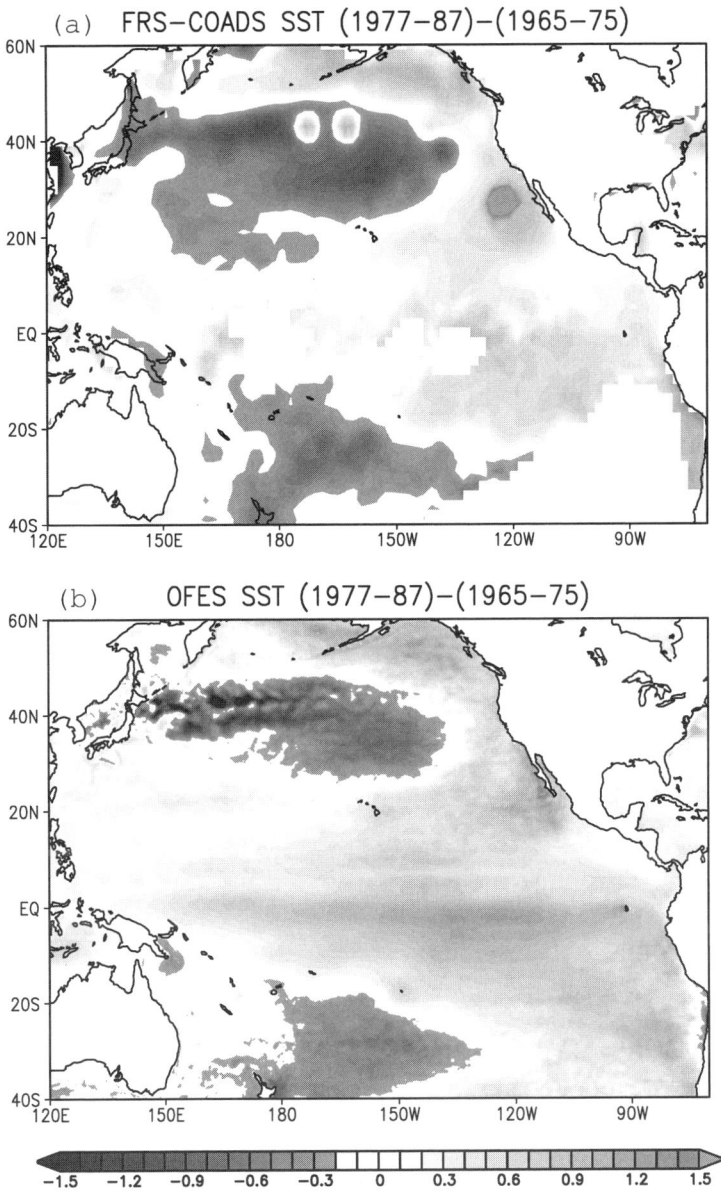

Fig. 10.14 Decadal sea surface temperature differences (°C) from 1965–1975 to 1977–1987 in the Pacific Ocean based on (**a**) the FRS-COADS data set (Tanimoto and Xie, 2002) and (**b**) the hindcast simulation. The FRS-COADS is a grid data constructed from quality-controlled ship and buoy measurements compiled in long marine reports in fixed length records (LMRF) of comprehensive ocean-atmosphere data set (COADS; Woodruff et al., 1987) with 10 days and 2° horizontal resolutions

A discrepancy between the observation and simulation appears in the region west of the dateline at around 20°N, where the simulated SST difference is positive while the observed data indicates negative values. This region is occupied by the subtropical front and the subtropical countercurrent, with relatively strong mesoscale eddy activity (see Figs. 10.2 and 10.3), whose meridional movements may affect the decadal variability in the SST difference.

To illustrate simulated subsurface temperature structure of the decadal variability, Figure 10.15 shows the horizontal distributions of the temperature differences at 50 and 400 m depths in the Kuroshio Extension region, in the similar manner to Deser et al. (1999). At the depth of 50 m, a horseshoe pattern of the negative temperature difference extends from the coast of Japan to around 140°W, with positive difference located at around 165°E, 35°N and near the eastern boundary. In contrast, at 400 m depth, the difference is concentrated within the western half of the basin, except for

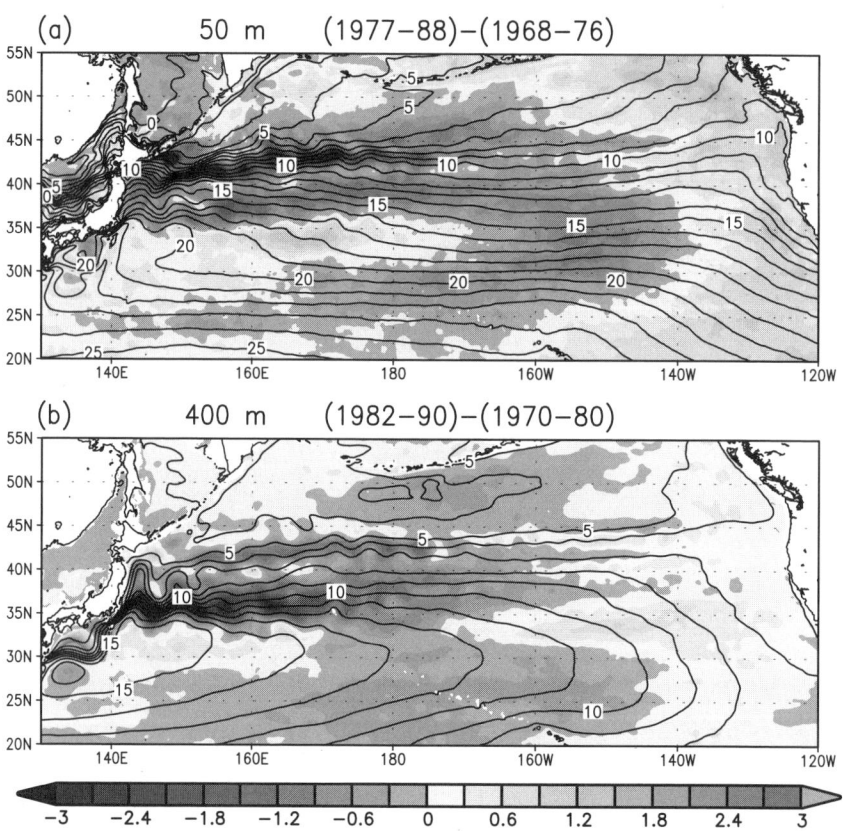

Fig. 10.15 Simulated decadal temperature differences (shadings; °C) and long-term mean temperature (contours; °C) in the subsurface layers in the Kuroshio Extension region. Contour intervals are 1°C. Following Deser et al. (1999), (**a**) the difference from 1968–1976 to 1977–1988 at 50 m depth and (**b**) that from 1970–1980 to 1982–90 at 400 m depth are plotted

the positive maximum near 150°W, 33°N. It is noteworthy that two zonally oriented bands of the large negative difference can be seen in the Kuroshio Extension region; one located at around 41°N, which has the larger signal at the depth of 50 m compared to the deeper layer, and the other at around 35°N, which is stronger at the deeper level at 400 m depth. These results are consistent with the observed structures of the subsurface temperature field (Deser et al., 1999). These successes in representations of the decadal subsurface temperature changes come from an ability to represent narrow structures and separation of the Kuroshio current from the coast of Japan at right latitude.

10.4.3.2 The Pan-Atlantic Decadal Oscillation

To examine how well the simulated results capture characteristics of the PADO (Xie and Tanimoto, 1998; Tanimoto and Xie, 2002), we define the PADO index by subtracting the area averaged SST over the northwestern box (65–75°W, 30–40°N) in the North Atlantic Ocean from the mean value of the two area averaged SST over the northern (45–55°W, 40–50°N) and the eastern (20–30°W, 10–20°N) boxes (see Fig. 10.17a for the locations of the boxes). Since the PADO appears clearly during the boreal winter, the SST averaged only during the winter is utilized. In general, the SST variability in the northern region is in phase with that in the eastern region and is out of phase to that in the northwestern region. Figure 10.16 shows the simulated and observed PADO indexes, smoothed with the 7-year running mean and normalized by their respective standard deviations. The standard deviations of the

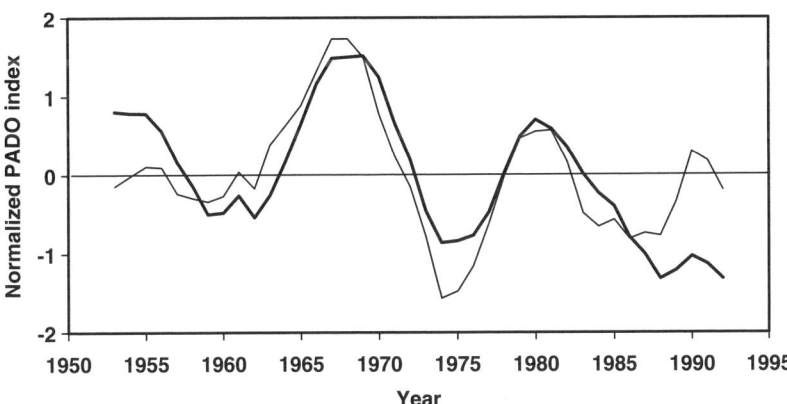

Fig. 10.16 The PADO indexes based on the hindcast simulation (*thick curve*) and the FRS-COADS data set (*thin curve*). The PADO index is defined by subtracting area mean SSTAs within the northwestern box (65–75°W, 30–40°N]) from average of two area mean values within the northern (45–55°W, 40–50°N) and the eastern (20–30°W, 10–20°N) boxes in boreal winter (January–March). The locations of the boxes are depicted in Fig. 10.17a. The indexes are smoothed by 7-year running mean and normalized by their respective standard deviations

smoothed SST in OFES (FRS-COADS) are 0.26°C (0.28°C) for the northwestern box, 0.35°C (0.30°C) for the northern box, and 0.25°C (0.14°C) for the eastern box, respectively. Relatively large difference of 0.11°C between the simulated and observed values for the eastern box results from the larger variance in OFES for the period from the late 1950s to the early 1960s, during which the observed variability is rather small. The observed PADO index indicates the positive extremes in 1967, 1980, and 1990, and the negative ones in 1974 and 1986. The OFES result captures these positive and negative extremes in the PADO index quite well, while the maximum in 1990 is significantly underestimated.

Figure 10.17 compares the simulated and observed distributions of SST differences in boreal winter between the positive and negative PADO years. The large-scale spatial structure of the SST differences associated with the PADO is well reproduced by the simulation. In addition, it is clearly seen that, in OFES, the SST differences in the northern and northwestern boxes are associated with the small-scale structures near the Gulf Stream, another indication of the correct separation latitude of the Gulf Stream along the east coast of the United States.

10.5 Concluding Remarks

We report an overview of simulated variations at various temporal scales reproduced in an eddy-resolving ocean general circulation model referred to as OFES, which has a horizontal grid spacing of $1/10°$ with 54 vertical levels and is driven by daily mean atmospheric reanalysis data for the period from 1950 to 2003. The simulated results nicely capture the variability ranging from the intraseasonal to the decadal timescales, for example, the Tropical Instability Waves, the El Niño and IOD events, the PDO, and the PADO, comparing with the observations.

Another eddy-resolving simulation of the quasiglobal ocean which incorporates with chlorofluorocarbons (CFCs) was performed (Sasai et al., 2004, 2005), following the spin-up integration. CFCs are absorbed at the sea surface and are carried by the oceanic circulation and mixing processes. Sasai et al. (2004) reported successful outcomes, representing the CFC-11 distribution in the Southern Ocean, where much of CFCs is entrained by the formation of the deep and bottom water around the Antarctic Continent. The model further indicates significant pathways of the newly formed Antarctic Bottom Water from the Weddell and Ross Seas. Figure 10.18 clearly shows the CFC-11 spreading with deep currents in the Weddell Sea. The successful simulation comes from realistic representations of model topography and the deep boundary currents in OFES.

Long-term high-resolution simulations of the global ocean such as the OFES simulations, on the one hand, provide us so huge output data sets, and it is necessary to consider data handling and visualization issues for them. The situation, though, has been improved by rapid developments in computational systems and visualization techniques (Uehara et al., 2006), which help us to extract efficiently information

10.5 Concluding Remarks

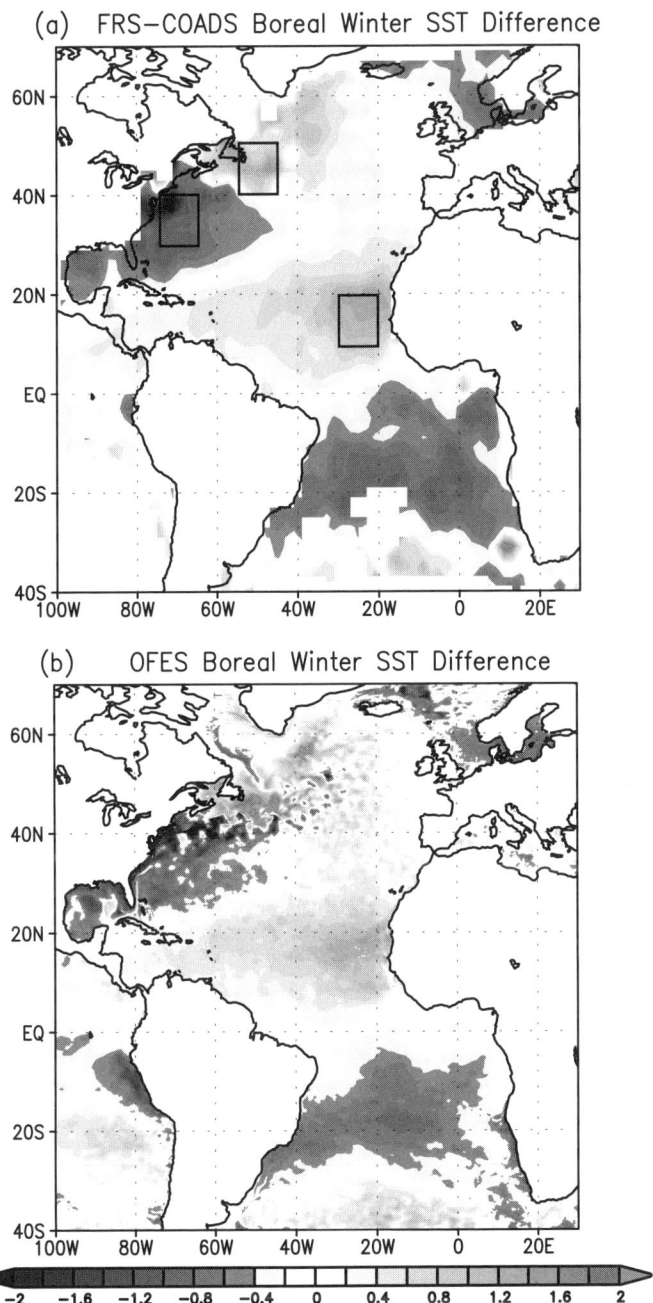

Fig. 10.17 Distributions of boreal winter (January–March) SST differences (°C) associated with the PADO. The differences between mean fields of six positive- (1968, 1969, 1970, 1979, 1980, 1981) and negative- (1972, 1973, 1974, 1984, 1985, 1986) phase years based on (**a**) the FRS-COADS data set and (**b**) the hindcast simulation are plotted

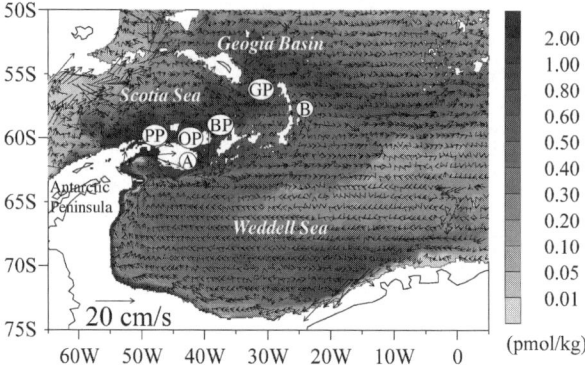

Fig. 10.18 Distribution of simulated CFC-11 concentration (pmol kg^{-1}) at 2,000 m depth in 1990 in the Weddell Sea. *Arrows* are last 10-year mean velocity (cm s^{-1}). Place names are: (A) South Scotia Ridge; (B) South Sandwich Trench; (PP) Philip Passage; (OP) Orkney Passage; (BP) Brace Passage; (GP) Georgia Passage

buried under a pile of the data. On the other hand, features represented in such simulations have so wide temporal and spatial ranges as summarized in this chapter, and simulation data set can be valuable for general oceanic studies. Then we have opened our OFES spin-up integration data set (http://www2.es.jamstec.go.jp/ofes/eng/), which may be able to contribute to research communities. We will also open a portion of the OFES hindcast data in the near future.

Acknowledgments We are grateful to OFES project members including Dr. H. Nakamura for their efforts in the model development. The OFES project using the Earth Simulator was originally initiated by Drs. T. Yamagata and H. Miyoshi. Our thanks are extended to researchers in Atmosphere and Ocean Simulation Research Group in Earth Simulator Center and Dr. Y. Tanimoto for valuable discussions, and to Messrs. Tsuda and Kitawaki for their tireless efforts to optimize the OFES for the Earth Simulator. Niño3 SSTA time series and PDO indexes were obtained from Climate Prediction Center in NOAA (http://cpc.ncep.noaa.gov/data/indices/) and Joint Institute for the Study of the Atmosphere and Ocean (JISAO, http://jisao.washington.edu/pdo/PDO.latest) respectively. Time series data from TAO/TRITON array were delivered from TAO Project Office (http://pmel.noaa.gov/tao). The SODA_1.4.2 reanalysis data were obtained from the University of Maryland Department of Meteorology's Data Storage (http://dsrs.atmos.umd.edu/DATA/SODA_1.4.2).

References

Antonov, J.I., S. Levitus, T.P. Boyer, M.E. Conkright, T. O'Brien, and C. Stephens, 1998a: World Ocean Atlas 1998 Vol. 1: Temperature of the Atlantic Ocean, *NOAA Atlas NESDIS* 27. U.S. Government Printing Office, Washington, DC.

Antonov, J.I., S. Levitus, T.P. Boyer, M.E. Conkright, T. O'Brien, and C. Stephens, 1998b: World Ocean Atlas 1998 Vol. 2: Temperature of the Pacific Ocean, *NOAA Atlas NESDIS* 28. U.S. Government Printing Office, Washington, DC.

References

Antonov, J.I., S. Levitus, T.P. Boyer, M.E. Conkright, T. O'Brien, C. Stephens, and T. Trotsenko, 1998c: World Ocean Atlas 1998 Vol. 3: Temperature of the Indian Ocean, *NOAA Atlas NESDIS* 29. U.S. Government Printing Office, Washington, DC.

Boyer, T.P., S. Levitus, J.I. Antonov, M.E. Conkright, T. O'Brien, and C. Stephens, 1998a: World Ocean Atlas 1998 Vol. 4: Salinity of the Atlantic Ocean, *NOAA Atlas NESDIS* 30. U.S. Government Printing Office, Washington, DC.

Boyer, T.P., S. Levitus, J.I. Antonov, M.E. Conkright, T. O'Brien, and C. Stephens, 1998b: World Ocean Atlas 1998 Vol. 5: Salinity of the Pacific Ocean, *NOAA Atlas NESDIS* 31. U.S. Government Printing Office, Washington, DC.

Boyer, T.P., S. Levitus, J.I. Antonov, M.E. Conkright, T. O'Brien, C. Stephens, and B. Trotsenko, 1998c: World Ocean Atlas 1998 Vol. 6: Salinity of the Indian Ocean, *NOAA Atlas NESDIS* 32. U.S. Government Printing Office, Washington, DC.

Boyer, T.P., C. Stephens, J.I. Antonov, M.E. Conkright, R.A. Locarnini, T.D. O'Brien, and H.E.Garcia, 2002: World Ocean Atlas 2001, Vol. 2: Salinity. S. Levitus, Ed., *NOAA Atlas NESDIS* 50, U.S. Government Printing Office, Washington, DC, 165 pp., CD-ROMs.

Carton, J.A., G.A. Chepurin, X. Cao, and B.S. Giese, 2000a: A Simple Ocean Data Assimilation analysis of the global upper ocean 1950–1995. Part I. Methodology, *J. Phys. Oceanogr.*, 30, 294–309.

Carton, J.A., G.A. Chepurin, and X. Cao, 2000b: A Simple Ocean Data Assimilation analysis of the global upper ocean 1950–1995. Part II. Results, *J. Phys. Oceanogr.*, 30, 311–326.

Chassignet, E.P. and Z.D. Garraffo, 2001: Viscosity parameterization and the Gulf Stream separation. In: P. Muller and D. Henderson, editors, *Aha Huliko'a Hawaiian Winter Workshop*. University of Hawaii, pp. 37–41.

Coward, A.C., B.A. de Cuevas, and D.J. Webb, 2002: Early results from a $1/12° \times 1/12°$ global ocean model. Poster from *WOCE Final Conference*, http://www.soc.soton.ac.uk/JRD/OCCAM/POSTERS.

Deser, C., M.A. Alexander, and M.S. Timlin, 1999: Evidence for a wind-driven intensification of the Kuroshio Current Extension from the 1970s to the 1980s. *J. Clim.*, 12, 1697–1706.

Ducet, N., P.Y.L. Traon, and G. Reverdin, 2000: Global high-resolution mapping of ocean circulation from TOPEX/Poseidon and ERS-1/2. *J. Geophys. Res.*, 105(C8), 19477–19498.

Feng, M. and G. Meyers, 2003: Interannual variability in the tropical Indian Ocean: A two-year time-scale of Indian Ocean Dipole. *Deep Sea Res. II*, 50, 2263–2284.

Gille, S.T., 1994: Mean sea surface height of the Antarctic circumpolar current from Geosat data: Method and application. *J. Geophys. Res.*, 99(C9), 18255–18273.

Gille, S.T., D.P. Stevens, R.T. Tokmakian, and K.J. Heywood, 2001: Antarctic circumpolar current response to zonally averaged winds. *J. Geophys. Res.*, 106(C2), 2743–2759.

Hughes, C.W. and E.R. Ash, 2001: Eddy forcing of mean flow in the Southern Ocean. *J. Geophys. Res.*, 106(C2), 2713–2722.

Hurlburt, H.E. and P.J. Hogan (2000): Impact of $1/8°$ to $1/64°$ resolution on Gulf Stream model-data comparisons in basin-scale subtropical Atlantic Ocean models. *Dyn. Atmos. Oceans*, 32, 283–329.

Jia, Y., 2000: Formation of an Azores Current due to Mediterranean overflow in a modeling study of the North Atlantic. *J. Phys. Oceanogr.*, 30, 2342–2358.

Kalnay, E., M. Kanamitsu, R. Kistler, W. Collins, D. Deaven, L. Gandin, M. Iredell, S. Saha, G. White, J. Woollen, Y. Zhu, M. Chelliah, W. Ebisuzaki, W. Higgins, J. Janowiak, K.C. Mo, C. Ropelewski, A. Leetmaa, R. Reynolds, and R. Jenne, 1996: The NCEP/NCAR 40-year reanalysis project. *Bull. Am. Meteorol. Soc.*, 77, 437–471.

Large, W.G., J.C. McWilliams, and S.C. Doney, 1994: Oceanic vertical mixing – A review and a model with a nonlocal boundary layer parameterization. *Rev. Geophys.*, 32, 363–403.

Legeckis, R., 1977: Long waves in the eastern equatorial Pacific Ocean: A view from a geostationary satellite. *Science*, 197, 1179–1181.

Lutjeharms, J.R.E. and D.J. Webb, 1995: Modelling the Agulhas Current system with FRAM (Fine Resolution Antarctic Model). *Deep Sea Res. I*, 42, 523–551.

Madden, R.A. and P.R. Julian, 1971: Description of a 40–50 day oscillation in the zonal wind in the tropical Pacific. *J. Atmos. Sci.*, 28, 702–708.

Maltrud, M.E. and J.L. McClean, 2005: An eddy resolving global $1/10°$ ocean simulation. *Ocean Model.*, 8, 31–54.

Mantua, N.J, S.R. Hare, Y. Zhang, J.M. Wallace, and R.C. Francis, 1997: A Pacific interdecadal climate oscillation with impacts on salmon production. *Bull. Am. Meteorol. Soc.*, 78, 1069–1079.

Masumoto, Y., H. Sasaki, T. Kagimoto, N. Komori, A. Ishida, Y. Sasai, T. Miyama, T. Motoi, H. Mitsudera, K. Takahashi, H. Sakuma, and T. Yamagata, 2004: A fifty-year eddy-resolving simulation of the world ocean: Preliminary outcomes of OFES (OGCM for the Earth Simulator). *J. Earth Simulator*, 1, 35–56.

Masumoto, Y., H. Hase, Y. Kuroda, H. Matsuura, and K. Takeuchi, 2005: Intraseasonal variability in the upper layer currents observed in the eastern equatorial Indian Ocean. *Geophys. Res. Lett.*, 32, L02607, doi:10.1029/2004GRL21896.

McPhaden, M.J., 1999: Genesis and evolution of 1997–1998 El Niño, *Science*, 283, 950–954.

McPhaden, M.J., 2004: Evolution of the 2002/03 El Niño. *Bull. Am. Meteorol. Soc.*, 85, 677–695.

Nakamura, H., G. Fin, and T. Yamagata, 1997: Decadal climate variability in the North Pacific during the recent decades. *Bull. Am. Meteorol. Soc.*, 78, 2215–2225.

Nonaka, M., H. Nakamura, Y. Tanimoto, T. Kagimoto, and H. Sasaki, 2006: Decadal variability in the Kuroshio–Oyashio Extension simulated in an eddy-resolving OGCM. *J. Clim.*, 19, 1970–1989.

Ohfuchi, W., H. Sasaki, Y. Masumoto, and H. Nakamura, 2005: Mesoscale resolving simulations of global atmosphere and ocean on the Earth Simulator. *EOS*, 86, 45–46.

Orsi, A.H., T. Whitworth III, and W.D. Nowlin, 1995: On the meridional extent and fronts of the Antarctic Circumpolar Current. *Deep Sea Res. I*, 42, 641–673.

Pacanowski, R.C. and A. Gnanadesikan, 1998: Transit response in a z-level ocean model that resolves topography with partial-cells. *Mon. Weather. Rev.*, 126, 3248–3270.

Pacanowski, R.C. and S.M. Griffies, 1999: The MOM 3 Manual, *GFDL Ocean Group Technical Report* No. 4, Princeton, NJ: NOAA/Geophysical Fluid Dynamics Laboratory, 680 pp.

Peterson, R.G., 1988: On the transport of the Antarctic Circumpolar Current through Drake Passage and its relation to wind. *J. Geophys. Res.*, 93(C11), 13993–14004.

Rao, S.A., S.K. Behera, Y. Masumoto, and T. Yamagata, 2002: Interannual subsurface variability in the tropical Indian Ocean with a special emphasis on the Indian Ocean Dipole. *Deep Sea Res. II*, 49, 1549–1572.

Rayhner, N.A., E.B. Horton, D.E. Parker, C.K. Folland, and R.B. Hachett, 1996: Version 2.2 of the global sea ice sea surface temperature data set. *Climate Research Technical Note* 74, UK Meteorological Office, Bracknell.

Rosati, A. and K. Miyakoda, 1988: A general circulation model for upper ocean circulation. *J. Phys. Oceanogr.*, 18, 1601–1626.

Saji, N.H., B.N. Goswami, P.N. Vinayachandran, and T. Yamagata, 1999: A dipole mode in the tropical Indian Ocean. *Nature*, 401, 360–363.

Sasai, Y., A. Ishida, Y. Yamanaka, and H. Sasaki, 2004: Chlorofluorocarbons in a global ocean eddy-resolving OGCM: Pathway and formation of Antarctic Bottom Water. *Geophys. Res. Lett.*, 31, L12305, doi:10.1029/2004GL019895.

Sasai, Y., A. Ishida, H. Sasaki, S. Kawahara, H. Uehara, and Y. Yamanaka, 2005: Spreading of Antarctic bottom water examined with the simulated CFC-11 distribution: Results of CFC-11 simulation in an eddy-resolving OGCM. *Polar Meteorol Glaciol*, 19, 15–27.

Saunders, P. and S.R. Thompson, 1993: Transport, heat and freshwater fluxes within a diagnostic model (FRAM). *J. Phys. Oceanogr.*, 23, 452–464.

Semtner, A.J. and R.M. Chervin, 1992: Ocean general circulation from a global eddy-resolving model. *J. Geophys. Res.*, 97(C4), 5493–5550.

Smith, R.D., M.E. Maltrud, F.O. Bryan, and M.W. Hecht, 2000: Numerical simulation of the North Atlantic Ocean at $1/10°$. *J. Phys. Oceanogr.*, 30, 1532–1561.

Stephens, C., J.I. Antonov, T.P. Boyer, M.E. Conkright, R.A. Locarnini, T.D. O'Brien, and H.E. Garcia, 2002: World Ocean Atlas 2001, Vol. 1: Temperature. S. Levitus, Ed., *NOAA Atlas NESDIS* 49, U.S. Government Printing Office, Washington, D.C., 167 pp., CD-ROMs.

Takayabu, Y. N., T. Iguchi, M. Kachi, A. Shibata, and H. Kanzawa, 1999: Abrupt termination of the 1997–98 El Niño in response to a Madden–Julian oscillation. *Nature*, 402, 279–282.

References

Tanimoto, Y. and S.P. Xie, 2002: Inter-hemispheric decadal variations in SST, surface wind, heat flux and cloud cover over the Atlantic Ocean. *J. Meteor. Soc. Jpn.*, 80, 1199–1219.

Uehara, H., S. Kawahara, N. Ohno, M. Furuichi, and A Kageyama, 2006: MovieMaker: A parallel movie-making software for large scale simulations. *J. Plasma Phys.*, 72, 841–844.

Wearn, R.B. and D.J. Baker, 1980: Bottom pressure measurements across the Antarctic Circumpolar Current and their relation to the wind. *Deep Sea Res.*, 27A, 875–888.

Webb, D.J., 2000: Evidence for shallow zonal jets in the South Equatorial Current region of the Southwest Pacific. *J. Phys. Oceanogr.*, 30, 706–720.

Whitworth, T. and R.G. Peterson, 1985: Volume transport of the Antarctic Circumpolar Current from bottom pressure measurements. *J. Phys. Oceanogr.*, 15, 810–816.

Woodruff, S.D., R.J. Sluts, R.L. Jenne, and P.M. Steurer, 1987: A comprehensive ocean-atmosphere dataset. *Bull. Am. Meteorol. Soc.*, 68, 521–527.

Xie, S.P. and Y. Tanimoto, 1998: A pan-Atlantic decadal climate oscillation. *Geophys. Res. Lett.*, 25, 2185–2188.

Chapter 11
Jets and Waves in the Pacific Ocean

Kelvin Richards, Hideharu Sasaki, and Frank Bryan

Summary Analysis of the output of high-resolution ocean models reveals the presence of a class of flow structures that is intermediate in scale between the large-scale circulation and the geostrophic eddy field. These structures are coherent over considerably long zonal distances and have a relatively small meridional scale of 300–500 km. They take the form of either quasipersistent multiple jets or long-crested Rossby waves with high-meridional and low-zonal wavenumbers. Here we discuss the properties of these features in the Pacific Ocean, possible mechanisms for their formation, and the potential impact on the transport of tracers.

11.1 Introduction

Our understanding of the circulation of the ocean has developed through a combination of theory, numerical experimentation, and observation (for a good overview, particularly of the theoretical aspects, see Pedlosky 1998). The classical view of the circulation at midlatitudes is one of swift flowing western boundary currents closing the more sluggish wind-driven gyres (Stommel 1948; Munk 1950). A major shift in thinking came about with the recognition that the wind and buoyancy forcing combine to shape the thermocline and the associated circulation (Luyten et al. 1983). The role of mesoscale eddies in setting the ocean circulation has always been in question since the pioneering observations of Crease and Swallow (cf. Swallow 1971) and subsequent recognition that the eddies contain a large fraction of the kinetic energy throughout the depth of the ocean (cf. Dickson 1983). Until recently, however, there has been a general perception that there is a clear separation between the eddy scale and that of the time-mean flow averaged over many eddy turnover times in such regions as the recirculating gyres. The change in thinking has come about through the results of high-resolution numerical models of the ocean, made possible through increases in computation power, and the analysis of remotely sensed data from satellites.

A robust feature of the flow field of high-resolution ocean models averaged over a few years is the presence of jet-like features, often alternating in direction, that are coherent over many degrees of longitude and have a relatively small meridional scale of 3–5° (Treguier et al. 2003; Nakano and Hasumi 2005; Galperin et al. 2004; Maximenko et al. 2005; Richards et al. 2006). These features have a scale which is intermediate between the eddy scale and that of the broader gyre flow. The presence of the jets should be no surprise. There are a number of mechanisms that will form jets in stratified, rotating flows (cf. Rhines 1994). The remarkable feature is that they dominate the flow in many parts of the model oceans throughout their full depth.

Direct observations of alternating jets in the interior of the ocean is scant. This is because of the heavy demands on both spatial and temporal sampling. An exception is in the deep Brazil basin where analysis of float data has revealed such flow structures (Hogg and Owens 1999). Indirect evidence comes from satellite altimetry. Maximenko et al. (2005) find alternating zonal jets in the anomaly of geostrophic velocity, derived from gridded sea surface height fields, when averaged over a few weeks. The same authors also find that the jets found in an ocean model have a strong surface signature which is similar to that found in the altimeter data.

Not all intermediate scale flow features found in the model flows are jets. In the subtropics long-crested Rossby waves, which have low-zonal and high-meridional wavenumbers, are found to dominate the flow. Here we discuss some of the characteristics of the jets and waves in high-resolution ocean model solutions. The implications of the presence of such features on the dispersion of tracers are briefly considered.

We focus on the flow in the Pacific Ocean and make use of the output from two high-resolution ocean models (both with nominally 1/10° horizontal resolution) that were run on the Japanese Earth Simulator (as considered in Richards et al. 2006). The first (which we refer to as POP) is a fully global implementation of the Parallel Ocean Program with 40 levels in the vertical and is very similar to that described in Maltrud and McLean (2005). The second (referred to as OFES) is based on MOM3 in a near global configuration (the North Pole is excluded), and with 54 levels in the vertical (Masumoto et al. 2004). The general characteristics of the zonal jets found in the results from the two models are very similar.

11.2 Jets

Figure 11.1 shows the zonal component of velocity at 380 m depth averaged over 3 years from a run of POP. The striking thing about this figure is the large zonal coherency of flow features. Although we see the imprint of the broad subtropical gyres in both the North and South Pacific the flow is dominated by much narrower jet-like structures that have a meridional scale of approximately 3–5°. A vertical meridional section along 180°E (Fig. 11.2) shows that these jets have a large vertical coherency which in places, such as the Southern Ocean and the northern Pacific, extends throughout the total depth of the ocean.

A wavelet transform (cf. Torrence and Compo 1998) of the zonal component of velocity at a given depth and along a given longitude tends to show three distinct peaks in the power spectrum (Fig. 11.3): in the ACC, in the tropics, and in the North Pacific. (In Fig. 11.3 the eastward jet at 18°S is associated with a topographic feature.) These three regions are also regions of strong eddy activity. The dominant

11.2 Jets 189

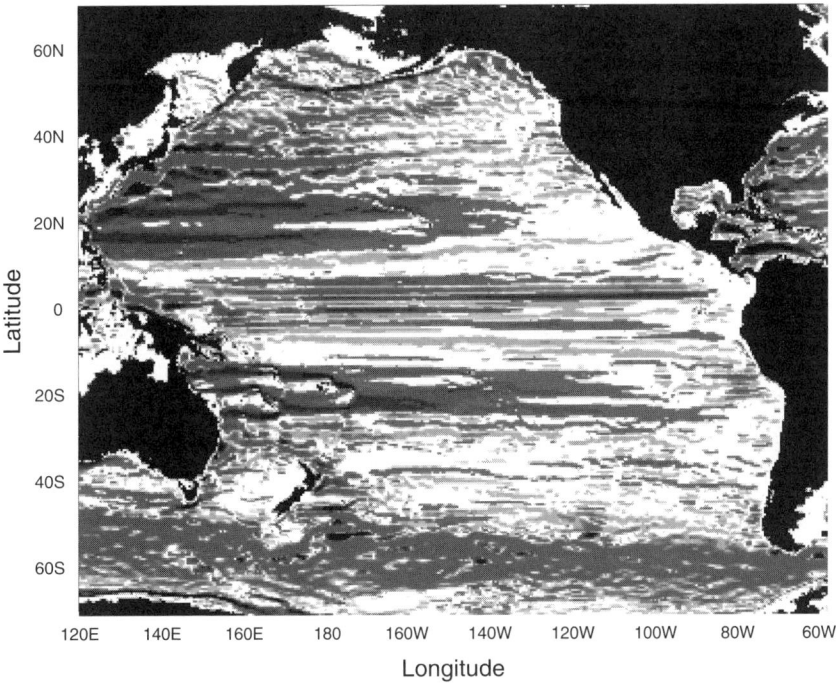

Fig. 11.1 Zonal component of velocity at 38 m depth averaged over 3 years from the climatological run of the POP model. Color saturates at -0.06 m s^{-1} (*blue*) and 0.06 m s^{-1} (*red*)

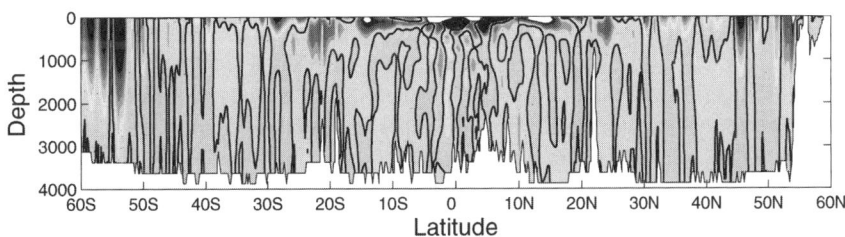

Fig. 11.2 Zonal component of velocity along 180°E averaged over 3 years from the climatological run of the POP model as a function of latitude and depth. Model run the same as in Fig. 11.1. Color saturates at -0.1 m s^{-1} (*blue*) and 0.1 m s^{-1} (*red*). Zero contour is given by *black line*

meridional wavelength is approximately constant with depth but varies with both longitude and latitude. Richards et al. (2006) show that the horizontal variation of the scale of the jets in these three regions is consistent with the variation in the Rhines scale given by $L_R = \sqrt{2u'/\beta}$, where u' is the r.m.s. eddy velocity and β the meridional gradient of the Coriolis parameter. This suggests that the jets in these regions may be formed through an eddy/mean flow interaction. Such a mechanism has been studied by a number of authors including Rhines (1975), Williams (1978),

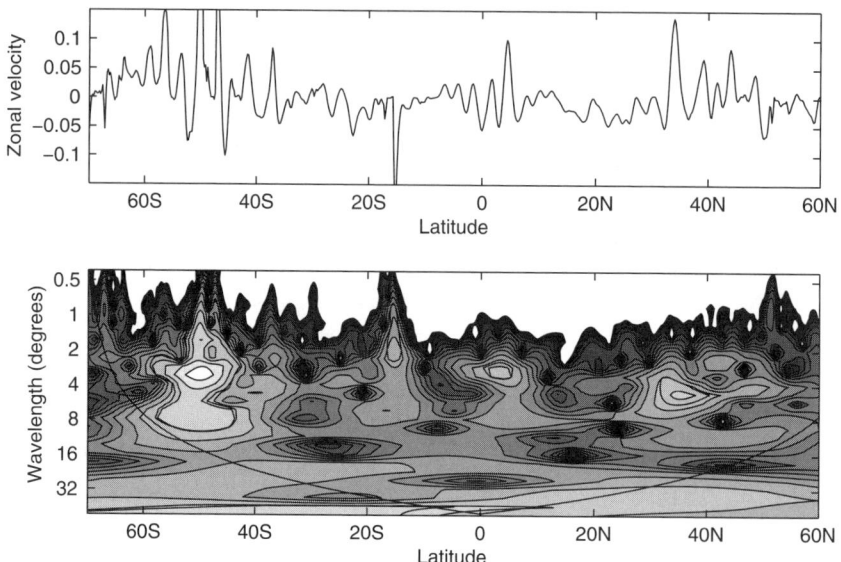

Fig. 11.3 *Upper panel*: Zonal component of velocity at 380 m depth along 180°E (from model run shown in Fig. 11.1). *Lower panel*: Wavelet power spectrum (normalized with respect to variance) of the zonal velocity shown in the upper panel as a function of meridional wavelength and latitude. Contours are at $\log_2$ intervals

Panetta (1993), Vallis and Maltrud (1993), Chekhlov et al. (1996), Huang and Robinson (1998), Sinha and Richards (1999), and Sukoriansky et al. (2002).

In other places topography plays a large role in shaping the flow, perhaps none more so than in the South West Pacific (Fig. 11.4 – see also Webb (2000) and Hughes (2002)). Here swift jets exist on the tips of many of the islands in the region. Water joining the northward and southward boundary currents along the eastern coast of Australia has negotiated a number of these jets. What impact this has on the propagation of anomalies through the region is unclear. It should be noted that the oncoming flow to the region itself is in the form of narrow jets. Relatively small meridional displacements in the position of these jets have the potential of feeding through to a significant change in the properties and bifurcation of the boundary currents.

11.3 Waves

Monthly averaged values of the zonal component of velocity at 380 m depth from OFES and averaged between 140 and 150°W are shown in Fig. 11.5 as a function of latitude and time. Polewards of 40°N and S there are features that are persistent in time. The tropics are dominated by the annual cycle. However a two-year running mean (see Richards et al. 2006) shows that individual features in the ACC, tropics,

11.3 Waves

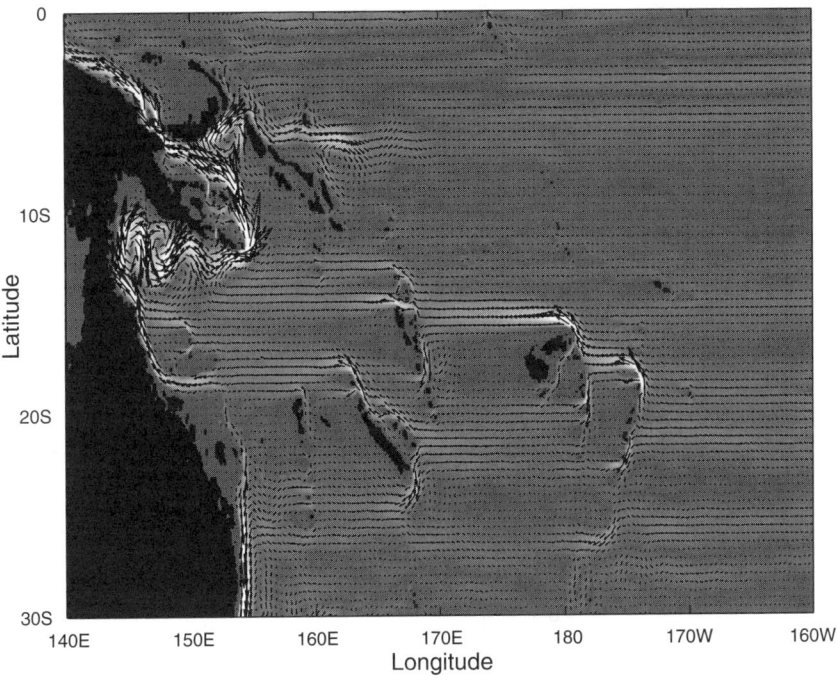

Fig. 11.4 Current vectors of the 3-year averaged flow at 380 m in the South West Pacific from POP. Shading indicates speed which saturates at 0.2 m s^{-1}

and northern Pacific generally persist for O(3–5 years) (sometimes much longer) with small deviations in their latitudinal position. The flow in the subtropics between 20 and 30°N and S shows a very different behavior. Here we see features propagating toward the equator over a number of degrees of longitude with a period of around 4 years.

Figure 11.6 shows the monthly averaged zonal component of velocity at 400 m depth for January 1980 in the South Pacific from OFES. Note the meridional scale has been exaggerated. The flow is dominated by long features that are oriented SW/NE. The features are surface intensified and extend below the thermocline. Similar features are found in the North Pacific at similar latitudes. An animation of the field shows the features to propagate westward with little change in shape with a phase speed between 0.04 and 0.05 m s^{-1}, and an equatorward phase speed 0.005 m s^{-1}. The zonal and meridional wavelengths are approximately 4,000 km and 500 km, respectively.

This equatorward phase propagation was first detected by Maximenko and Bang (2005) both in satellite altimeter sea level anomaly data and in the OFES model run forced by the NCEP climatology. They estimated the meridional phase speed to be close to 0.45 cm s^{-1} at a dominant time period of 3.5 years in the midlatitudes

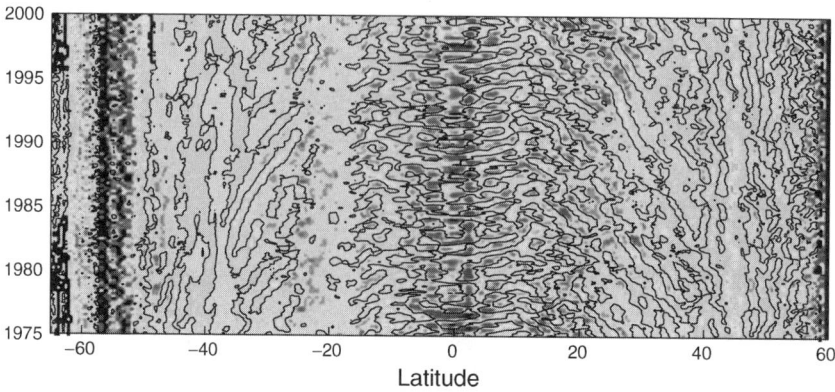

Fig. 11.5 Monthly averaged zonal component of velocity at 380 m depth averaged between 140°W and 150°W, as a function of latitude and time, from OFES. Color saturates at -0.2 m s^{-1} (*blue*) and 0.2 m s^{-1} (*red*). Zero contour is given by *black line*

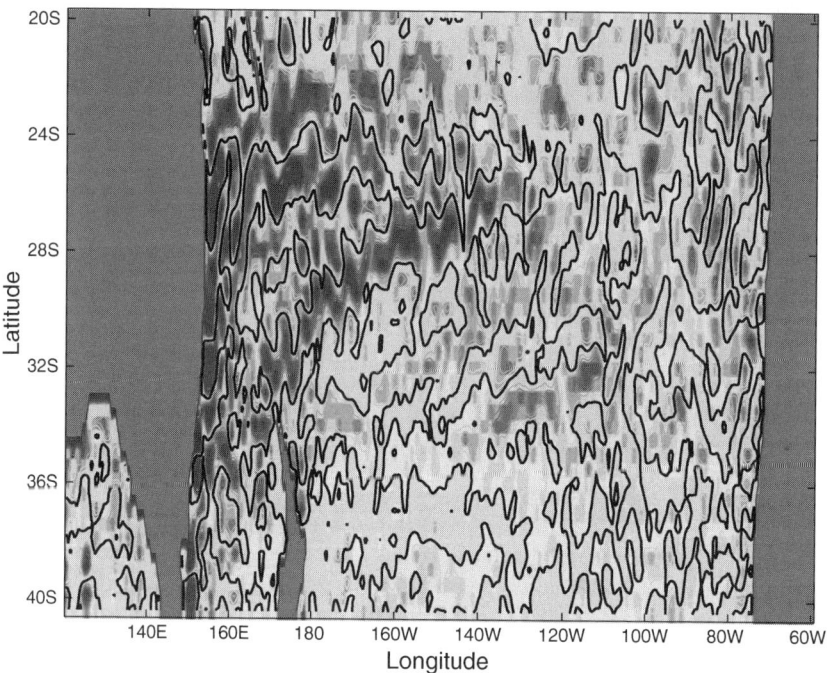

Fig. 11.6 Monthly averaged zonal component of velocity in the South Pacific for January 1980 at 400 m depth, from OFES. Color saturates at -0.2 m s^{-1} (*blue*) and 0.2 m s^{-1} (*red*). Zero contour is given by black line

of both hemispheres of the Pacific Ocean. On a horizontal plane, the jet-like structures resembled a pattern of linear waves with almost meridional orientation of wave vector.

The westward phase speed is consistent with what we would expect for a long (in the zonal sense) Rossby wave at approximately 30° latitude (Killworth and Blundell 2003). The equatorward phase speed implies a poleward group velocity. A notable feature is that these long-crested waves dominate the flow in the subtropical regions. It is only in the long term time-mean that we see again the jet-like structures oriented approximately east/west as seen in Fig. 11.1. (There are some differences in the structure of the jets in the long-term means of OFES and POP. In particular the eastward jet at around 30°S in OFES does not penetrate quite as deep as that in POP and is weaker at 400 m. These differences may be caused by the different forcing used for the two model runs.)

The tilt of the waves in the model is inconsistent with what we would expect if they had been produced on the American coast and were propagating toward the west (see, e.g., Qiu et al. 1997). Because of the decrease in phase speed with latitude the wave would be tilted in the opposite direction to that found. The phase speed of long Rossby waves at 25° latitude is approximately twice that at 35°. Superimposed on the large wave-like structure shown in Fig. 11.6 smaller scale features are seen to propagate at a speed similar to the Rossby wave phase speed appropriate for their latitude. What produces the largescale waves with small meridional scale and what sets the meridional and zonal scales is unclear. Radiation from low latitudes is one possibility. Glazman and Weichman (2005) also find Rossby waves with a relatively high meridional wavenumber in their spectral analysis of Topex altimeter data in the subtropical Pacific. However they consider waves with periods of less than two years in $10° \times 10°$ areas. It would be useful to apply their methodology to the longer time and space scales we consider here.

11.4 Impact on the Transport of Tracers

The presence of relatively high horizontal shears associated with the zonally coherent intermediate scale structures in the flow will significantly alter the dispersion characteristics of tracers in the model ocean through shear dispersion (Young et al. 1982; Bartello and Holloway 1991) or chaotic advection by modulated Rossby waves (Pierrehumbert 1991; Dupont et al. 1998). We therefore may expect a fundamentally different behavior for tracer transport and dispersion in the flows considered here as compared to that in coarser more diffuse ocean models.

It is unclear how the Eulerian properties, as shown here, relate to the tracer transport, or Lagrangian properties, of the flow. A study of the transport properties of the intermediate flow structures using particle tracking techniques is planned for the near future. As a very crude estimate we can use the effective longitudinal dispersion coefficient, K_E, appropriate for a zonal flow which varies sinusoidally in the meridional direction with magnitude u' and inverse wavenumber L, and which varies *slowly* with time, given by Young et al. (1982). If the dispersion is averaged for *sufficiently* long, then K_E is given by

$$K_E = \frac{(u'L)^2}{4K_d}, \tag{11.1}$$

where K_d is the background cross-stream diffusion coefficient. Young et al. (1982) discuss when the above limits apply. Taking $u' = 0.02$ m s^{-1}, $L = 50$ km, and $K_d = 100$ m^2 s^{-1} then $K_E = 2,500$ m^2 s^{-1}. K_E should be viewed as an upper bound, but the implication is that dispersion in the zonal direction is likely to be significantly enhanced by the sheared flow. The likely importance of zonally coherent structures in the flow is also indicated by the work of Speer et al. (2002), who study dispersion away from the Mid-ocean Ridge using particle trajectories calculated from the flow of the ocean model reported by Maltrud and McLean (2005). They find the time taken for the particle dispersion to become *Fickian*-like, rather than ballistic (roughly the Lagrangian integral time) is around 200 days. This is considerably longer than the timescale of individual eddies (typically of order 10 days – cf. LaCase and Speer (1999)), and suggests a longer time coherency to the flow.

11.5 Conclusions

Intermediate scale features with a meridional scale of O(300–500 km), which are persistent for a few years, are found to be a robust feature of the flow in high-resolution/low-dissipation ocean models. The features take the form of jets and propagating long-crested Rossby waves and in many places they dominate the zonal component of the flow. There is growing observational evidence that such features exist in the real ocean, principally from sea surface height, but also from a few in situ measurements.

What sets the amplitude and spatial and temporal scales of the jets and waves? In some locations such as the South West Pacific, the presence of islands and seamounts obviously plays a dominant role in the production of the multiple jet-like flow. In others the meridional scale of the jets is consistent with the Rhines scale hinting that perhaps they are formed through an eddy/mean flow interaction on a background potential vorticity gradient. However we are yet to determine the momentum and vorticity dynamics of the jets to show if this is indeed correct. As pointed out by Richards et al. (2006), although the meridional scale of the jets is relatively constant with depth, the strength of the eddying activity decreases markedly with depth. Whether or not the jet scale is set by the eddying activity above the thermocline or by the strength of the barotropic motions is unclear. The situation in the tropics is further complicated by the presence of the near surface system of zonal flows and tropical instability waves that are a major contributor to the eddy kinetic energy. Careful analysis of the results from high-resolution ocean models and numerical experimentation with more idealized models is required to address these issues.

Although the presence of intermediate flow structures in high-resolution ocean models is intriguing the bottom line is whether or not they matter. Should you worry that the ocean component of your global coupled model does not contain such features? Do they affect the interannual, decadal, or longer characteristics of the coupled system? The short answer is that we do not know. As suggested in Sect. 11.4 these features have the potential of affecting the transport and diffusive properties of the ocean. Hopefully answers will be forthcoming as we learn more about the many scale interactions in both the ocean and the atmosphere.

Acknowledgments The POP and OFES model runs were performed on the Earth Simulator (ES) of Japan. We would like to thank Nikolai Maximanko (IPRC) for many discussions on the subject of jets in data and model results. This research was supported in part by JAMSTEC through its sponsorship of the IPRC. IPRC/SOEST contribution nos. 415/6965.

References

Bartello, P., and G. Holloway (1991) Passive scalar transport in beta-plane turbulence. J. Fluid Mech., 223, 521–536.
Chekhlov, A., S. Orszag, S. Sukoriansky, B. Galperin, and I. Staroselsky (1996) The effect of small-scale forcing on large-scale structries in two-dimensional flows. Phys. D, 98, 321–334.
Dickson, R. R. (1983) Global summaries and intercomparisons: flow statistics from long-term current meter moorings. In *Eddies in Marine Science*, Robinson, A. R. (ed.), Springer, Berlin.
Dupont, F., R. I. McLachlan, and V. Zeitlin (1998) On a possible mechanism of anomalous diffusion by Rossby waves. Phys. Fluids, 10, 3185–3193.
Galperin, B., H. Nakano, H.–P. Huang, and S. Sukoriansky (2004) The ubiquitous zonal jets in the atmospheres of giant planets and the Earth's oceans. Geophys. Res. Lett., 31, L13303, doi:10.1029/2004GL019691.
Glazman, R. E., and P. B. Weichman (2005) Meridional component of oceanic Rossby wave propagation. Dynamics Atoms. Oceans, 38, 173–193.
Hogg, N., and B. Owens (1999) Direct measurement of the deep circulation within the Brazil basin. Deep-Sea Res., 46, 335–353.
Huang, H.-P, and W. Robinson (1998) Two-dimensional turbulence and persistent zonal jets in a global barotropic model. J. Atmos. Sci., 55, 611–632.
Hughes, C. W. (2002) Zonal jets in and near the Coral Sea, seen by satellite altimetry. Geophys. Res. Lett., 29, 1330, doi: 10.1029/2001GL014002.
Killworth, P. D., and J. Blundell (2003) Long extratropical planetary wave propagation in the presence of slowly varying mean flow and bottom topography. Part II: ray propagation and comparison with observations. J. Phys. Oceanogr., 33, 802–821.
LaCase, J. H., and K. G. Speer (1999) Lagrangian statistics in unforced barotropic flows. J. Mar. Res., 57, 245–274.
Luyten, J. R., J. Pedlosky, and H. Stommel (1983) The ventilated thermocline. J. Phys. Oceanogr., 13, 292–309.
Maltrud, M. E., and J. L. McLean (2005) An eddy resolving global 1/10 degree ocean simulation. Ocean Modell., 8, 31–54.
Masumoto, Y., et al. (2004) A fifty-year eddy-resolving simulation of the world ocean: preliminary outcomes from OFES (OGCM for the Earth Simulator). J. Earth Simul. 1, 35–56.
Maximenko N. A., and B. Bang (2005) Alternating zonal jets observed in the upper ocean. IPRC Annual Symposium, Univeristy of Hawaii.
Maximenko N. A., B. Bang, and H. Sasaki (2005) Observational evidence of alternating zonal jets in the world ocean. Geophs. Res. Lett., 32, L12607, doi:10.1029/2005GL022728.
Munk, W. H. (1950) On the wind-driven ocean circulation. J. Meteorol., 7, 79–93.

Nakano, H, and H. Hasumi (2005) A series of zonal jets embedded in the broad zonal flows in the Pacific obtained in eddy-permitting ocean general circulation models. J. Phys. Oceanogr., 35, 474–488.

Panetta, R. L. (1993) Zonal jets in wide baroclinically unstable regions: persistence and scale selection. J. Atmos. Sci., 50, 2073–2106.

Pedlosky J. (1998) *Ocean Circulation Theory*, Springer Berlin. 453 pp.

Pierrehumbert R. T. (1991) Chaotic mixing of a tracer and vorticity by modulated traveling Rossby waves. Geophys. Astrphys. Fluid Dynamics, 58, 285–320.

Qiu B., W. Miao, and P. Muller (1997) Propagtion and decay of forced and free baroclinic Rossby waves in off-equatorial oceans. J. Phys. Oceanogr., 27, 2405–2417.

Rhines, P. B. (1975) Waves and turbulence on a beta-plane, J. Fluid Mech., 69, 417–443.

Rhines, P. B. (1994) Jets. Chaos, 4, 417–443.

Richards, K. J., N. A. Maximenko, F. O. Bryan, and H. Sasaki (2006) Zonal jets in the Pacific Ocean. Geophys. Res. Lett., 33, L03605, doi:10.1029/2005GL024645.

Sinha B., and K. J. Richards (1999) Jet structure and scaling in Southern Ocean models. J. Phys. Oceanogr., 29, 1143–1155.

Speer K. G., M. E. Maltrud, and A. M. Thurnherr (2002) A global view of dispersion above the Mid-ocean Ridge. In *Energy and Mass Transfer in Marine Hydrothermal Systems*, Halbach, P. E. V. Tunnicliffe, and J. R. Hein (eds.), Dahlem Workshop Report, 89, Dahlem Univierisity Press, Berlin.

Stommel., H. (1948) The westward intensification of wind–driven ocean currents, Trans. Am. Geophys. Union, 99, 202–206.

Swallow, J. C. (1971) The Aries current measurements in the Western North Atlantic. Phil. Trans. R. Soc. A, 270, 460–470.

Sukoriansky, S., B. Galperin, and N. Dikovskaya (2002) Universal spectrum of two–dimensional turbulence on a rotating sphere and some basic features of atmospheric circulation on giant planets. Phys. Rev. Lett., 89, 124 501.

Torrence, C., and G. P. Compo (1998) A practical guide to wavelet analysis. Bull. Am. Meteor. Soc., 79, 61–78.

Treguier, A. M., N. G. Hogg, M. Maltrud, K. Speer, and V. Thierry (2003) The origin of deep zonal flows in the Brazil Basin. J. Phys. Oceanogr., 33, 580–599.

Vallis, G., and M. Maltrud (1993) Generation of mean flows and jets on a beta plane over topography. J. Phys. Oceanogr., 33, 1346–1362.

Williams G. (1978) Planetary circulations: 1. Barotropic representation of Jovian and terrestrial turbulence. J. Atmos. Sci., 35, 1399–1426.

Webb, D. J. (2000) Evidence for shallow zonal jets in the south equatorial current region of the Southwest Pacific. J. Phys. Oceanogr., 30, 706–720.

Young W. R., P. B. Rhines, and C. J. R. Garrett (1982) Shear-flow dispersion, internal waves and horizontal mixing in the ocean. J. Phys. Oceanogr., 12, 515–527.

Chapter 12
The Distribution of the Thickness Diffusivity Inferred from a High-Resolution Ocean Model

Yukio Tanaka, Hiroyasu Hasumi, and Masahiro Endoh

Summary Using a high-resolution ocean model ($1/8 \times 1/12$, 85 levels) with realistic geometry and forcing, the diffusion coefficient of the isopycnal thickness is evaluated in the Southern Ocean. It is found that the coefficient is large in the Argentine Basin, the Agulhas Retroflection region, and the east of the Kerguelen Pleatue at 2000 m depth. The large coefficient regions coincide with high growth rate regions of the baroclinic instability.

12.1 Introduction

Satellite observation data (Nerem et al. 1994; Wunsch and Stammer 1995; Ducet et al. 2000) show that the large amplitude mesoscale eddies are located in the Antarctic circumpolar current (ACC) and western boundary currents (Kuroshio, Gulf Stream, Malvinas, Agulhas, and East Australian). The large amplitude regions correspond to high growth rate regions of the baroclinic instability (Stammer, 1998; Treguier, 1999). The time scale of the eddies is of order 5–50 days and agrees qualitatively with that of the baroclinic instability. The length scale is of order 10–500 km and agrees with the length scale of the instability theory (Stammer, 1998). These results suggest that the baroclinic instability is a primary source of the mesoscale eddies in the ocean.

The mesoscale eddies affect the larger space and longer timescales of the ocean flow. Gent and McWilliams (1990) have proposed a parameterization scheme (GM parameterization) to account for this eddy effect. The parameterization mixes isopycnal layer thickness along isopycnal surfaces. The mixing is expressed by diffusion of the layer thickness and the coefficient of this diffusion is called thickness diffusivity. This parameterization improves the performance of the coarse resolution ocean models substantially in several climatically important aspects of the ocean circulation (Danabasoglu et al., 1994; Hirst and McDougall, 1996). Following its implementation in a coarse resolution ocean circulation model (Danabasoglu et al., 1994), the scheme is widely used and implemented in the majority of ocean circulation models which utilize height coordinates.

The value of thickness diffusivity is not determined in the process of the derivation of the parameterization. Danabasoglu et al. (1994) employed the constant value of 1,000 m² s⁻¹ in their coarse resolution ocean model simulation. Hirst and McDougall (1996) also employed the constant value of 1,000 m² s⁻¹. Visbeck et al. (1997) proposed to use the spatial variable thickness diffusivity to improve the performance of coarse resolution ocean models.

By using eddy-resolving ocean models, the capability of the parameterization and the distribution of the thickness diffusivity can be evaluated. With simple idealized geometry and forcing, several studies have been carried out to evaluate the capability of the parameterization and distribution of the thickness diffusivity. Lee et al. (1997) carried out an eddy-resolving simulation with a periodic current channel. The authors concluded that the obtained results broadly support the approach of the GM parameterization. The authors also suggest that the parameterization would be improved by employing potential vorticity rather than thickness.

Treguier (1999) carried out an eddy-resolving simulation with a zonally periodical channel model. The author calculated the distribution of the eddy density flux divergence and showed that away from the surface layers the distribution of the eddy density flux divergence is consistent with the advection of density by the eddy-induced velocities of the GM parameterization. The author obtained a vertical structure of the thickness diffusivity. The diffusivity value is about 200 m² s⁻¹ between the surface and 2,500 m depth and increases rapidly to about 2,500 m² s⁻¹ at around 3,000 m depth.

Solovev et al. (2002) studied the ability of the GM and other parameterizations by performing an eddy-resolving simulation in an idealized ocean basin. The authors evaluated the divergence of the eddy heat fluxes and compared it with the GM and other parameterizations. The authors concluded that the parameterizations have some skill but does not identify any one scheme as being superior to the others.

Visbeck et al. (1997) carried out three different eddy-resolving simulations and attempted to fit the solutions to noneddy-resolving model results with parameterizations. The authors concluded that the best fit could be obtained by using the GM parameterization with the following spatial variable thickness diffusivity

$$\kappa = c_e l^2 \tau^{-1}, \tag{12.1}$$

where l is length scale of the energy-containing eddies, τ^{-1} is the Eady growth rate of the baroclinic instability, and c_e is a constant of proportionality. This form of the diffusivity is based on the suggestion by Green (1970) and Stone (1972) which is derived from the study of baroclinic eddies in the troposphere. The same form of the thickness diffusivity was also obtained by Larichev and Held (1995) by performing homogeneous geostrophic turbulence simulation.

The results of these idealized simulations should be tested by performing more realistic numerical simulations. There has been limited number of these studies up to now because realistic simulations which resolve mesoscale eddies require a large amount of computer resources. Rix and Willebrand (1996) investigated the value of the thickness diffusivity in the eastern part of the North Atlantic and estimated it as approximately $1{,}000\,\mathrm{m^2\,s^{-1}}$. Bryan et al. (1999) evaluated the distribution of the eddy-induced velocities using a world ocean model result and compared it with the GM parameterization as well as others. The authors concluded that the GM parameterization with the thickness diffusivity of the form of (12.1) represented the effect of eddies in the model better than other parameterizations.

The aim of this paper is to evaluate the spatial distribution pattern of the thickness diffusivity by using one of the highest resolution ocean model up to now with a realistic geometry and forcing. The evaluations are performed by using output data obtained from the ocean model with 53 year time integration. The resolution used in this study is $1/8° \times 1/12°$ (longitude × latitude) with 85 vertical levels which is one of the highest resolution up to date to investigate the eddy forcing term with realistic geometry and forcing. The evaluation using the resolutions of $1/4° \times 1/6°$ with 85 vertical levels are also carried out to investigate sensitivity of the eddy effect to horizontal resolution.

The simulation is performed in the region limited to the Southern Ocean. The effect of the eddies is important in western boundary currents and in the Southern Ocean. In this chapter we limited the simulation area to the Southern Ocean and use available computer resources to increase the horizontal resolution, which is critical to resolve the mesoscale eddies.

The structure of the chapter is as follows. In Sects. 12.2 and 12.3, the GM parameterization and the numerical model are described, respectively. In Sect. 12.4, the calculated distribution of the eddy buoyancy flux convergence and the thickness diffusivity are described. A summary and discussion is described in Sect. 12.5.

12.2 GM Parameterization

The averaged buoyancy equation in the adiabatic ocean interior is expressed as

$$\frac{\partial \overline{b}}{\partial t} + \overline{\mathbf{u}} \cdot \nabla \overline{b} = -\nabla \cdot (\overline{b'\mathbf{u}'}), \tag{12.2}$$

where b is buoyancy and $\mathbf{u}$ is velocity. The overbar represents the spatial and temporal mean and the prime deviations from it. The averaging scales are larger than those of mesoscale eddies. The buoyancy is defined as $b = -g(\rho - \rho_0)/\rho_0$, where g is gravity acceleration, ρ is potential density, and ρ_0 is reference density ($= 10^3\,\mathrm{kg\,m^{-3}}$). The term $-\nabla \cdot (\overline{b'\mathbf{u}'})$ is the eddy forcing term of the equation and called the eddy buoyancy flux convergence (EBFC) hereafter.

Since the eddy forcing term consists of the correlation of the velocity and buoyancy of the eddies, the characteristics of the eddy forcing terms depend on the motion of the eddies. In the adiabatic ocean interior, there is little mixing across the isopycnals and the volume between any two isopycnals is conserved. In addition, the baroclinic instability, which is a primary source of the mesoscale eddies, tends to release the potential energy. The eddy forcing term, which represents the mesoscale eddies, should have these characteristics.

In order to represent these characteristics, the density coordinates are used and the averaging is carried out on isopycnals. The averaged equation of the isopycnal thickness can be expressed as

$$\frac{\partial \overline{h}}{\partial t} + \overline{\mathbf{u}} \cdot \nabla_\rho \overline{h} = -\nabla_\rho \cdot (\overline{h'\mathbf{u}'}), \tag{12.3}$$

where h is isopycnal layer thickness, $\mathbf{u}$ is velocity, and ∇_ρ is the differential operator of the density coordinates. The averaging scales are larger than these of mesoscale eddies. The right-hand side represents the eddy forcing term of this equation. Gent and McWilliams (1990) parameterize the eddy term as the downgradient of the averaged thickness, i.e.,

$$\nabla_\rho \cdot (\overline{h'\mathbf{u}'}) = -\nabla_\rho \cdot \kappa \nabla_\rho \overline{h}, \tag{12.4}$$

where κ is the thickness diffusivity. The eddy forcing term of the height coordinates can be obtained from (12.4) and expressed as

$$\nabla \cdot (\overline{b'\mathbf{u}'}) = \nabla \cdot \mathbf{F}_{\text{GM}} = -\nabla_H \cdot \kappa \nabla_H \overline{b} + \frac{\partial}{\partial z}\left[\kappa(\nabla_H \overline{b})^2 \Big/ \frac{\partial \overline{b}}{\partial z}\right], \tag{12.5}$$

$$\mathbf{F}_{\text{GM}} = -\kappa \left(\nabla_H \overline{b}, -(\nabla_H \overline{b})^2 \Big/ \frac{\partial \overline{b}}{\partial z}\right), \tag{12.6}$$

where ∇_H is horizontal Laplacian operator in height coordinates (Gent and McWilliams, 1990). It should be noted that the overbar in (12.4) represents the temporal and spatial averaging on an isopycnal but the overbar in (12.5) represents those averaging at a fixed height.

It is readily seen that the flux $\mathbf{F}_{\text{GM}}$ (GM flux) is along the mean isopycnal (Griffies, 1998) and the parameterization can be rewritten in the following advection form

$$\nabla \cdot \mathbf{F}_{\text{GM}} = \mathbf{v}^* \cdot \nabla_H \overline{b} + w^* \frac{\partial \overline{b}}{\partial z}, \tag{12.7}$$

where $\mathbf{v}^* = \frac{\partial}{\partial z}\kappa(\nabla_H \overline{b}/\frac{\partial \overline{b}}{\partial z})$ and $w^* = -\nabla_H \cdot \kappa(\nabla_H \overline{b}/\frac{\partial \overline{b}}{\partial z})$ are the eddy-induced velocities (Gent et al., 1995).

12.3 Model Description

The model used is the Center for Climate System Research (CCSR) Ocean Component model (COCO) (Hasumi, 2000) and solves the primitive equations with the explicit free surface method. This model is previously applied to the eddy-permitting modeling of the Pacific Ocean (Nakano and Hasumi, 2005).

There are 85 levels in the vertical. The grid spacing is constant at 50 m from the sea surface to 2,000 m depth, then is linearly increased up to 200 m at 4,000 m depth. From 4,000 to 5,500 m, a constant 200-m grid spacing is used. For horizontal grids, two different horizontal resolutions of $1/4° \times 1/6°$ and $1/8° \times 1/12°$ (longitude × latitude) are used and called SO6 and SO12, respectively.

In the tracer equations, a high-accuracy tracer advection scheme (Hasumi and Suginohara, 1999) and biharmonic diffusion scheme are employed. The biharmonic diffusivities are 10^9 and $5 \times 10^7\,\mathrm{m^4\,s^{-1}}$ for SO6 and SO12, respectively. For the vertical diffusivity, a constant of $5 \times 10^{-5}\,\mathrm{m^2\,s^{-1}}$ is used.

In the momentum equations, an enstrophy conserving scheme (Ishizaki and Motoi, 1999) and biharmonic friction with a Smagorinsky-like viscosity (Griffies and Hallberg, 2000) are employed. This viscosity depends on both the horizontal grid spacing and horizontal deformation rate of the flow and controlled by one parameter C. We used the same value of $C = 2.5$ for SO6 and SO12 cases. This viscosity is large in the eddy-rich regions and the typical value is about $10^9\,\mathrm{m^4\,s^{-1}}$ in the Argentine Basin for SO12 case. For the vertical viscosity, a constant value of $10^{-4}\,\mathrm{m^2\,s^{-1}}$ is used.

The simulation area is the Southern Ocean from 20°S to 75°S. The model has a realistic world-ocean bathymetry based on the ETOPO2 data.

The initial value of the model is calculated by integrating it for 1 year from a state of rest with constant temperature and salinity. The temperature and salinity fields are relaxed toward annual-mean climatology of the World Ocean Atlas 98 (WOA98, Antonov et al. 1998; Boyer et al. 1998) with relaxation time of 30 days throughout the water columns. During the integration, the model is forced by the monthly mean climatology of the sea surface wind stress compiled by Röske (2001), which is based on the ECMWF reanalysis data.

From the initial condition, the model is integrated for 53 years with the sea surface wind stress described above. The sea surface temperature and salinity are restored to the monthly mean of the climatology WOA98 with relaxation time of 30 days. The temperature and salinity from 20°S to 23°S and those of the southernmost grid points are restored to the annual mean of the climatology with relaxation time of 180 days. A rigid wall boundary is used at 20°S with a no-slip condition applied to the tangential flow.

12.4 Results

12.4.1 The Distribution of the EBFC

The EBFC of two different horizontal resolutions (i.e., SO6 and SO12) at 2,000 m depth are shown in Fig. 12.1 a, b, respectively. The output data for the last 3 years are collected once every 10 days and are used to evaluate the EBFC. The obtained EBFC is spatially smoothed with $4° \times 2°$ length scales using a method employed by Bryan et al. (1999).

Although there are differences in detail between the calculated EBFCs of SO6 and SO12, the overall spatial structures and strengths are similar. It is anticipated from these results that model results with horizontal resolution higher than the SO12 case

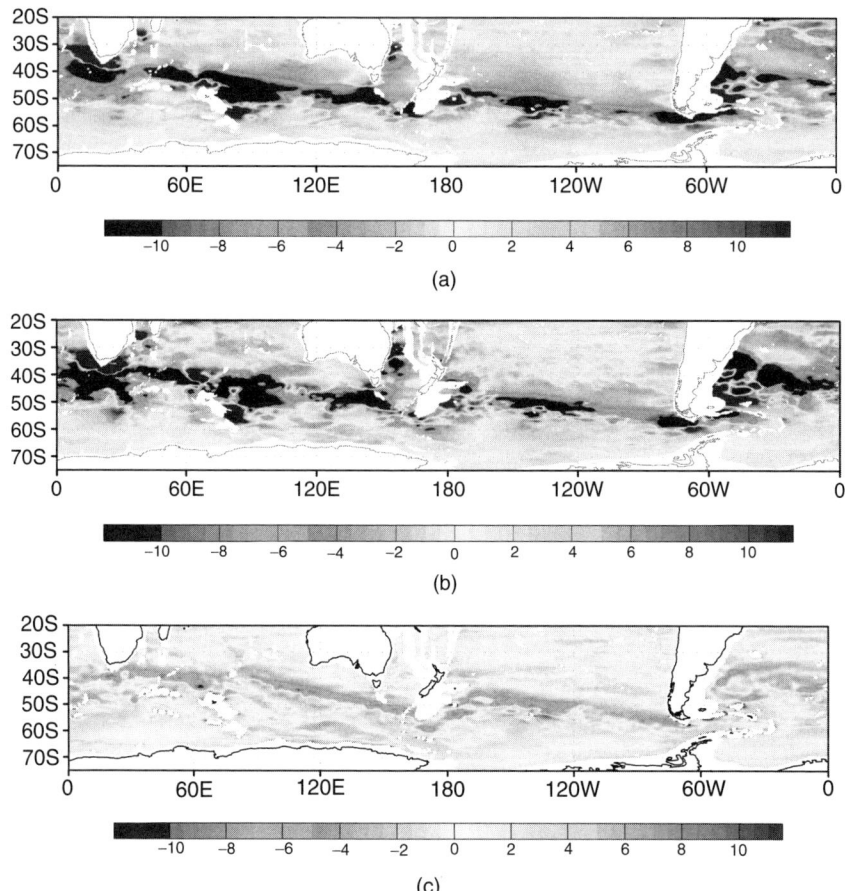

Fig. 12.1 The EBFC of (**a**) the SO6, (**b**) SO12, and (**c**) the Gent–McWilliams flux convergence (unit: 4×10^{-13} m s^{-3}) at 2,000 m depth

may not have very different spatial structure and intensity. We, therefore, use the output data of the SO12 for the further analyses.

The EBFC distribution of the SO12 in Fig. 12.1 b shows that the intensity is strong in the ACC and the Agulhas Retroflection region. In the ACC, the EBFC is positive on the poleward flank, where eddies decrease the density, and negative on the equatorward flank, where eddies increase the density. In the Drake Passage and Argentine Basin, the negative EBFC regions are located on the north of the Subantarctic Front and the positive regions are located on the south of the Polar Front. In the east of the Kerguelen Plateau, the strong negative (positive) region exists on the north (south) of the Polar Front. In the Agulhas Retroflection region, the EBFC is positive on the north side of the Agulhas Current and negative on the south side of it. Figure 12.2 a shows that the horizontal potential density distribution of SO12. Overall, the eddies decrease (increase) the density on the heavy (light) side of the current jets and work to flatten the isopycnals.

12.4.2 The Distribution of the Thickness Diffusivity

The capability of the GM parameterization is investigated by comparing the calculated EBFC with the GM flux convergence (GMFC), $-\nabla \cdot \mathbf{F}_{GM}$, in (12.5). Figure 12.1 c shows the GMFC calculated with a constant thickness diffusivity $\kappa = 400 \, \text{m}^2 \, \text{s}^{-1}$. The time-averaged buoyancy is spatially averaged with length scales of $4° \times 2°$ in the horizontal and 100 m in the vertical to obtain $\overline{b}$ which is used to calculate the GMFC. Although there are many differences in detail between the GMFC and EBFC, the GMFC captures the essential features of the EBFC; the GMFC and EBFC are positive on the poleward flank of the ACC and negative on the equatorward flank. In the Agulhas Retroflection region, they are positive on the north side of the Agulhas Current and negative on the south side of it.

The spatial variability of the thickness diffusivity is estimated. In the above comparison, the constant thickness diffusivity is used. However, in the Agulhas Retroflection region and Argentine Basin, the intensity of the EBFC shown in Fig. 12.1 b is higher than that of the GMFC and an appropriate diffusivity value seems to be larger than $400 \, \text{m}^2 \, \text{s}^{-1}$. On the other hand, in the north of the ACC, the strength of the EBFC is comparable with that of the GMFC. The spatial variability of the diffusivity can be determined by solving (12.5) for κ and this equation may be modified as

$$\nabla \cdot (\overline{b'\mathbf{u}'}) + \kappa \left(\nabla_H^2 \overline{b} - \frac{\partial}{\partial z} \left[(\nabla_H \overline{b})^2 / \frac{\partial \overline{b}}{\partial z} \right] \right)$$
$$+ \nabla_H \kappa \cdot \nabla_H \overline{b} - \frac{\partial \kappa}{\partial z} \left[(\nabla_H \overline{b})^2 / \frac{\partial \overline{b}}{\partial z} \right] = 0. \quad (12.8)$$

Instead of solving this equation, we apply the least-square fitting for κ in (12.8) with neglecting the third and fourth terms, i.e., a coefficient $\overline{\kappa}$ which minimizes the following quantity

$$\sum_{k,l} \left[\nabla \cdot (\overline{b'\mathbf{u}'})_{k,l} + \overline{\kappa}_{i,j} \left(\nabla_H^2 \overline{b} - \frac{\partial}{\partial z} \left[(\nabla_H \overline{b})^2 / \frac{\partial \overline{b}}{\partial z} \right] \right)_{k,l} \right]^2 \quad (12.9)$$

is calculated. The summation $\sum_{k,l}$ is taken for grids which are in the area of $10° \times 5°$ (longitude $\times$ latitude) with the (i, j) point is at the center. The validity of neglecting third and fourth terms in (12.9) will be described in Sect. 12.5.

Figure 12.2b shows the calculated coefficient $\overline{\kappa}$ at 2,000 m depth. Although there are negative value regions, the positive value regions dominate. The large positive value regions are located in the Argentine Basin, Agulhas Retroflection region, and the east of the Kerguelen Plateau and the values exceed $1,000\,\mathrm{m}^2\,\mathrm{s}^{-1}$. The values of the coefficient are relatively small outside the ACC region.

The effect of the summation area size on the calculated coefficient is investigated. The coefficient calculated without the summation area has high spatial variability with negative values (not shown). The high spatial variability is due to the position difference between the spatial structure of the EBFC and GMFC. The least-square fitting with the summation area can reduce this high spatial variability because it uses all data in the summation area to determine the coefficient.

The coefficient $\bar{\kappa}$ obtained by using the summation area size much larger than mesoscale eddies such as $10° \times 5°$ would be appropriate coefficient as the thickness diffusivity coefficient of the GM parameterization. Since the GM parameterization represents the mesoscale eddy effects on the ocean flow whose spatial scale is much larger than that of mesoscale eddies, the higher spatial frequency component of the coefficient calculated without the summation area does not seem to be important to represent the eddy effect in the parameterization. The least-square fitting with relatively large summation area size can remove the higher frequency components and the calculated coefficient would be appropriate for the parameterization.

The thickness diffusivity formulation of (12.1) is examined. Figure 12.2 shows that the locations of the large $\bar{\kappa}$ region are roughly in accordance with the locations of the strong current flow one. Since the growth rate of the baroclinic instability is large where the current flow is strong, it suggests that (12.1) would be useful to represent the distribution of the thickness diffusivity. In order to evaluate validity of (12.1), the domain is divided into eight regions; it is divided into four in longitude from 0° with 90° width and into two in latitude from 70°S with 20° width. The averaged value of the diffusivity in each region is used to evaluate (12.1). The τ^{-1} in (12.1) is evaluated as $V\lambda^{-1}$, where V and λ are the vertical mean velocity and the radius of deformation, respectively. This evaluation of τ^{-1} is based on Larichev and Held (1995) and Treguier et al. (1997), who assumed that the flow is very unstable and produces a substantial inverse energy cascade. The length scale of the energy-containing eddies, l, is evaluated as λ. The calculated $\bar{\kappa}$ and $l^2\tau^{-1}$ are plotted in Fig. 12.3. It shows a correlation between $\bar{\kappa}$ and $l^2\tau^{-1}$ and the proportionality constant obtained by the regression line is $c_e = 0.044$.

12.5 Summary and Discussion

The distribution of the EBFC, which represents the effect of mesoscale eddies on the climate timescale current flow, is obtained by using one of the highest resolution ocean model up to now with realistic geometry and forcing. It is found that the EBFC intensity is strong in the Argentine Basin, Agulhas Retroflection region, and the east of the Kerguelen Plateau. The horizontal distribution of the EBFC clearly shows that the EBFC works to decrease the slope of the isopycnals.

12.5 Summary and Discussion

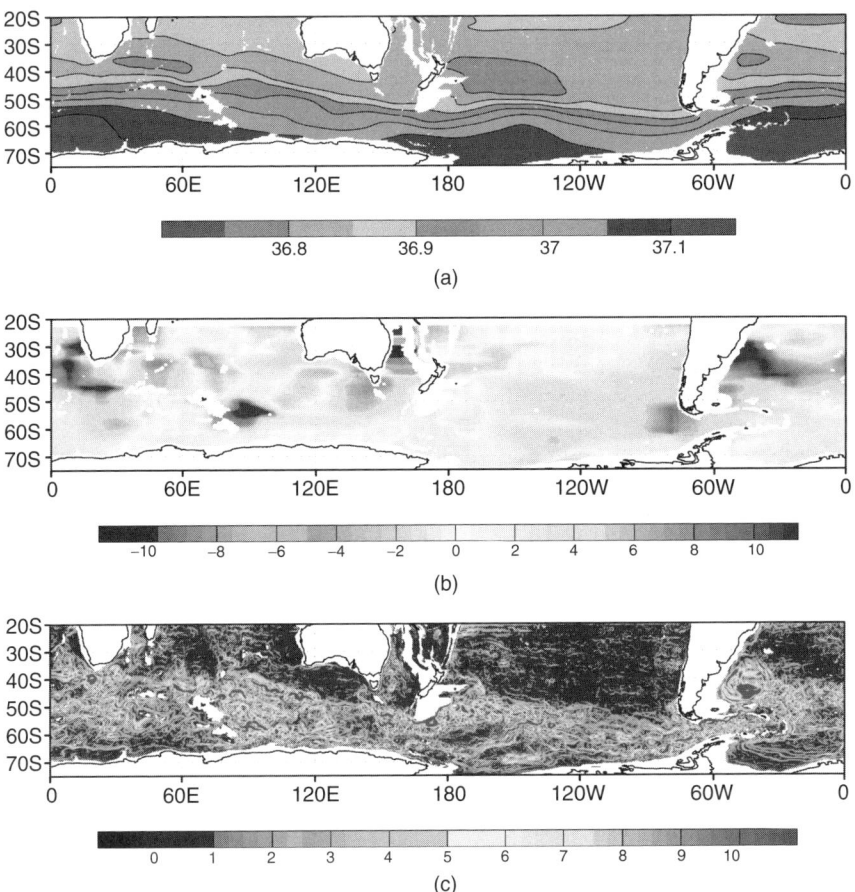

Fig. 12.2 (a) The potential density distribution (σ_2), (b) the coefficient $\overline{\kappa}$ calculated with the summation area size of $10° \times 5°$ (unit : 10^2 m^2 s^{-1}), and (c) the current strength at 2,000 m depth (unit : 10^{-2} m s^{-1})

In the least-square fitting of determining the coefficient $\overline{\kappa}$ in (12.9), the last two terms in (12.8)

$$\nabla_H \overline{\kappa} \cdot \nabla_H \overline{b} - \frac{\partial \overline{\kappa}}{\partial z} \left[(\nabla_H \overline{b})^2 / \frac{\partial \overline{b}}{\partial z} \right]$$

are not evaluated. If these terms are small compared with the EBFC, it is rationalized to drop these terms in (12.9). The order of the second term would be estimated as

$$| \frac{\partial \overline{\kappa}}{\partial z} || (\nabla_H \overline{b})^2 / \frac{\partial \overline{b}}{\partial z} | \sim | \frac{\partial \overline{\kappa}}{\partial z} || \nabla_H \overline{b} || s_\rho | \ll | \nabla_H \overline{\kappa} || \nabla_H \overline{b} |,$$

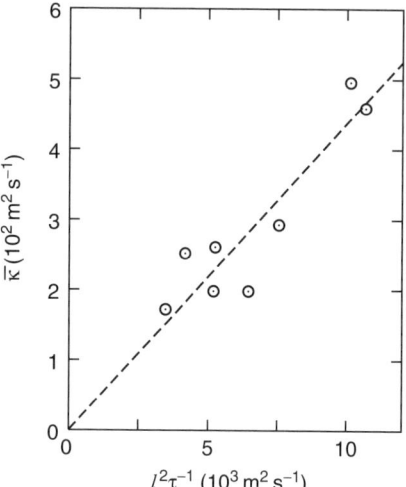

Fig. 12.3 The area averaged coefficient $\bar{\kappa}$ vs. $l^2\tau^{-1}$ in (12.1). The inclination of the regression line gives an estimate of the proportional constant and the resulting value is $c_e = 0.044$

where $s_\rho \sim 0.01$ represents the slope of the isopycnals and $|\frac{\partial \bar{\kappa}}{\partial z}| \leq |\nabla_H \bar{\kappa}|$ is employed on the assumption that the vertical dependency of the coefficient is small. The second term is, therefore, would be negligible. The horizontal spatial derivative of the $\bar{\kappa}$ is estimated as $\nabla_H \bar{\kappa} \sim 10^{-4}$ m s^{-1} by using the distribution of the coefficient in Fig. 12.2 b and that of the buoyancy is estimated as $\nabla_H \bar{b} \sim 10^{-9}$ s^{-2} by using the distribution in Fig. 12.2 a. Then, $\nabla_H \bar{\kappa} \cdot \nabla_H \bar{b} \sim 10^{-13}$ m s^{-3} is smaller than the EBFC in most regions.

Acknowledgments We thank all members of the Global Environment Modeling Research Program, Frontier Research Center for Global Change. We also thank the reviewer for providing helpful suggestions to improve the manuscript. The numerical experiments were performed on the Earth Simulator. The graphics were produced with the GrADS and the Gnuplot.

References

Antonov, J. I., S. Levitus, T. P. Boyer, M. E. Conkright, T. D. O'Brien, and C. Stephens, 1998: *World Ocean Atlas 1998 Vol. 1: Temperature of the Atlantic Ocean*. NOAA Atlas NESDIS 27, US Government Printing Office, Washington, DC

Boyer, T. P., S. Levitus, J. I. Antonov, M. E. Conkright, T. D. O'Brien, and C. Stephens, 1998: *World Ocean Atlas 1998 Vol. 4: Salinity of the Atlantic Ocean*. NOAA Atlas NESDIS 30, US Government Printing Office, Washington, DC

Bryan, K., J. Dukowicz, and R. Smith, 1999: On the mixing coefficient in the parameterization of bolus velocity. *J. Phys. Oceanogr.*, **29**, 2442–2456.

Danabasoglu, G., J. McWilliams, and P. Gent, 1994: The role of mesoscale tracer transports in the global ocean circulation. *Science*, **264**, 1123–1126.

References

Ducet, N., P. Le Traon, and G. Reverdin, 2000: Global high-resolution mapping of ocean circulation from TOPEX/Poseidon and ERS-1 and -2. *J. Geophys. Res.*, **105(C8)**, 19477–19498.

Gent, P. and J. McWilliams, 1990: Isopycnal mixing in ocean circulation models. *J. Phys. Oceanogr.*, **20**, 150–155.

Gent, P., J. Willebrand, T. McDougall, and J. McWilliams, 1995: Parameterizing eddy-induced tracer transports in ocean circulation models. *J. Phys. Oceanogr.*, **25**, 463–474.

Green, J., 1970: Transfer properties of the large-scale eddies and the general circulation of the atmosphere. *Q. J. R. Meteorol. Soc.*, **96**, 157–185.

Griffies, S., 1998: The Gent–McWilliams skew flux. *J. Phys. Oceanogr.*, **28**, 831–841.

Griffies, S. and W. Hallberg, 2000: Biharmonic friction with a Smagorinsky-like viscosity for use in large-scale eddy-permitting ocean model. *Mon. Weather Rev.*, **128**, 2935–2946.

Hasumi, H., 2000: *CCSR Ocean Compoment Model (COCO) version 2.1*. CCSR Rep. 13, Univ. of Tokyo, Tokyo, 68 pages.

Hasumi, H. and N. Suginohara, 1999: Sensitivity of a global ocean general circulation model to tracer advection schemes. *J. Phys. Oceanogr.*, **29**, 2730–2740.

Hirst, A. and T. McDougall, 1996: Deep-water properties and surface buoyancy flux as simulated by a z-coordinate model including eddy-induced advection. *J. Phys. Oceanogr.*, **26**, 1320–1343.

Ishizaki, H. and T. Motoi, 1999: Reevaluation of the Takano–Onishi scheme for momentum advection on bottom relief in ocean models. *J. Atmos. Oceanic. Technol.*, **16**, 1994–2010.

Larichev, V. and I. Held, 1995: Eddy amplitudes and fluxes in a homogeneous model of fully developed baroclinic instability. *J. Phys. Oceanogr.*, **25**, 2285–2297.

Lee, M.-M., D. Marshall, and R. Williams, 1997: On the eddy transfer of tracers:Advective or diffusive? *J. Mar. Res.*, **55**, 483–505.

Nakano, H. and H. Hasumi, 2005: A series of zonal jets embedded in the broad zonal flows in the Pacific obtained in eddy-permittimg ocean general circulation models. *J. Phys. Oceanogr.*, **35**, 474–488.

Nerem, R., E. Schrama, C. Koblinsky, and B. Beckly, 1994: A preliminary evaluation of ocean topography from the TOPEX/POSEIDON mission. *J. Geophys. Res.*, **99(C12)**, 24565–24583.

Rix, N. and J. Willebrand, 1996: Parameterization of mesoscale eddies as inferred from a high-resolution circulation model. *J. Phys. Oceanogr.*, **26**, 2281–2285.

Röske, F., 2001: An atlas of surface fluxes based on the ECMWF re-analysis – A climatological dataset to force global ocean general circulation models. *Max-Planck-Institut für Meteorologie, Hamburg*, Report No. 323.

Solovev, M., P. Stone, and P. Malanotte-Rizzoli, 2002: Assessment of mesoscale eddy parameterizations for a sigle-basin coarse-resolution ocean model. *J. Geophys. Res.*, **107(C9)**, 3126, doi:10.1029/2001JC001032.

Stammer, D., 1998: On eddy characteristics, eddy transports, and mean flow properties. *J. Phys. Oceanogr.*, **28**, 727–739.

Stone, P., 1972: A simplified radiative-dynamical model for the static stability of rotating atmospheres. *J. Atmos. Sci.*, **29**, 405–418.

Treguier, A., 1999: Evaluating eddy mixing coefficients from eddy-resolving ocean models: A case study. *J. Mar. Res.*, **57**, 89–108.

Treguier, A., I. Held, and V. Larichev, 1997: On the parameterization of the quasigeostrophic eddies in primitive equation ocean models. *J. Phys. Oceanogr.*, **27**, 567–580.

Visbeck, M., J. Marshall, T. Haine, and M. Spall, 1997: Specifications of eddy trasfer coefficients in coarse-resolution ocean circulation models. *J. Phys. Oceanogr.*, **27**, 181–402.

Wunsch, C. and D. Stammer, 1995: The global frequency–wavenumber spectrum of oceanic variability estimated from TOPEX/POSEIDON altimetric measurements. *J. Geophys. Res.*, **100(C12)**, 24895–24910.

Chapter 13
High Resolution Kuroshio Forecast System: Description and its Applications

Takashi Kagimoto, Yasumasa Miyazawa, Xinyu Guo, and Hideyuki Kawajiri

Summary We have developed a forecast system for the Kuroshio large meander with a high horizontal resolution (approximately 10 km). Using the system, we succeeded in predicting the path transitions of the Kuroshio from the nearshore nonlarge meander path to the offshore nonlarge meander path in 2003, and from the nearshore nonlarge meander path to the typical large meander path in 2004 as well as the occurrence of its triggering small meander south of Kyushu Island. We have also been developing a higher resolution forecast model for coastal oceans and bays south of Japan, where physical and biological states of the ocean are much affected by the path variation of the Kuroshio. This model, although the development is still under way, represents tides and tidal currents in two bays south of Japan in a realistic way.

13.1 Introduction

The Kuroshio, one of the strongest surface currents in the world oceans, flows northeastward or eastward near Japan as a part of the subtropical gyre in the North Pacific. Because it carries enormous amounts of heat to the high latitude (e.g., 0.76 Peta Watt (PW) by Bryden et al.'s (1991) observation and 0.25–0.46 PW by Ichikawa and Chaen's (2000)), it plays a crucial role in maintaining and changing the global climate system. In particular the heat flux by the geostrophic advection of the Kuroshio has a substantial contribution to the local heat balance in the Kuroshio Extension region of the central North Pacific (Qiu and Kelly, 1993). Changes in the sea surface temperature associated with the Kuroshio path variations cause the surface wind to change (Nonaka and Xie, 2003), leading to some influence on the Pacific storm track (Inatsu et al., 2002). Thus in order to know climate variations in the North Pacific, it is important to grasp how the Kuroshio is and will be.

In addition the Kuroshio also affects coastal (local) areas because of its peculiar path variations. For example, the 100 m-depth temperature in Sagami Bay[1] is

---

[1] Reader may be unfamiliar to the names of those places. The locations of Enshu-nada Sea, Shikoku, Kyushu and Hachijo-jima Islands, and the Kii Peninsula are denoted by E, S, Q, H, and K, respectively, in the upper left panel of Fig. 13.6, that of Sagami Bay is by Sa in the left panel of Fig. 13.12, and that of the Izu Peninsula is in Fig. 13.14.

remarkably higher during the typical large meander path of the Kuroshio than other paths (Kawabe and Yoneno, 1987). The large meander of the Kuroshio may enhance the onshore intrusion of the warm water into the shelf region of the Enshu-nada Sea[1] and thereby contribute toward increasing the temperature in spawning habitats of sardine, leading to increase of postlarval sardine but to decrease of postlarval Japanese anchovy there (Nakata et al., 2000). Recent observations based on the long-term current measurements (Iwata and Matsuyama, 1989) and the HF radars (Hinata et al., 2005) indicate that the warm water intrusion into Sagami Bay is enhanced when the Kuroshio is located near the Izu Peninsula,[1] corresponding to either the nearshore nonlarge meander path or the typical large meander path. In addition to the physical and biological aspects, the location of the Kuroshio may have an influence on engineering activities such as the determination of optimal routes for ferries and cargo vessels. Thus the path variation of the Kuroshio is a general concern of many people as well as oceanographers.

In order to respond to their concerns it is necessary to predict the path variations of the Kuroshio. Several previous studies on development of a nowcast/forecast system for the Kuroshio have been accomplished. Komori et al. (2003) examine a capability for short-range forecast of the Kuroshio path variations south of Japan using a 1.5 layer primitive equation model with 1/12° resolution into which an objective analysis data of the sea surface height (Kuragano and Shibata, 1997) is assimilated with a variational method. The analysis fields obtained from the model are quite reasonable because the root-mean-square (RMS) differences with regard to the Kuroshio axis location south of the Enshu-nada Sea and the sea level at Hachijo-jima Island[1] between the reanalysis and the observations are less than the observed variances. The prediction errors are kept within the variances during 75-day prediction period, suggesting that the model has a predictability skill up to 75 days for the Kuroshio path variations. Ishikawa et al. (2004) point out, using the same forecast system, that an appropriate assimilation of a shoaling of the thermocline associated with a small meander of the Kuroshio south of Shikoku Island[1] into the model is of great importance in predicting the transition from the nearshore nonlarge meander path to the large meander path. Kamachi et al. (2004a) develop an operational ocean data assimilation system for the Kuroshio region using a more sophisticated ocean general circulation model with 1/4° resolution but with the simpler assimilation scheme of a time-retrospective nudging method than Komori et al.'s, and confirm that the reanalysis represents the real ocean reasonably well by comparing with individual observation data. The system, which has been operated in the Japan Meteorological Agency since January 2001, can predict the Kuroshio axis in 30- to 80-day prediction period in a manner that depends upon the ocean state in the initial condition (Kamachi et al., 2004b).

Previous studies have indicated several important factors in predicting the path transition of the Kuroshio: anticyclonic mesoscale eddies that appear in the subtropical frontal zone and move all the way to the south of Japan along the Kuroshio (Akitomo and Kurogi, 2001; Miyazawa et al., 2004), mesoscale eddies propagating westward from the Kuroshio Extension region to the southeast of Kyushu Island[1] (Ichikawa and Imawaki, 1994; Ebuchi and Hanawa, 2000; Akitomo and Kurogi,

2001; Mitsudera et al., 2001), the baroclinic instability process leading to the development of the small meander and the subsequent typical large meander (Endoh and Hibiya, 2000; Qiu and Miao, 2000), and a small seamount and associate abyssal circulation (Hurlburt et al., 1996; Endoh and Hibiya, 2000; Miyazawa et al., 2004). All of them suggest that numerical models are required to have a fine horizontal grid spacing so as to represent those factors reasonably well. Moreover, only a numerical model with a fine horizontal resolution can represent the realistic vertical structure of the velocity and the associated density structure in the Kuroshio region (Guo et al., 2003).

On the basis of the implications of previous studies, we developed a high resolution Kuroshio forecast system, called the Japan Coastal Ocean Predictability Experiment (JCOPE) ocean forecast system. The JCOPE project[2] has begun in October 1997 under an initiative of the Frontier Research Center for Global Change (FRCGC) / Japan Agency for Marine-Earth Science and Technology (JAMSTEC). As a part of the project, we have carried out fundamental simulation studies. Guo et al. (2003) examined effectiveness of a one-way nesting technique for the simulation of the Kuroshio. Miyazawa et al. (2004) pointed out the role of mesoscale eddies in the path variations of the Kuroshio using an eddy-resolving POM. Recently Miyazawa et al. (2005) indicated using a prototype of the forecast system that an initialization process with mesoscale eddies taken into account is crucial in predicting the path variations of the Kuroshio accurately. The present JCOPE forecast system finally reaches a stage of near-operational level using almost all available data obtained from satellites, ARGO (the Array for Real-time Geostrophic Oceanography) floats and ships, and its outcome is available on the Web site of the JCOPE project.[2] In this chapter we will describe details of the JCOPE ocean forecast system that is currently operated.

This chapter is organized as follows. In Sect. 13.2, we describe the numerical model, surface forcing data, and data assimilation scheme that we use for the forecast system in great detail. A couple of examples to show the forecast of the Kuroshio path transition are shown in Sect. 13.3. We also describe the sensitivity of the forecast to the parameters used in the data assimilation scheme. In Sect. 13.4, we briefly introduce our recent study toward downscaling of the Kuroshio forecast for coastal oceans and bays. Section 13.5 is devoted to summary and exemplification of some applications of the JCOPE outcome.

13.2 Description of Forecast System

13.2.1 Numerical Model

The numerical model that we adopt for the Kuroshio forecast system is a three-dimensional primitive-equation ocean general circulation model based on the POM with a generalized vertical coordinate (Mellor et al., 2002). The forecast system

---
[2] See http://www.jamstec.go.jp/frcgc/jcope/

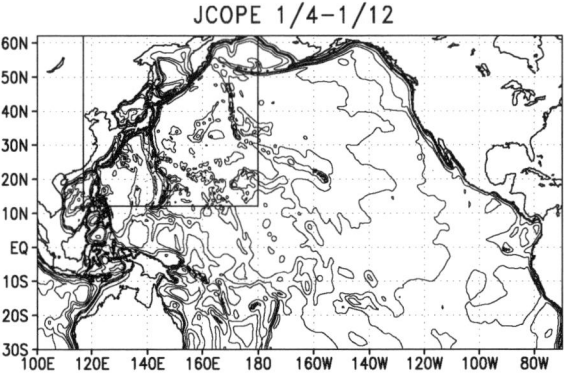

Fig. 13.1 Bottom topography used in the coarse model. Contour interval is 1,000 m. The small rectangle region delineated by solid straight lines is the computational domain for the fine model

consists of two models with different computational domains and resolutions: one is for the North Pacific basin with the variable horizontal resolution (1/4° near Japan and 1/2° in the east of basin) and 21 vertical levels; and the other is for the western North Pacific basin with 1/12° resolution and 45 levels (Fig. 13.1). The former model (hereafter coarse model) provides the boundary conditions for all prognostic variables described below to the latter (hereafter fine model), but the latter does not affect the former (no feedback).

The prognostic variables in the governing equations of the model are the surface elevation (η), the zonal and meridional components of the vertically averaged velocity ($\bar{u}, \bar{v}$), the zonal and meridional velocity (u, v), potential temperature (T), salinity (S), twice the turbulent kinetic energy (q^2), and turbulent length scale (l). A set of the equations for the latter two variables is a turbulent closure submodel (Mellor and Yamada, 1982), which is used for the estimation of the vertical eddy mixing coefficients. Mellor (2001) modified the treatment of the turbulent dissipation term in the submodel in such a way that the surface mixed layer did not become shallow overly in summer. With this correction Ezer (2000) succeeded in simulating the realistic vertical temperature profile in the North Atlantic, but Kagimoto (2003) pointed out that the correction caused overly mixing in the surface equatorial region where the vertical shear was strong and thereby the Equatorial Undercurrent almost disappeared. Hence we do not employ this modification. The horizontal eddy mixing coefficient is, on the other hand, evaluated by Smagorinsky's (1963) nonlinear formulation with constant coefficients of 0.1 for the coarse model and 0.075 for the fine model.[3] The coefficient for tracers is half of that for momentum, meaning that the turbulent Prandtl number is two.

The horizontal coordinate system is a conventional longitude–latitude grid, whereas the vertical one is a generalized "z-plus-sigma" grid. The depth of layer

---

[3] Those values correspond approximately to the so-called Smagorinsky's constants of 0.2659 and 0.23, respectively.

$1 \leq k \leq kb$ at location (x, y) and time t is defined by $z = \eta(x, y, t) + s(x, y, k, t)$, where kb is the number of vertical levels. A z-level system is $s^{(z)} = \sigma(k)[H_{\max} + \eta(x, y, t)]$, and a sigma coordinate system is $s^{(\sigma)} = \sigma(k)[H(x, y) + \eta(x, y, t)]$, where $H_{\max}$ is the maximum depth and the relative distribution of vertical layers is $0 \leq \sigma(k) \leq -1$ (Ezer and Mellor, 2004). In our formulation for the vertical grid, z-level grids are aligned on sigma grids. The boundary between the z-level system and the sigma-level system is located at k_e, the minimum value of indices that satisfy the following inequality,

$$|s^{(z)}(x, y, k_e, t)| > 0.5 H(x, y) \tag{13.1}$$

Note that this formulation keeps surface layers from being over-resolved particularly in shallow areas (Mellor et al., 2002) and then prevents the sea surface temperature (SST) from being overestimated there. We also note that the bottom topography can be represented with the terrain following grid system more realistically than the z-level grids. In fact, the bottom relief is crucial to the representation of the Kuroshio because the bottom pressure torque plays a major role in the vorticity budget over the continental slope region at which the main axis of the Kuroshio is located (Kagimoto and Yamagata, 1997).

The bottom topography used in the model is based on the global DTM5 (Digital Terrain Models) with 1/12° resolution. For the fine model, the topography data with 500 m resolution archived by Hydrographic and Oceanographic Department, Japan Coastal Guard is also used for the area near Japan. In order to reduce the error in computing the pressure gradient force near steep topography, the topography data is smoothed out using two kinds of smoothing filters: the Gaussian smoothing with the 1/8° length scale, and the filter based on Mellor et al. (1994). The latter filter is continuously applied to the data until the depth satisfies the inequality $\delta H / \bar{H} \leq 0.35$, where δH is the depth difference between two neighboring grid points and $\bar{H}$ is the average depth between them. Note that the former filter is useful particularly for the variable grid system, which is adopted in our coarse model, because an effect of the filter that is not based on the local depth gradient but depth "difference" is inhomogeneous in space, thereby the bottom relief in areas with coarser resolutions is heavily smoothed out rather than with finer ones. The depths near the boundary of the fine model are calculated from the smoothed topography data used in the coarse model by using a bilinear interpolation in such a way that solutions of the fine model may be connected smoothly with those of the coarse model.

In the original POM, a partial slip condition is applied to the tangential component of the velocity at the boundary. We change it to the no-slip boundary condition, because the positive vorticity supply may be important to the development of the Kuroshio large meander (Akitomo et al., 1991). The temperature and salinity near the artificial boundaries in the coarse model are restored to the monthly mean climatology (Levitus and Boyer, 1994; Levitus et al., 1994) in such a way that fictitious waves neither propagate along nor reflect at the boundaries. Since sea ice processes are not taken into account in the forecast system, both the temperature and salinity of the coarse model in the Sea of Okhotsk and the Bering Sea are also restored to the monthly mean climatology. In the fine model, they are restored to the climatology only in the Sea of Okhotsk.

The initial condition is a state of rest with the 1/4° annual mean climatological temperature and salinity fields (Boyer and Levitus, 1997). The coarse model is spun up for 10 years from the initial state with the use of the monthly mean surface forcings described below. Then it is further driven by 6 hourly surface forcings covering from September 1999 to December 2001. The fine model is also forced from the initial state with the monthly mean surface forcings but for 5 years, and is continued to be forced with the 6 hourly forcings from September 1999 to December 2001. The data assimilation scheme described in Sects. 13.2.5 and 13.2.7 is introduced from September 1999. Then the forecast experiment together with the data assimilation is started in December 2001.

13.2.2 Nesting

As described earlier, the coarse model provides the boundary values of the prognostic variables to the fine model. All prognostic variables except for two turbulent properties in the coarse model are linearly interpolated on to the boundary grid points of the fine model. In order to reduce noise caused by a mismatch between two models, we modify the vertically averaged velocity normal to the boundary with the use of the following radiation condition (Flather, 1976; Guo et al., 2003),

$$\bar{u}_f = \bar{u}_c - c_{rf}\sqrt{\frac{g}{H}}(\eta_f - \eta_c) \tag{13.2}$$

where $\bar{u}_c$ and η_c are, respectively, the interpolated values of the normal component of the velocity and the surface elevation, g is the gravity constant, η_f is the surface elevation of the fine model at a grid inside the open boundary, and c_{rf} is an adjustable parameter ($0 < c_{rf} \leq 1$), for which we select 0.1. The turbulent properties, q^2 and $q^2 l$, are set to be a constant of 10^{-10}, meaning that the turbulent energy advected from the outside is ignored.

13.2.3 Surface Forcings

The zonal and meridional components of the momentum flux (τ_x, τ_y) are parameterized with the following bulk formula,

$$(\tau_x, \tau_y) = \rho_a C_D \sqrt{u_{10m}^2 + v_{10m}^2}(u_{10m}, v_{10m}) \tag{13.3}$$

where $\rho_a = 1.225$ kg m^{-3} is the air density at sea level, u_{10m} and v_{10m} are, respectively, the zonal and meridional components of wind velocity at 10 m above sea level

13.2 Description of Forecast System

(m s^{-1}), and C_D is the drag coefficient based on Large and Pond's (1981) formula, that is,

$$10^3 \times C_D = \begin{cases} 1.14, & (\sqrt{u_{10m}^2 + v_{10m}^2} < 10) \\ 0.49 + 0.067\sqrt{u_{10m}^2 + v_{10m}^2}, & (10 \le \sqrt{u_{10m}^2 + v_{10m}^2} < 26) \\ 0.49 + 0.065 \times 26, & (26 \le \sqrt{u_{10m}^2 + v_{10m}^2}) \end{cases} \quad (13.4)$$

For u_{10m} and v_{10m}, we adopt the QuikSCAT Near-Realtime data products (Liu et al., 1998), whose spatial and temporal resolutions are half a degree and 12 h, respectively. Although the resolutions are coarser than the intrinsic ones in the NASA/JPL's Sea-Winds Scatterometer, the products have the advantage of less sampling error along the satellite track than others.

The turbulent heat flux $\overline{w'T'}$ is evaluated using the following equation,

$$\rho_a C_p \overline{w'T'} = (Q_L + Q_E + Q_H) - \frac{dQ}{dT}(T_s - T_{obs}) \quad (13.5)$$

where $C_p = 1.005 \times 10^3$ J kg^{-1} K^{-1} is the specific heat capacity, Q_L is the long-wave radiation, Q_E is the latent heat flux, Q_H is the sensible heat flux, $dQ/dT = -35$ W m^{-2} K^{-1} is the coupling coefficient, and T_s and T_{obs} are the model and observed SST, respectively. The last term of (13.5) is the correction term to reduce SST biases. The first three terms are parameterized with the bulk formulas,

$$Q_L = \epsilon \sigma T_{obs}^4 (0.39 - 0.05\sqrt{e_a})(1 - 0.8C) + 4\epsilon \sigma T_{obs}^3 (T_{obs} - T_a) \quad (13.6)$$

$$Q_E = \rho_a C_E L \sqrt{u_{2m}^2 + v_{2m}^2} (q_{sat}(T_{obs}) - q_a) \quad (13.7)$$

$$Q_H = \rho_a C_p C_H \sqrt{u_{2m}^2 + v_{2m}^2} (T_{obs} - T_a) \quad (13.8)$$

where $\epsilon = 0.97$ is the emissivity of the ocean, $\sigma = 5.67 \times 10^{-8}$ W m^{-2} K^{-4} is the Stefan-Boltzmann constant, C is the fraction cloud cover, T_a is the air temperature, $L = 2.501 \times 10^6$ J kg^{-1} is the latent heat of fusion, $q_{sat}(T_{obs})$ is the saturated specific humidity at T_{obs}, and q_a is the specific humidity. The saturated specific humidity is related to the saturated vapor pressure, which is approximately a polynomial function of temperature (Tetens, 1930). The vapor pressure, e_a, is calculated from q_a using the same relationship. The zonal and meridional wind velocities at 2 m above sea level, (u_{2m}, v_{2m}), are estimated from u_{10m} and v_{10m} using the logarithmic law. The turbulent exchange coefficients, C_E and C_H, are based on Kondo's (1975) formulas.

The incoming shortwave radiation at the sea surface, Q_s, is estimated with an empirical formula,

$$Q_s = Q_s^{sky}(0.77 - 0.5C^2)(1 - \alpha) \quad (13.9)$$

where Q_s^{sky} is the solar radiation incident on the ocean under clear skies, and $\alpha = 0.08$ is the ocean surface albedo. Equation (13.9) is different from numerous empirical relations proposed in previous studies (e.g., Reed, 1977). The factor,

$(0.77 - 0.5C^2)$, is found out with trial and error in such a way that the climatological model SST becomes close to the observed one. The penetration of the shortwave radiation into the ocean is taken into account on the basis of Paulson and Simpson's (1977) formulation,

$$I(z) = Q_s \left(\gamma \exp\left(\frac{z}{\zeta_1}\right) + (1-\gamma) \exp\left(\frac{z}{\zeta_2}\right) \right) \qquad (13.10)$$

where $I(z)$ is the penetrated solar radiation flux at depth z, ζ_1 and ζ_2 are, respectively, the e-folding depths of the long (ζ_1), and short visible and ultra violet (ζ_2) wavelengths, and γ is a fraction for the former part. We select $\gamma = 0.58$, $\zeta_1 = 0.35\,\text{m}$, and $\zeta_2 = 23.0\,\text{m}$, corresponding to the type I water of Jerlov's (1976) classification.[4]

Those fluxes require the observed SST, surface air temperature, specific humidity, the fraction cloud cover, and the solar radiation incident on the ocean under clear skies. We adopt the 6 hourly NCEP/NCAR reanalysis product (Kalnay et al., 1996) for them.

The sea surface salinity, S_s, is restored to the monthly climatology (Levitus et al., 1994), S_{obs}, in the form of the salinity flux,

$$\overline{w's'} = \gamma_s (S_s - S_{\text{obs}}) \qquad (13.11)$$

where $\gamma_s = 10\,\text{m month}^{-1}$ is the restoring speed.

13.2.4 Observational Data

The observations assimilated into the model are the satellite altimeter data obtained from the TOPEX/POSEIDON and ERS-1 during September 1999 to June 2002 and from the Jason-1 and the GeoSat Follow-On during June 2002 to the present, the sea surface temperature from the advanced very high resolution radiometer/multi-channel sea surface temperature (AVHRR/MCSST) Level 2 products, and vertical profile data of the temperature and salinity provided by the Global Temperature-Salinity Profile Program (GTSPP). From these data, we build up 7 days interval data on 1/4° grids using the optimal interpolation technique described later. The sampling period for the satellite data is 7 days (before and after 3 days of an analysis time), while that for the temperature and salinity profile is 31 days. The spatial extent of sampling data is 360 km.

The gridded anomaly of a variable, a_g, is calculated using a linear combination of the observed anomaly of the variable, a_o,

$$a_g = W^T a_o \qquad (13.12)$$

---

[4] The type II or IA is better for the water off Japan, but the use of the e-folding depths corresponding to the waters resulted in a warm bias of the SST. Hence we chose the type I water.

13.2 Description of Forecast System

where a_g and a_o are, respectively, the column vectors with N and M elements, W is a $M \times N$ matrix of weights,[5] M is the number of observation points, and N is the number of grid points. The superscript of W, T, denotes the transpose of the matrix. The matrix, W, is found by minimizing the variance of interpolation errors, and obtained from solving the algebraic equation (Gandin, 1963),

$$(P^{oo} + \lambda I)W = P^{og} \tag{13.13}$$

where I is the unit matrix. The $M \times M$ matrix of the autocorrelation of the observed values, P^{oo}, is parameterized with a Gaussian model,

$$\begin{aligned} P^{oo} &= (P^{oo}_{i,j}) \\ &= P^{oo}(\boldsymbol{x}_i; \boldsymbol{x}_j) \\ &= C_0(x_i, y_i, z_i) \exp\left(-\left(\tfrac{\Delta x - c_x \Delta t}{L_x}\right)^2 - \left(\tfrac{\Delta y}{L_y}\right)^2 - \left(\tfrac{\Delta t}{L_t}\right)^2\right) \end{aligned} \tag{13.14}$$

where $\boldsymbol{x}_i = (x_i, y_i, z_i, t_i)$ and $\boldsymbol{x}_j$ are two observation points, C_0 is the autocorrelation at lag 0, $\Delta x = x_i - x_j$, $\Delta y = y_i - y_j$, $\Delta t = t_i - t_j$, c_x is a phase speed in the zonal direction, and L_x, L_y, and L_t, respectively, denote the zonal, meridional, and temporal decorrelation scales. The $M \times N$ matrix of the autocorrelation between the observed and interpolated values, P^{og}, is also calculated from (13.14), but $\boldsymbol{x}_j$ is replaced with $\boldsymbol{x}_g$, which denotes a grid point to which the data are interpolated. The signal-to-noise ratio, λ in (13.13), is calculated with C_0,

$$\lambda = \frac{1 - C_0}{C_0} \tag{13.15}$$

The parameters, C_0, L_x, L_y, L_t, and c_x, required by the optimal interpolation are calculated from the following procedures using the observation data listed in Table 13.1.

1. Create a mean field during a period
2. Calculate the autocorrelation at observation points and average the correlation matrices within $5° \times 5°$ regions

Table 13.1 Data used to estimate the parameters in (13.14)

Data	Instruments	Distributor	Period	Mean
SSH	TOPEX/POSEIDON, ERS-2	CCAR	Sep 1999–Jun 2002	None
SST	AVHRR	JPL, NAVO-CEANO	Oct 2001–Aug 2002	Mean during the period
T & S (0–400 m)	CTD, XBT, MBT, etc.	GTSPP	Mar 1990–Mar 2002	Boyer and Levitus (1997)

---

[5] For the sake of convenience, (13.12) is expressed with the matrix form, but in the numerical calculation W is evaluated at each grid point. Thus in practice a_g is a scalar, W is an M elements column vector.

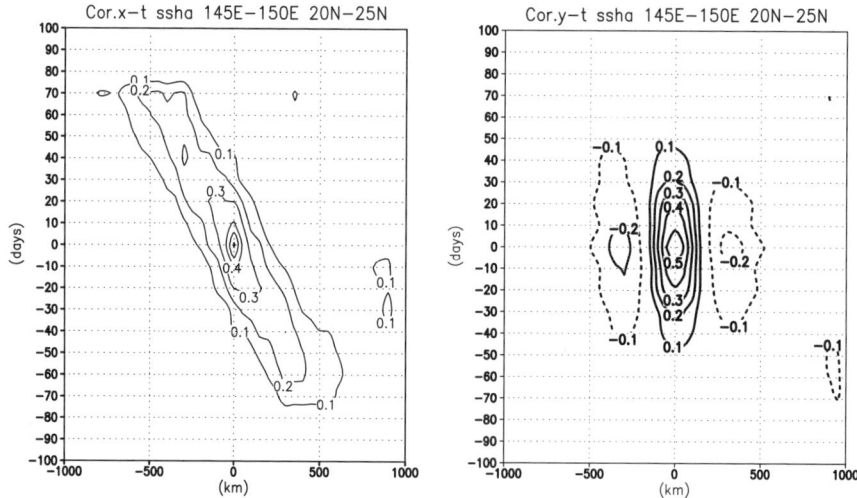

Fig. 13.2 Autocorrelation of the sea surface height (SSH) anomaly in the region, 20°N–25°N, 145°E–150°E. (*Left*) Time-longitude lag-correlation. (*Right*) Time-latitude lag-correlation. Contour interval is 0.1. *Solid line* denotes positive values and *dashed line* denotes negative values

3. Estimate the parameters by fitting the Gaussian model described earlier
4. Interpolate/extrapolate the parameters on to 1/4° grids
5. Smooth out the parameters using the Gaussian filter with 3° horizontal scale

Figure 13.2 shows an example of the autocorrelation of the sea surface height (SSH) anomaly based on the altimetry data in the region, 20°N–25°N, 145°E–150°E, obtained from procedure 2. By fitting the autocorrelation, in a least square sense, to the Gaussian model of (13.14) (procedure 3), we obtain the parameters,

$$(C_0, c_x, L_x, L_y, L_t) = (0.626, -0.07 \text{ m s}^{-1}, 152 \text{ km}, 108 \text{ km}, 53 \text{ days}) \quad (13.16)$$

The approximate autocorrelation function with the above parameters is shown in Fig. 13.3. Both the time-longitude and time-latitude correlation functions are good approximations, although the negative correlation at 300 km apart from the specific point is not represented in the latter (right panels of Figs. 13.2 and 13.3).

13.2.5 Data Assimilation

The SSH anomaly, $\tilde{\eta}^o$, and the vertical profile of the temperature and salinity, $T^o(z_1)$, $T^o(z_2), \cdots, T^o(z_M), S^o(z_1), S^o(z_2), \cdots, S^o(z_M)$, gridded in the way described in the previous subsection are combined with a forecast data, leading to an analysis data on the 1/12° horizontal grids and the "z-plus-sigma" vertical grids, which is used for the increment analysis update (IAU) process described in Sect. 13.2.7. In our

13.2 Description of Forecast System

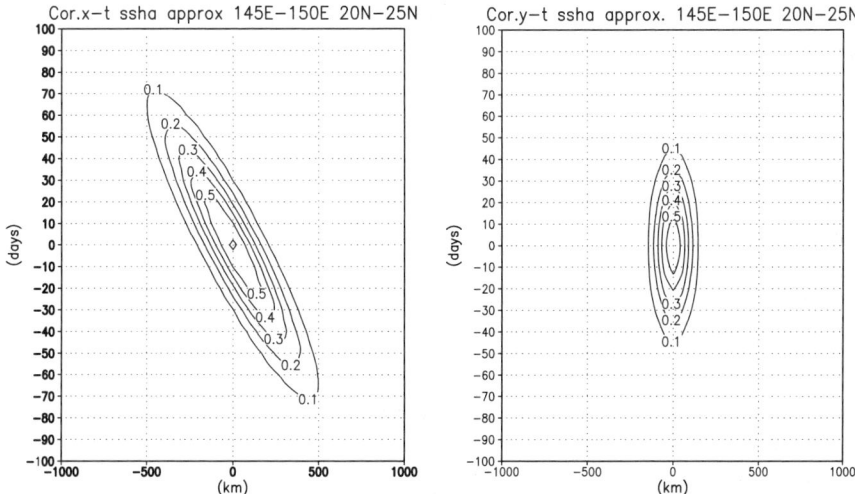

Fig. 13.3 Approximate autocorrelation function of the SSH anomaly in the region, 20°N–25°N, 145°E–150°E fitted to the Gaussian model, (13.14). (*Left*) Time-longitude lag-correlation. (*Right*) Time-latitude lag-correlation. Contour interval is 0.1

system, the number of vertical levels for the gridded observation data, M, is 6, and the corresponding depths, $z_k (k = 1, \cdots, M)$, are 0, 50, 100, 200, 300, and 400 m.

For the data assimilation process, we use a multivariate optimal interpolation scheme described later (Lorenc, 1981; Ezer and Mellor, 1997). Before the interpolation, we subtract basin-wide seasonal variations such as the thermal expansion from the SSH anomaly, that is,

$$\tilde{\eta}' = \tilde{\eta} - \sum_{\text{latitude}} \tilde{\eta} \qquad (13.17)$$

This enables us to assimilate the data of the SSH variations associated with mesoscale eddies effectively into the model (Y. Wakata, personal comm., 1999).

Define the analysis state, X^a, as a $2N + 1$ elements column vector,

$$X^a = (\tilde{\eta}'^a, T_1^a, T_2^a, \cdots, T_N^a, S_1^a, S_2^a, \cdots, S_N^a) \qquad (13.18)$$

where N denotes the number of vertical "z-plus-sigma" levels. The forecast state, X^f, and the observation value, y^o, are also defined in the same way,

$$X^f = (\tilde{\eta}'^f, T_1^f, T_2^f, \cdots, T_N^f, S_1^f, S_2^f, \cdots, S_N^f) \qquad (13.19)$$

$$y^o = (\tilde{\eta}'^o, T_1^o, T_2^o, \cdots, T_M^o, S_1^o, S_2^o, \cdots, S_M^o) \qquad (13.20)$$

The analysis state is calculated from the best linear invariant estimation,

$$X^a = X^f + PH^T(HPH^T + R)^{-1}(y^o - HX^f) \qquad (13.21)$$

where P is the error covariance matrix of the forecast state with $(2N+1) \times (2N+1)$ elements, H is the observation matrix with $(2M + 1) \times (2N + 1)$ elements that is a linear operator mapping the forecast state into the observed data, and R is the observation error covariance matrix with $(2M+1) \times (2M+1)$ elements. The Kalman gain matrix, $PH^T(HPH^T + R)^{-1}$, acts as interpolation and/or extrapolation of the innovation, $y^o - HX^f$, to the model state. The error covariance matrix is the function of time and space that depends upon the model physics. For simplicity, however, we assume it to be constant in time, and estimate it in advance.

Suppose, for a simple explanation, that $M = N = 1$. In this case, the observation matrix becomes the 3×3 unit matrix, and the matrix, PH^T is expressed as follows,

$$PH^T = \begin{pmatrix} \langle(\delta\tilde{\eta}'^f)^2\rangle & \langle\delta\tilde{\eta}'^f\delta T^f\rangle & \langle\delta\tilde{\eta}'^f\delta S^f\rangle \\ \langle\delta T^f\delta\tilde{\eta}'^f\rangle & \langle(\delta T^f)^2\rangle & \langle\delta T^f\delta S^f\rangle \\ \langle\delta S^f\delta\tilde{\eta}'^f\rangle & \langle\delta S^f\delta T^f\rangle & \langle(\delta S^f)^2\rangle \end{pmatrix} \qquad (13.22)$$

where δx and $\langle x \rangle$ denote, respectively, the error and expectation of a variable, x. Assuming that each element of the matrix, PH^T, can be approximately estimated from the variance of the variables obtained from the model (Mellor and Ezer, 1991).

$$PH^T \approx \begin{pmatrix} \overline{(\Delta\tilde{\eta}'^f)^2}E^f_{\eta'} & \overline{\Delta\tilde{\eta}'^f\Delta T^f}E^f_T & \overline{\Delta\tilde{\eta}'^f\Delta S^f}E^f_S \\ \overline{\Delta T^f\Delta\tilde{\eta}'^f}E^f_{\eta'} & \overline{(\Delta T^f)^2}E^f_T & \overline{\Delta T^f\Delta S^f}E^f_S \\ \overline{\Delta S^f\Delta\tilde{\eta}'^f}E^f_{\eta'} & \overline{\Delta S^f\Delta T^f}E^f_T & \overline{(\Delta S^f)^2}E^f_S \end{pmatrix}$$

$$= \begin{pmatrix} 1 & F_{\eta'T} & F_{\eta'S} \\ F_{T\eta'} & 1 & F_{TS} \\ F_{S\eta'} & F_{ST} & 1 \end{pmatrix} \begin{pmatrix} E^f_{\eta'} & 0 & 0 \\ 0 & E^f_T & 0 \\ 0 & 0 & E^f_S \end{pmatrix}$$

$$\times \begin{pmatrix} \overline{(\Delta\tilde{\eta}'^f)^2} & 0 & 0 \\ 0 & \overline{(\Delta T^f)^2} & 0 \\ 0 & 0 & \overline{(\Delta S^f)^2} \end{pmatrix} \qquad (13.23)$$

where $\overline{x}$ and Δx denote the time mean of a variable, x, and the deviation from it, respectively. The regression coefficient between variables, x and y, denoted by F_{xy}, is expressed as follows,

$$F_{xy} = \frac{\overline{\Delta x \Delta y}}{\overline{\Delta y^2}} \qquad (13.24)$$

In this way, the estimation of the forecast error covariance matrix results in the estimation of the nondimensional parameters, $E^f_{\eta'}$, E^f_T, and E^f_S, which are the forecast error variances normalized by the variances in the forecast data. Note that the spatial pattern of the regression coefficients depends strongly upon the model performance, suggesting that the improvement of the model performance is crucial to refinement of the regression coefficients.

Assuming that an observation error of a variable is not correlated to that of another, R becomes a diagonal matrix,

$$R \approx \begin{pmatrix} E^o_{\eta'} & 0 & 0 \\ 0 & E^o_T & 0 \\ 0 & 0 & E^o_S \end{pmatrix} \begin{pmatrix} \overline{(\Delta\tilde{\eta}'^f)^2} & 0 & 0 \\ 0 & \overline{(\Delta T^f)^2} & 0 \\ 0 & 0 & \overline{(\Delta S^f)^2} \end{pmatrix} \qquad (13.25)$$

13.2 Description of Forecast System

where $E_{\tilde{\eta}'}^o$, E_T^o, and E_S^o are the observation error variances normalized by the variances in the forecast data.

If the probability distributions of the observation and forecast errors are independent of each other, we obtain the innovation covariance matrix,

$$\langle (y^o - HX^f)(y^o - HX^f)^T \rangle = HPH^T + R \tag{13.26}$$

Comparison of the diagonal components between both sides of (13.26) gives a simple relation of the observation and forecast errors to the model statistics for each variable,

$$\begin{aligned}
(E_{\tilde{\eta}'}^f + E_{\tilde{\eta}'}^o)\langle (\Delta \tilde{\eta}'^f)^2 \rangle &= \langle (\tilde{\eta}'^f - \tilde{\eta}'^o)^2 \rangle \\
(E_T^f + E_T^o)\langle (\Delta T^f)^2 \rangle &= \langle (T^f - T^o)^2 \rangle \\
(E_S^f + E_S^o)\langle (\Delta S^f)^2 \rangle &= \langle (S^f - S^o)^2 \rangle
\end{aligned} \tag{13.27}$$

Thus the amplitude of the observation and forecast errors are not determined independently, but the ratio between them is. Note that the ratio strongly affects forecast skills, because when E_x^o is bigger (smaller) than E_x^f the forecast (observation) data are set above the observation (forecast) in determining the analysis data used in the IAU process described in Sect. 13.2.7. In fact we realize that the prediction of the Kuroshio large meander is sensitive particularly to the error ratio associated with the SSH anomaly. Now we introduce an adaptive assimilation scheme (Fox et al., 2000) in order to estimate the adequate forecast error variance in (13.27),[6]

$$\begin{aligned}
E_{\tilde{\eta}'}^f G[\langle (\Delta \tilde{\eta}'^f)^2 \rangle] &= G[\langle (\tilde{\eta}'^f - \tilde{\eta}'^o)^2 \rangle - \alpha_{\eta'} e_i^2(\tilde{\eta}^o)] \\
E_{\tilde{\eta}'}^o G[\langle (\Delta \tilde{\eta}'^f)^2 \rangle] &= G[\alpha_{\eta'} e_i^2(\tilde{\eta}^o)]
\end{aligned} \tag{13.28}$$

$$\begin{aligned}
E_{T_0}^f G[\langle (\Delta T_0^f)^2 \rangle] &= G[\langle (T_0^f - T_0^o)^2 \rangle - \alpha_{T_0} e_i^2(T_0^o)] \\
E_{T_0}^o G[\langle (\Delta T_0^f)^2 \rangle] &= G[\alpha_{T_0} e_i^2(T_0^o)]
\end{aligned} \tag{13.29}$$

where $G[\cdot]$ is the spatial filter of the Gaussian shape with a zonal scale of 1.4° (the equator)–0.7° (60°N) and a meridional scale of 0.4° (the equator)–0.2° (60°N), and T_0 is the SST. The interpolation error covariance of a variable x, $e_i^2(x)$, is defined by

$$e_i^2 = V_s^2(1 - W^T P^{og}) \tag{13.30}$$

where V_s^2 is a signal variance of the variable. The difference between the original adaptive assimilation scheme proposed by Fox et al. (2000) and ours is the existence of the adjustable parameters, $\alpha_{\eta'}$ and α_{T_0}, which are found out from the sensitivity experiments in such a way that the predicted Kuroshio large meander is as close to the observation as possible. If there are sufficient data to interpolate on to the model grids, α_x should be unity. In our case, however, the amount of satellite data is insufficient particularly on the onshore side of the Kuroshio, and then the interpolation error

---

[6] Because the satellite data has much information due to high spatio-temporal resolutions, the error variances of the SSH and SST obtained from the adaptive assimilation scheme are more reliable than those of the temperature and salinity from hydrographic observations. Hence we apply the scheme to the SSH and SST data, although the forecast is not so sensitive to the choice of the SST error variance.

covariance has a large uncertainty. Hence we introduce the adjustable parameter in (13.28) and (13.29). Note that there is a constraint on the sensitivity experiments. Since the distribution of the following variable,

$$(y^o - HX^f)^T (HPH^T + R)^{-1} (y^o - HX^f)$$

asymptotically approaches to the normal distribution with mean M and variance $2M$, as M goes to infinity (Lupton, 1993), we need to confirm that the sum of the above variable over the computational domain is close to the unity (Menard and Chang, 2000),

$$\frac{1}{LM} \sum_{i=1}^{L} (y_i^o - HX_i^f)^T (HP_i H^T + R_i)^{-1} (y_i^o - HX_i^f) \approx 1 \qquad (13.31)$$

where L is the number of horizontal grid points.

Several parameters listed in Table 13.2 are introduced into the data assimilation scheme described above. The adaptive assimilation scheme is applied only in the Kuroshio region of isadpt $\leq$ longitude $\leq$ ieadpt and jsadpt $\leq$ latitude $\leq$ jeadpt, because our main target to predict is the Kuroshio path variations. Hence the normalized error variances for the SSH and SST out of the region are set to be constants. Those for the temperature below the surface and the salinity are also assumed to be constants because their effect on the forecast skill is not so large. The values of E_x^o and E_x^f for a variable x are, respectively, listed in the lines of vobsadj0 and vmodeladj0 of Table 13.2. The parameter, vobsieadj, corresponds to α_x for a variable x.

The data assimilation (the estimation of the analysis values) does not take place in the whole of the computational domain. If the depth at a grid point is out of range of hup to hlow or less than hassim, the observation data is not assimilated, that is, the Kalman gain in (13.21) is set to be zero. The parameter, vobsiemax0, is

Table 13.2 Parameters used for the data assimilation

Parameters	SSH	SST	T	S
hup	0 m	0 m	50 m	0 m
hlow	3 000 m	200 m	3 000 m	3 000 m
hassim	200 m	10 m	50 m	10 m
vobsiemax0	0.9	0.9	0.8 [7]	0.8 [7]
vobsadj0	0.48	0.01	0.192	0.192
vmodeladj0	0.036	0.01	0.048	0.048
vcorimpt	1.0	1.0	1.0	0.0
vcorimps	0.1	0.0	0.0	1.0
isadpt	117°E	117°E	-	-
ieadpt	141°E	180°E	-	-
jsadpt	12°N	12°N	-	-
jeadpt	44°N	62°N	-	-
vobsieadj	0.6	1.0	-	-

---

[7] The value of 0.6 is used for the temperature at 50 and 100 m and the salinity at 0, 50, and 100 m.

the acceptable maximum interpolation error, implying that if e_i/V_s is larger than the value the observation data is not assimilated. Additionally all data except for the SST are not unconditionally assimilated in the region of 117°E to 180° and 44°N to 62°N. In fact the data assimilation in this area is not significant, partly because the temperature and salinity are restored to the monthly mean climatology in the Okhotsk Sea described in Sect. 13.2.1, partly because the number of hydrographic observation data in the northern North Pacific is quite few, and partly because the relationship between the SSH anomaly and the in-situ temperature and salinity is unclear there, meaning that the regression coefficient is small.

In the data assimilation process, the regression coefficients between the temperature at the depth k and a variable x, $F_{T_k x}$, and between the salinity at the depth k and the variable, $F_{S_k x}$, are factorized by the adjustable parameters, vcorimpt and vcorimps, respectively, because we realize that the forecast skill got worse when the temperature and salinity are assimilated with the same weight as each other. These parameters are empirically determined.

13.2.6 Quality Control

Both the raw and gridded data of the observation are quality controlled in a simple way. They are required to be within the range of valmin to valmax, and the absolute value of their anomaly to be less than valamax. If the data value is out of range, it is converted to be a missing value. Even if within the range, the raw data at the depth shallower than hobsmin are excluded in the optimal interpolation process. Additionally the gridded data at the depth shallower than hmin are discarded before the subsequent multivariate optimal interpolation process is carried out. The exclusion process is due to the worse accuracy of the satellite altimetry data particularly in shallow coastal areas and marginal seas than in open oceans. The parameters used in the data quality control process for each variable are listed in Table 13.3.

The analysis data obtained from the multivariate optimal interpolation scheme is also quality controlled. The analysis value of the temperature, T^a, and the salinity, S^a, is required to be within the range,

$$|T^a - T^f| \leq 8, \quad -1 \leq T^a \leq 32 \qquad (13.32)$$

Table 13.3 Parameter values for the data quality control

	SSH	SST	T	S
hmin	200 m	10 m	10 m	10 m
hobsmin	100 m	10 m	10 m	10 m
valmin	−1 m	−2°C	−2°C	0 psu
valmax	1 m	32°C	32°C	36 psu
valamax	1 m	10°C	7°C	1 psu

$$|S^a - S^f| \leq 1, \quad 25 \leq S^a \leq 36 \tag{13.33}$$

where T^f and S^f are the forecast values of the temperature and salinity, respectively.

13.2.7 Incremental Analysis Update

The analysis data estimated using an optimal interpolation scheme is generally used as the initial condition of the next forecast experiment (e.g., Mellor and Ezer 1991; Ezer and Mellor 1994). The estimated data, however, does not satisfy governing equations of the model because the optimal interpolation scheme doest not take account of them. Hence the forecast result might be infected by fictitious inertiogravity waves. Mellor and Ezer (1991) modified the velocity field through a diagnostic calculation in such a way that it is geostrophically balanced with the density field based on the analysis data.

Another simple initialization scheme is nudging technique. In the scheme, the forecast data is continuously restored to the analysis data,

$$X^{n+1} = X^n + \frac{X^a - X^n}{\tau_{nudge}} \Delta t \tag{13.34}$$

where X^{n+1} and X^n are values of a variable at time level $n+1$ and n, respectively, X^a is an analysis value of the variable at an analysis time, τ_{nudge} is e-folding time scale during which the forecast value approaches to the analysis value, and Δt is the model time step. It is well known that the nudging scheme tends to impose more fictitious inertio-gravity waves on the model than the optimal interpolation scheme. Moreover the response of the model to the nudging, as shown in the end of this subsection, depends strongly upon the e-folding time scale.

In our forecast system we introduce, instead of the above schemes, an incremental analysis update (IAU) scheme (Bloom et al., 1996). The formulation is similar to the nudging scheme,

$$X^{n+1} = X^n + \frac{X^a - X^f}{\tau_{IAU}} \Delta t \tag{13.35}$$

where X^f is a forecast data at the same analysis time as for X^a and $\tau_{IAU} = 7$ days is a period of the IAU cycle. Equation (13.35) indicates that the analysis increments that are differences between the analysis and the forecast data are gradually incorporated into the model over the period. The increments are constants during each IAU cycle.

Figure 13.4 illustrates the flow of the IAU scheme. (1) A previous IAU cycle provides an initial condition of a forecast. (2) The forecast experiment is conducted for $\tau_{IAU}/2$ (=3.5 days).[8] (3) The output data of the forecast process is provided to the

---

[8] The forecast experiment is, in fact, conducted for 60 days in each cycle as shown in Fig. 13.4 in order to provide the information about the Kuroshio path variation for public people. Miyazawa et al. (2005) suggest that the model has a predictability skill up to approximately 2 months for the variation.

13.2 Description of Forecast System

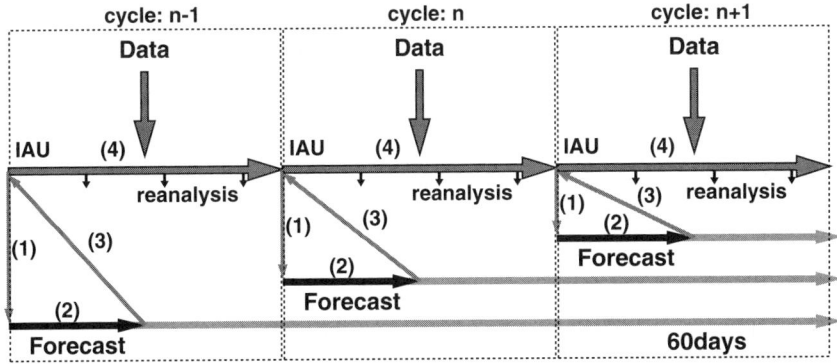

Fig. 13.4 Flow chart for the IAU and forecast. Numbers in the parentheses indicate the order of procedures

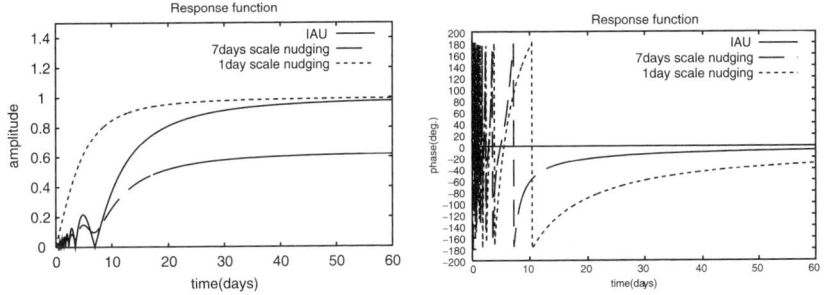

Fig. 13.5 Comparison of the amplitude (*left*) and phase (*right*) response of filters. *Solid*: The IAU with τ_{IAU}=7 days. *Dashed*: The nudging with τ_{nudge}=7 days. *Dotted*: The nudging with τ_{nudge}=1 day

IAU scheme, and the analysis increment is determined. (4) The increment is incorporated into the model by following (13.35). During a cycle of the IAU process, model results are stored at 2 days interval as shown in Fig. 13.4. We call these output the reanalysis. Then these procedures are continuously carried out.

Finally we will show the superiority of the IAU scheme to the nudging scheme. Assuming that the governing equations are linear, we can examine the response characteristics of the filters such as the IAU and the nudging (Bloom et al., 1996). After the IAU with τ_{IAU}=7 days is applied, phenomena on the 20 days time scale will lose 20% of their amplitude, but the decrease of the signal amplitude asymptotically approaches to zero as the time scale of phenomena gets longer (left panel of Fig. 13.5). Moreover the phase error is almost null for phenomena with time scale longer than 7 days (right panel of Fig. 13.5). On the other hand, although the phase error due to the nudging with the same time scale as the IAU is quite small, the amplitude response is terribly bad. Phenomena with the 60 days time scale will lose 40% of its amplitude because of the nudging. The nudging with the e-folding scale of

1 day is superior to the others in the amplitude response, but inferior in the phase response. This comparison implicates that the IAU is a better low-pass filter than the nudging.

13.3 Predictions of the Kuroshio Large Meander

We have carried out the forecast experiment of the Kuroshio large meander since December 2001. The forecast lead time is 2 months, which is determined in such a way that RMS errors of ensemble forecast experiments do not exceed the magnitude of the model climatic variation, and those obtained from nonassimilated simulations and persistence (Miyazawa et al., 2005). Forecast results and the reanalysis data described in the previous section are distributed from the web site http://www.jamstec.go.jp/frsgc/jcope/, and updated on every Saturday. In this section, we will take a couple of examples from the recent forecasts of the Kuroshio path variations. We will also describe the sensitivity of the July–October 2004 forecast to a couple of parameters used in the data assimilation scheme.

13.3.1 Case for April to June 2003

In April 2003 the Kuroshio took the nearshore nonlarge-meander path with a small offshore displacement south of the Kii Peninsula moving approximately 250 km east from south of Shikoku Island. On the offshore side of the Kuroshio south of the peninsula a westward surface current is evident between 30.5°N and 32.5°N, suggesting the existence of an anticyclonic eddy (upper left panel of Fig. 13.6). Existence of another small anticyclonic eddy around 140°E on the offshore side of the Kuroshio is also suggested.

The forecast model is initialized with the IAU output on April 26, 2003. The upper right panel of Fig. 13.6 shows a nowcast of the SSH, indicating that the Kuroshio flowing along the Japanese coast and the existence of two major anticyclonic eddies south of the Kii Peninsula are represented well.

In the May 2003 forecast (1-month lead time forecast), the model predicts the development of the offshore displacement (small meander) of the path southeast of the Kii Peninsula (middle right panel of Fig. 13.6), in perfect accord with the observation on its zonal and meridional extents: approximately 450 km and 250 km. In the meander, a cyclonic eddy is formed. On the onshore side of the eddy, the speed of the westward coastal current exceeds 0.5 m s^{-1} at places based on the one-time observation (middle left panel of Fig. 13.6), while that in the 2-day mean forecast is at most 0.25 m s^{-1}. The underestimate is partly because the observation fields are instantaneous, and partly because the model needs more horizontal resolution to resolve such local phenomena.

13.3 Predictions of the Kuroshio Large Meander

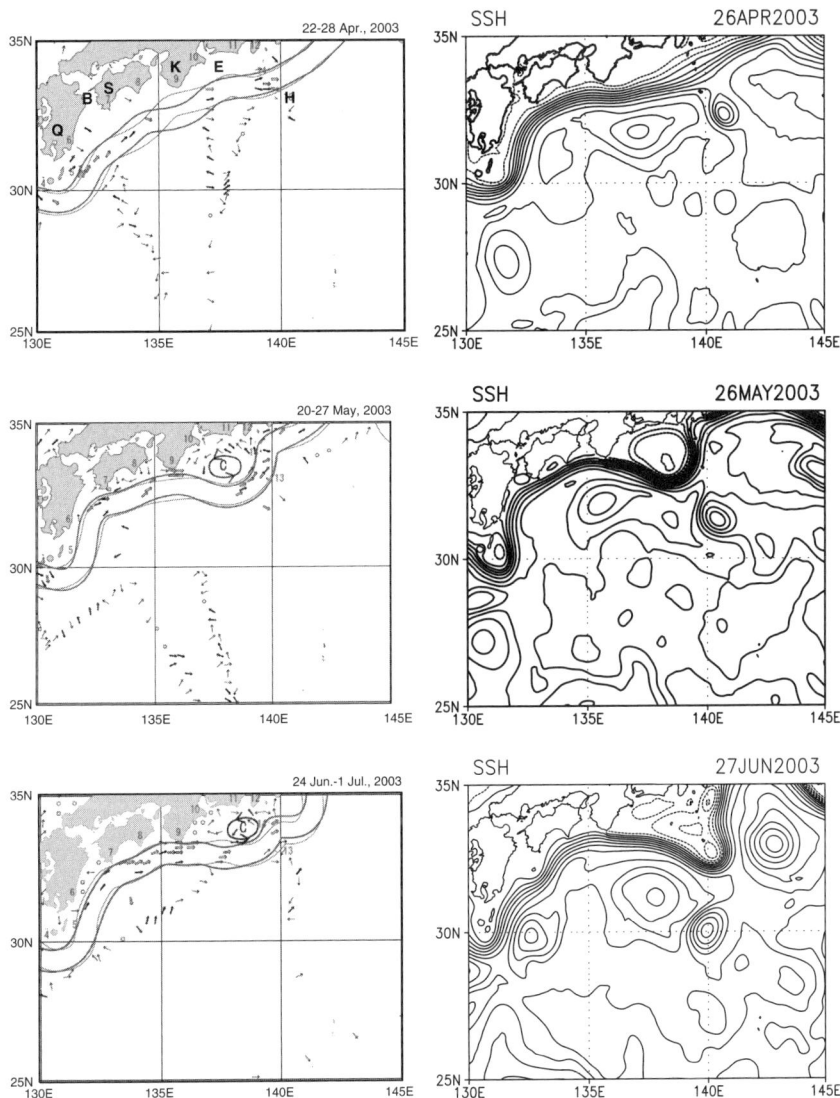

Fig. 13.6 *Left*: Weekly mean observed Kuroshio path during April 22, 2003 to July 1, 2003 reported in the Quick Bulletin Ocean Conditions provided by Hydrographic and Oceanographic Department, Japan Coastal Guard. The *thin line* delineates the Kuroshio path of a week ago and the thick is the current path. *Arrows* indicate the surface velocity at 0.15–0.5 m s^{-1} (*thin*), 0.5–1.0 m s^{-1} (*thick*) and 1.1–2.5 m s^{-1} (*open*). *Right*: Model SSH in 2-day mean of the forecast initialized on April 26, 2003. Contour interval is 0.05 m. Solid contours correspond to positive values of the SSH, and dotted contours to negative values. In the *upper left panel*, K, E, S, Q, H, and B denote the Kii Peninsula, Enshu-nada Sea, Shikoku, Kyushu, and Hachijo-jima Islands, and the Bungo Channel, respectively

In the June 2003 forecast (2-month lead time forecast), the small meander moves eastward, leading to the offshore nonlarge meander path, which is close to the observation, although the observed path is distorted in the W-shaped pattern (lower panels of Fig. 13.6). Furthermore, our forecast model initialized in the beginning of June predicts the subsequent eastward movement of the meander, eventually causing the Kuroshio to take the nearshore nonlarge meander path in the end of July (not shown). Note that the model seems to predict well the anticyclonic eddy southeast of Kyushu Island, which is barely perceptible from the spatially sparse observation (lower left panel of Fig. 13.6).

13.3.2 Case for May to July 2004

A small meander of the Kuroshio occurs southeast of Kyushu in the beginning of 2004 (not shown), stays there with growing its size until June 2004 (middle left panel of Fig. 13.7). It starts to move eastward slowly in the second week of June 2004, and grows into a large meander at the end of July 2004.

In 1-week lead time forecast initialized on May 1, 2004, the small meander moves to the east and its head reaches the south tip of the Kii Peninsula (upper right panel of Fig. 13.7), showing a pattern similar to the observation on June 1–8, 2004 or in a week before delineated by the thin line in the middle left panel of Fig. 13.7.

In the June 2004 forecast (1-month lead time forecast), the center of the meander reaches the south of the peninsula in a way similar to the observation in the end of July 2004. As the meander moves to the east, the anticyclonic eddies in the east and west of it develop (middle right panel of Fig. 13.7).

In the July 2004 forecast (2-month lead time forecast), the Kuroshio large meander reaches its mature state. The anticyclonic eddy east of the meander moves westward along the path and is merged into another, then the eddy west of the meander further develops (lower right panel of Fig. 13.7).

This is the first success in predicting the Kuroshio large meander as well as the occurrence of its triggering meander southeast of Kyushu Island (not shown) using both the numerical model and available observation data, although the model has a bias in the eastward movement speed of the small meander along the Japanese coast as described earlier. In the model the small triggering meander takes approximately 4 months to grow into the large meander after its appearance, while it does approximately 7 months in the real ocean. Previous hydrographic observations, however, indicate that the small meander usually takes approximately 4 months to grow into the large meander after its appearance southeast of Kyushu Island (Kawabe, 1986), suggesting that its eastward movement speed in the model is consistent with those in past events. This means that the model will perhaps be able to predict the path transition of the Kuroshio observed in the more past in a relatively accurate way compared with the 2004 forecast, but, at the same time that the model perhaps misses some mechanism to explain why the small meander takes approximately 7 months to grow into the large meander after its appearance.

13.3 Predictions of the Kuroshio Large Meander

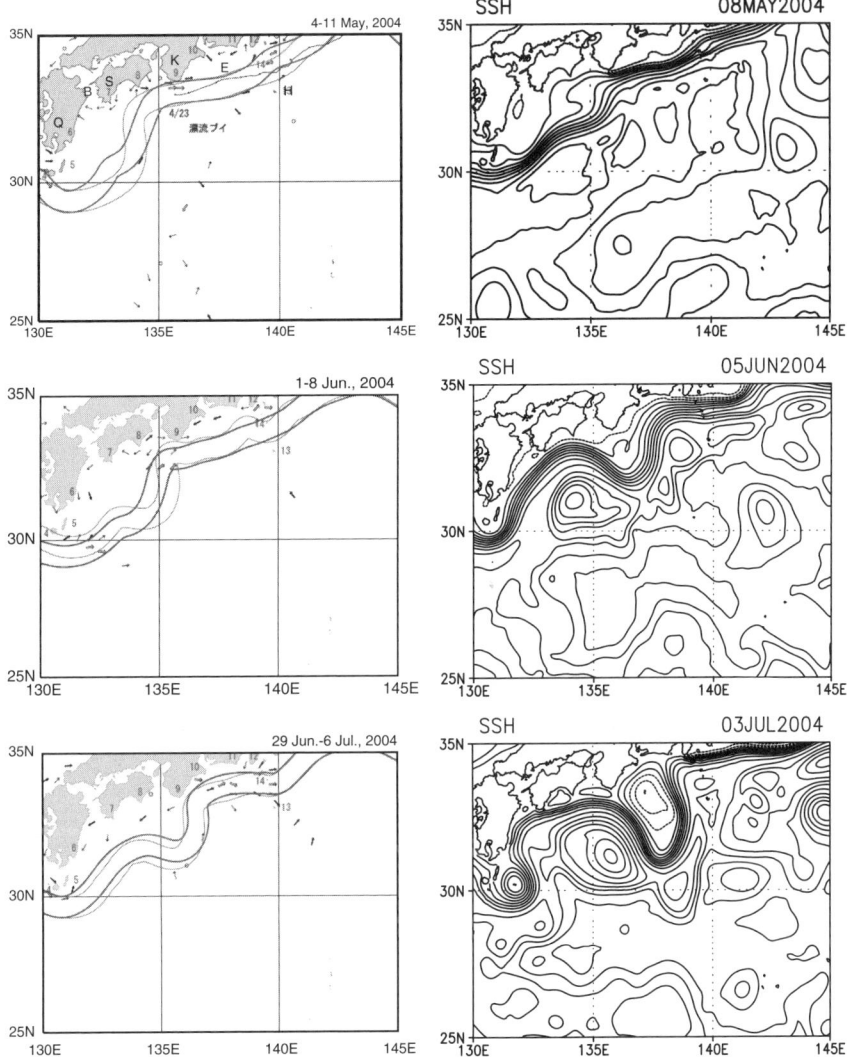

Fig. 13.7 Same as Fig. 13.6 except for the period during May 4, 2004 to July 6, 2004 and the forecast initialized on May 1, 2004

13.3.3 Sensitivity of the Forecast to Parameters

Since the forecast is quite sensitive to parameters used in the data assimilation scheme (Table 13.2), we have conducted lots of trial-and-error experiments in such a way that the forecast of the path transition of the Kuroshio that occurred in the past becomes reasonable. In this subsection, we will take an example to show how

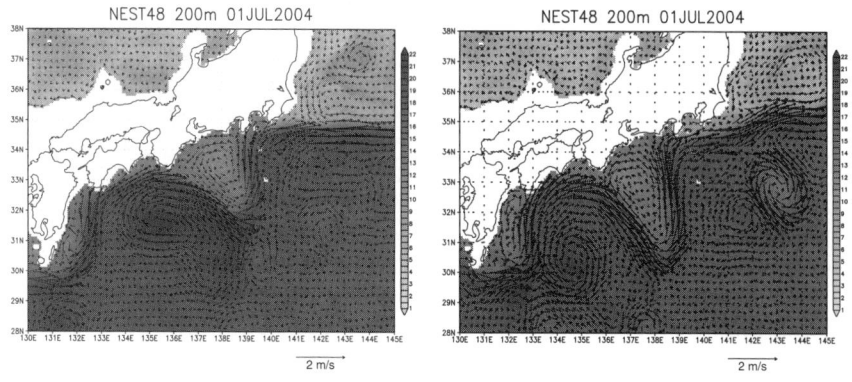

Fig. 13.8 Comparison of the 2-month lead time forecasts initialized on May 6, 2004 obtained with two different sets of parameters. Shading and vectors denote the temperature (°C) and velocity (m s^{-1}) at 200 m, respectively. *Left*: With parameters listed in Table 13.2. *Right*: Same as the left panel except that vcorimps and vobsieadj for the SSH are 1.0 and 0.01, respectively

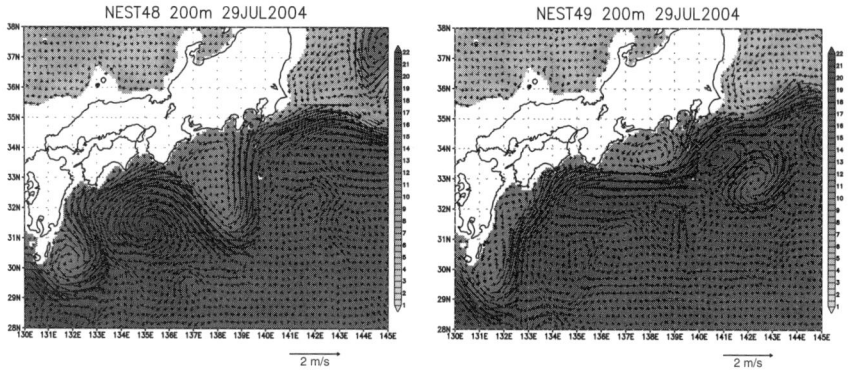

Fig. 13.9 Same as Fig. 13.8 except for the forecast initialized on June 3, 2004

sensitive the model is to a couple of parameters. The experiment with parameter values listed in Tables 13.2 and 13.3 is called the standard case, while that with the same values except for vcorimps and vobsieadj for the SSH is the test case. The former one is the adjustable parameter by which the regression coefficient between the salinity at the depth k and the SSH, and the latter is α'_η in (13.28).

The standard case shows that the large meander path is stable in a sense of its meridional extent, although the path slightly vibrates in the zonal direction (left panels of Figs. 13.8–13.11). On the other hand, in the test case where vcorimps and vobsieadj for the SSH are set to be 1.0 and 0.01, respectively, the meridional amplitude of the large meander changes drastically in a manner that depends upon the initialization date. These changes are mainly due to the detachment of anticyclonic eddies from the Kuroshio (right panel of Fig. 13.9). Numbers of pinched-off eddies suggest that the Kuroshio might be under the baroclinically unstable condition.

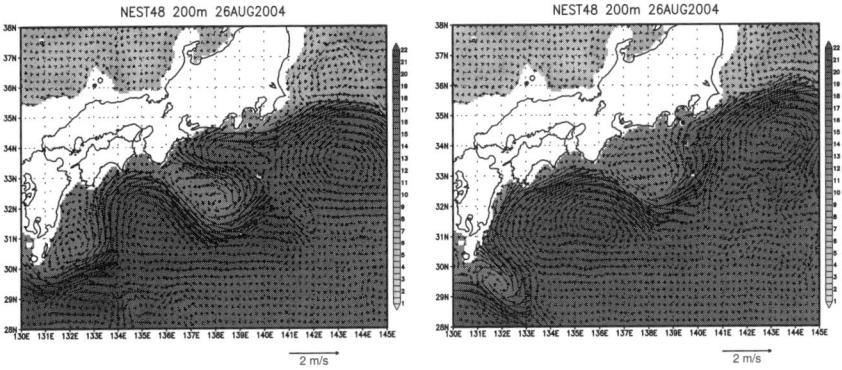

Fig. 13.10 Same as Fig. 13.8 except for the forecast initialized on July 1, 2004

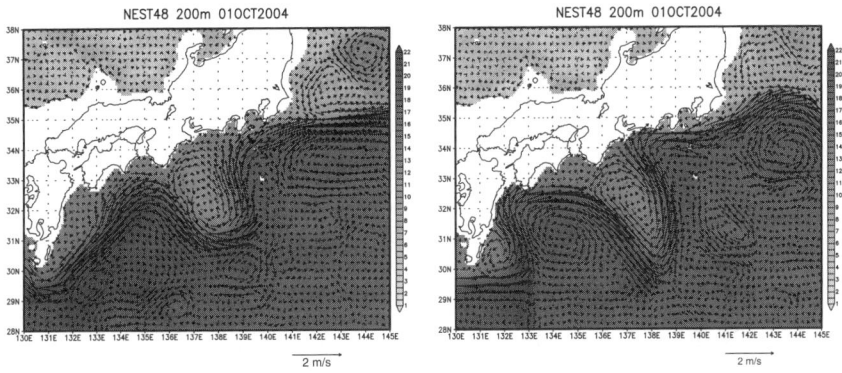

Fig. 13.11 Same as Fig. 13.8 except for the forecast initialized on August 6, 2004

The above example indicates that even a couple of parameters may drastically change a dynamical balance in the model, thus that a careful choice of the parameters is required in such a way that a reasonable forecast is accomplished. The parameters that we use for the operational forecast of the Kuroshio path variation are not perfect, but better at this moment. We still now explore the parameter sensitivity.

13.4 Toward the Kuroshio Forecast Downscaling for Coastal Oceans and Bays

As described in Introduction, the coastal waters around Japan are strongly influenced by the Kuroshio. The variation in its path such as the large meander has been well known to affect the water temperature in some semienclosed bays along the southern coast of Japan (Kawabe and Yoneno, 1987) and a biological state there (Nakata et al., 2000). Even during the period of the nearshore nonlarge meander path, a short-term variation in the Kuroshio also affects hydrographic conditions in the enclosed bays.

For example, it has been reported that an abnormally strong warm current (the Kyucho) originating from the Kuroshio intrudes into the Bungo Channel[9] (Takeoka et al. 1993) and the Sagami Bay (Uda 1953; Matsuyama and Iwata 1977), and results in a steep temperature rise there. Such local phenomena, however, cannot be resolved by the present Kuroshio forecast model described in the previous sections, thus we need to develop a further fine resolution forecast model for coastal oceans and bays south of Japan in order to understand and predict the influence of the Kuroshio on the coastal waters.

As the first step of such efforts, which are extensions of the present Kuroshio forecast system, we chose the Suruga Bay, Sagami Bay, and Tokyo Bay as our target areas (see Fig. 13.12). These bays are surrounded by many important industrial areas in Japan and several big cities such as Tokyo and Yokohama. As a research site for oceanography, they are one of the most active places in Japan. At present, in addition to the physical problems such as the Kyucho and internal tides, the primary production and the fisheries are also being deeply concerned, for which several big research projects have been released by the national and local governments and universities. Therefore, our modeling efforts will provide a framework for the future interdisciplinary researches.

In this section, we will introduce some results of a tidal simulation. Other simulations forced by either the geostrophic currents or the river runoff, or by all of them including the tides are still under way, and the results will be reported in other papers. We note that the basic idea beneath our activities is that the correct prediction of an oceanic current intrusion into a coastal water depends upon not only the correct open boundary conditions but also a good background field inside the coastal water. The former has been partly guaranteed by the present Kuroshio forecast system but the latter still needs more efforts.

The numerical model that we use for the tidal simulation is based on the two-dimensional version of the POM. The horizontal resolution is $1/108° \times 1/108°$ (approximately 0.854 km in the zonal and 1.031 km in the meridional direction). The bottom topography data with 500 m resolution archived by Hydrographic and Oceanographic Department, Japan Coastal Guard is used after smoothed out with the filter based on Mellor et al. (1994), as described in Sect. 13.2.1. The bottom drag coefficient is set to be a constant (=0.0026). The external forcing is only the barotropic tide listed in the Table 13.4. The tidal constants (amplitude and phase lag) for each tidal constituent are provided by Matsumoto et al.'s (2000) model output

Table 13.4 Four principal components of the tide used as the forcing

Abbreviation	Character	Period [h]
M_2	Principal lunar semidiurnal	12.4206
S_2	Principal solar semidiurnal	12.0000
O_1	Lunar diurnal	25.8193
K_1	Lunisolar diurnal	23.9345

---

[9] The location of the Bungo Channel is denoted by B in the upper left panel of Fig. 13.6.

with 1/12° resolution, into which the available harmonic constants obtained from the tide gauges around Japan and those over the open ocean calculated directly from TOPEX/POSEIDON altimetry are assimilated. The tide calculated with the spatially interpolated tidal constants is given at three lateral open boundaries, but the tidal potential forcing is not taken into account. The barotropic velocity normal to the open boundary is modified in the same way as described in Sect. 13.2.2 except that η_c in (13.2) is the tide, $c_{rf} = 1$ and $\bar{u}_c$ is the barotropic velocity at a grid inside the open boundary. The model is integrated for 7 days from a state of rest with the time step of 0.5 s, and the output from the last 2 days is used for harmonic tidal analysis.

Figures 13.12 and 13.13 show the corange and cotidal lines obtained from the model forced with M_2 and O_1 tides, respectively. Both tidal waves propagate from east to west. Whereas the amplitude of M_2 tide increases from east to west, that of O_1 tide from the offshore side to the onshore side, suggesting a boundary trapped wave. In fact O_1 tide does not have a feature of Poincaré wave, but that of Kelvin wave because its frequency ($=0.818f$) is less than the inertial frequency, f. Comparison between the tide gauge observation and the model result indicates that their differences with regard to both the amplitude and phase are within a reasonable range (the amplitude error of M_2 tide is less than 5% of the observed amplitude in most of places) at all observation points except for those around the Tokyo Bay. The large difference in the Tokyo Bay may be due to the lack of the horizontal resolution compared to the least width of the bay mouth (approximately 7 km).

The tidal current is generally parallel to the coast line, as shown by flat ellipses in Fig. 13.14. It is, however, intriguing that M_2 tidal current indicates an obvious

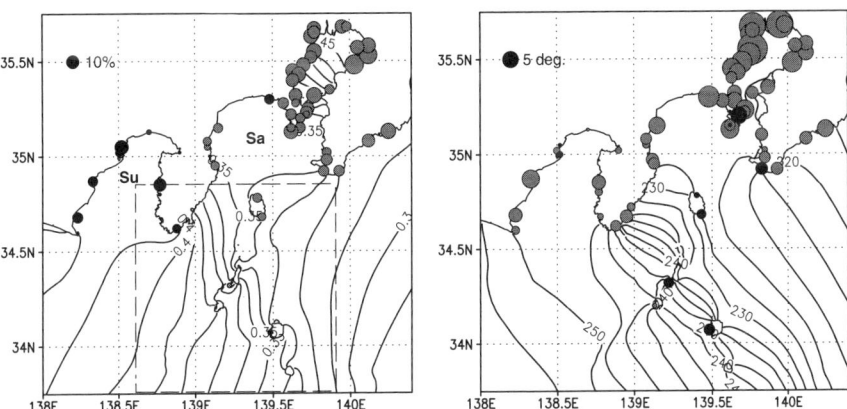

Fig. 13.12 Comparison of M_2 tide amplitude (*left*) and phase (*right*) between the model and observation based on the tide gauge. The area of circles is proportional to the difference between them. *Black (Gray) circles* indicates that the model amplitude is larger (smaller) than the observation, and that the model (observation) lags the observation (model). *Contour lines* delineate the corange in m (left) and cotide in degree (right) obtained from the model. Contour interval is 0.01 m for corange and 2° for cotide. Su and Sa denote Suruga Bay and Sagami Bay, respectively. *Dashed line* denotes the area in which the tidal ellipse is delineated (Fig. 13.14)

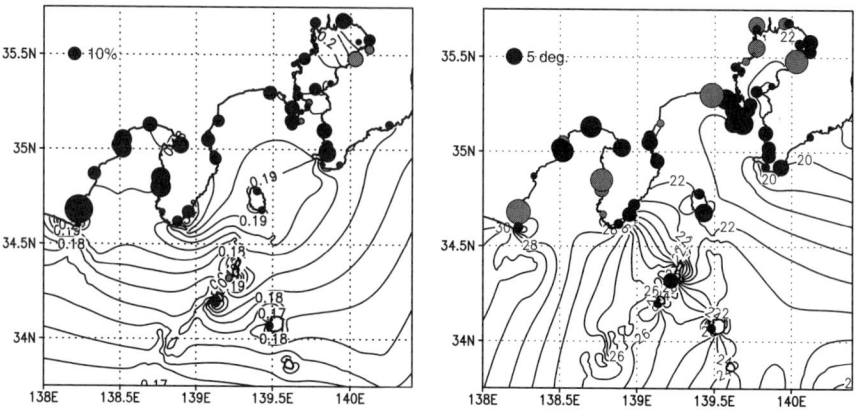

Fig. 13.13 Same as Fig. 13.12 except for the O_1 tide. Contour interval is 0.002 m for corange and 0.5° for cotide

clockwise rotation over the Izu-Ogasawara Ridge, particularly near the Izu Islands, whereas the tidal current rotates anticlockwise in the south of Japan apart from the ridge (not shown). Although the observed basic features of the tidal current are represented well in the model, the tidal current speed is underestimated particularly in the head of Sagami Bay and Suruga Bay. This is because of the presence of strong internal tide in these bays (Ohwaki et al. 1991; Takeuchi and Hibiya 1997). At present, we have finished simulations of the internal tide and confirmed the reproduction of the observed strong tidal currents in Sagami Bay and Suruga Bay with the use of our coastal model. Readers can refer to Kawajiri et al. (2006) for the details about the reproduction of internal tide as well as other aspects of the tidal simulation such as effects of tidal potential forcing, problems related to the open boundary conditions, and an interaction between four tidal constituents listed in Table 13.4.

13.5 Summary

We have described the detail of the Kuroshio forecast system with approximately 10 km resolution, called the JCOPE (Japan Coastal Ocean Predictability Experiment) ocean forecast system, specifically with regard to the spatio-temporal interpolation of the observation data and the data assimilation technique including the incremental analysis update scheme. Although the system is, as described in Sect. 13.3.3, not perfect particularly in a sense of parameter setting, we believe that this documentation is useful for readers to develop an ocean forecast system.

Two examples of the prediction of the Kuroshio path variations using the system are introduced. The system successfully predicts the path transitions from the nearshore nonlarge meander path to the offshore nonlarge meander path in 2003, and

13.5 Summary 235

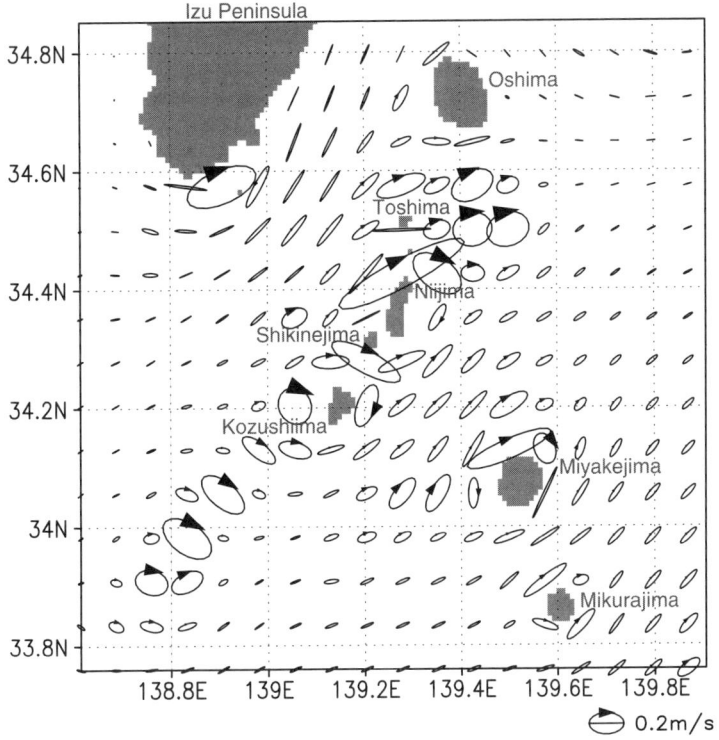

Fig. 13.14 M_2 tidal current ellipses around the Izu Islands

from the nearshore nonlarge meander path to the typical large meander path in 2004 as well as the occurrence of its triggering small meander south of Kyushu Island. Note that the success of the 2004 forecast is attributed to the recent establishment of observation network (satellites, ships, and ARGO floats) as well as the sophistication of both the numerical model and the data assimilation technique, suggesting that they are just like the wheels of a car. We also note that it enables us to deeply understand how the path transitions of the Kuroshio occur.

For the purpose of the Kuroshio forecast downscaling for coastal oceans and bay areas, we have, for the first step, conducted the tidal simulation with a state-of-the-art model. Comparison between the tide gauge observation and the model result indicates that their differences with regard to both amplitude and phase are within a reasonable range. It is quite intriguing that M_2 tidal current rotates clockwise only over the Izu-Ogasawara Ridge in our analysis area. Moreover the strong tidal current at that place, particularly near the Kuroshio path, suggests its nonlinear interaction with the Kuroshio. It is, however, out of the scope of this manuscript, hence we leave them as future subjects.

More accurate forecasts require further reduction of model biases shown in Sect. 13.3. Comparing the JCOPE reanalysis data and the individual hydrographic observation that are not assimilated into the model, we realize that the model temperature tends to have a positive bias in the onshore side of the Kuroshio and a negative bias in the upper 50 m and the thermocline layer of its offshore side. The representation of the density field particularly in the onshore side of the Kuroshio is very crucial in predicting the path transition of the Kuroshio accurately, because the eastward movement and development of the small trigger meander are related to the baroclinic instability. Hence we have been investigating causes of the biases through numerous sensitivity experiments to parameters and numerical schemes, and also developing, on the basis of Hukuda and Guo (2004), a two-way nesting forecast model which consists of the current fine model and a regional model for the Kuroshio area with 1/36°resolution and thereby may represent a sharp density front in the onshore side of the Kuroshio more accurately than the current model.

Outputs of the current JCOPE ocean forecast system can be utilized as pseudo-data for many purposes. Because of the fineness of the horizontal resolution, the JCOPE reanalysis data have been used for the lateral boundary condition of a regional ocean model (e.g., Suzuki et al. 2004; Uchimoto et al. 2006) and for the bottom boundary condition of a regional atmospheric model (e.g., Yamaguchi et al. 2005). In the latter case, the short-term variations of the rainfall and surface wind associated with the typhoon passing are represented consistently with the observation by a hindcast experiment with the SST obtained from the JCOPE model compared to another experiment with the surface skin temperature from NCEP Global Forecast System, suggesting that the predicted SST by the JCOPE system may be useful to local severe weather forecasting and notification. Recently the Fisheries Research Agency has started to investigate dispersal of eggs and larvae of anchovy and horse mackerel using the velocity field of the JCOPE reanalysis data to attain the goal of evaluating fisheries resources and predicting fishing grounds. The JAMSTEC has tentatively utilized the velocity data for the operation of the Deep Sea Drilling Vessels "CHIKYU" because the strong current has a particularly serious impact on the ability of a drilling rig to maintain its position. We expect that the JCOPE ocean forecast system may contribute to as many societies as possible because of its wide availability.

Acknowledgments We are greatly indebted to Prof. Toshio Yamagata and Dr. Hirofumi Sakuma for their encouragement and the management of our research activity. The JCOPE project is supported by the Frontier Research Center for Global Change/JAMSTEC. The QuikSCAT data is obtained from the NASA/NOAA sponsored data system Seaflux, at JPL through the courtesy of Dr. W. Timothy Liu and Dr. Wenqing Tang. The comparison study between the JCOPE output and the hydrographic data and the sensitivity study are supported by Dr. Kosei Komatsu, Dr. Takashi Setou, and many collaborators in the Fisheries Research Agency.

References

Akitomo, K., T. Awaji, and N. Imasato, 1991: Kuroshio path variation south of Japan 1. Barotropic inflow-outflow model. *J. Geophys. Res.*, **96**, 2549–2560.
Akitomo, K. and M. Kurogi, 2001: Path transition of the Kuroshio due to mesoscale eddies: A two-layer, wind-driven experiment. *J. Oceanogr.*, **57**, 735–741.

References

Bloom, S. C., L. L. Takacs, A. M. da Silva, and D. Ledvina, 1996: Data assimilation using Increment Analysis Updates. *Mon. Wea. Rev.*, **124**, 1256–1271.

Boyer, T. P. and S. Levitus, 1997: Objective Analyses of Temperature and Salinity for the World Ocean on a 1/4° Grid. Noaa/nesdis atlas 11, U.S. Gov. Printing Office, Washington, D.C.

Bryden, H. L., D. H. Roemmich, and J. A. Church, 1991: Ocean heat transport across 24°N in the Pacific. *Deep-Sea Res.*, **38**, 297–324.

Ebuchi, N. and K. Hanawa, 2000: Mesoscale eddies observed by TOLEX-ADCP and TOPEX/POSEIDON altimeter in the Kuroshio recirculation region south of Japan. *J. Oceanogr.*, **56**, 43–57.

Endoh, T. and T. Hibiya, 2000: Numerical study of the generation and propagation of trigger meanders of the Kuroshio south of Japan. *J. Oceanogr.*, **56**, 409–418.

Ezer, T., 2000: On the seasonal mixed layer simulated by a basin-scale ocean model and the Mellor-Yamada turbulence scheme. *J. Geophys. Res.*, **105**, 16843–16855.

Ezer, T. and G. L. Mellor, 1994: Continuous assimilation of Geosat altimeter data into a three-dimensional primitive equation Gulf Stream model. *J. Phys. Oceanogr.*, **24**, 832–847.

— 1997: Data assimilation experiments in the Gulf Stream region: How useful are satellite-derived surface data for nowcasting the subsurface fields? *J. Atmos. Oceanic Technol.*, **14**, 1379–1391.

— 2004: A generalized coordinate ocean model and a comparison of the bottom boundary layer dynamics in terrain-following and in z-level grids. *Ocean Model.*, **6**, 379–403.

Flather, R. A., 1976: A tidal model of the northwest European continental shelf. *Memories da la Societe Roylale des Sicences de Liege*, **6**, 141–164.

Fox, A. D., K. Haines, B. A. de Cuevas, and D. J. Webb, 2000: Altimeter assimilation in the OCCAM global model Part II: TOPEX/POSEIDON and ERS-1 assimilation. *J. Mar. Systems*, **26**, 323–347.

Gandin, L. S., 1963: Objective analysis of meteorological field. Gidrometeorologicheskoe Izdate'stvo., 287 pp.

Guo, X., H. Hukuda, Y. Miyazawa, and T. Yamagata, 2003: A triply nested ocean models – Roles of horizontal resolution on JEBAR. *J. Phys. Oceanogr.*, **33**, 146–169.

Hinata, H., T. Yanagi, T. Takao, and H. Kawamura, 2005: Wind-induced Kuroshio warm water intrusion into Sagami Bay. *J. Geophys. Res.*, **110**, doi:10.1029/2004JC002300.

Hukuda, H. and X. Guo, 2004: Application of a two-way nested model to the seamount problem. *J. Oceanogr.*, **60**, 893–904.

Hurlburt, H. E., A. J. Wallcraft, W. J. Schmitz, Jr., P. J. Hogan, and E. J. Metzger, 1996: Dynamics of the Kuroshio/Oyashio current system using eddy-resolving models of the North Pacific Ocean. *J. Geophys. Res.*, **101**, 941–976.

Ichikawa, H. and M. Chaen, 2000: Seasonal variation of heat and freshwater transports by the Kuroshio in the East China Sea. *J. Mar. Systems*, **24**, 119–129.

Ichikawa, K. and S. Imawaki, 1994: Life history of a cyclonic ring detached from the Kuroshio Extension as seen by the Geosat altimeter. *J. Geophys. Res.*, **99**, 15953–15966.

Inatsu, M., H. Mukougawa, and S.-P. Xie, 2002: Tropical and extratropical SST effects on the midlatitude storm track. *J. Meteor. Soc. Jpn.*, **80**, 1069–1076.

Ishikawa, Y., T. Awaji, N. Komori, and T. Toyoda, 2004: Application of sensitivity analysis using an adjoint model for short-range forecasts of the Kuroshio path south of Japan. *J. Oceanogr.*, **60**, 293–301.

Iwata, S. and M. Matsuyama, 1989: Surface circulation in Sagami Bay: The response to variations of the Kuroshio axis. *J. Oceanogr. Soc. Jpn.*, **45**, 310–320.

Jerlov, N. G., 1976: Marine Optics. Elsevier Sci. Pub. Co., Amsterdam, 231 pp.

Kagimoto, T.: 2003, Sensitivity of the Equatorial Undercurrent to the vertical mixing parameterization. *Proc. 2003 Terrain-Following Ocean Models Users Workshop in Seattle*, available from http://www.aos.princeton.edu/WWWPUBLIC/htdocs.pom/SIG03/SEATTLE03.pdf.

Kagimoto, T. and T. Yamagata, 1997: Seasonal transport variations of the Kuroshio: An OGCM simulation. *J. Phys. Oceanogr.*, **27**, 403–418.

Kalnay, E., M. Kanamitsu, R. Kistler, W. Collins, D. Deaven, L. Gandin, M. Iredell, S. Saha, G. White, J. Woollen, Y. Zhu, M. Chelliah, W. Ebisuzaki, W. Higgins, J. Janowiak, K. C. Mo,

C. Ropelewski, J. Wang, A. Leetmaa, R. Reynolds, R. Jenne, and D. Joseph, 1996: The NCEP/NCAR 40-Year Reanalysis Project. *Bull. Am. Meteorol. Soc.*, **77**, 437–471.

Kamachi, M., T. Kuragano, H. Ichikawa, H. Nakamura, A. Nishina, A. Isobe, D. Ambe, M. Arai, N. Gohda, S. Sugimoto, K. Yoshita, T. Sakurai, and F. Uboldi, 2004a: Operational data assimilation system for the Kuroshio south of Japan: Reanalysis and validation. *J. Oceanogr.*, **60**, 303–312.

Kamachi, M., T. Kuragano, S. Sugimoto, K. Yoshita, T. Sakurai, T. Nakano, N. Usui, and F. Uboldi, 2004b: Short-range prediction experiments with operational data assimilation system for the Kuroshio south of Japan. *J. Oceanogr.*, **60**, 269–282.

Kawabe, M., 1986: Transition processes between the three typical paths of the Kuroshio. *J. Oceanogr. Soc. Jpn.*, **41**, 307–326.

Kawabe, M. and M. Yoneno, 1987: Water and flow variations in Sagami Bay under the influence of the Kuroshio path. *J. Oceanogr. Soc. Jpn.*, **43**, 283–294.

Kawajiri, H., X. Guo, T. Kagimoto, and Y. Miyazawa. Simulations of barotropic and baroclinic tides in coastal waters south of Tokyo, Japan. *J. Geophys. Res.*, (to be submitted).

Komori, N., T. Awaji, Y. Ishikawa, and T. Kuragano 2003: Short-range forecast experiments of the Kuroshio path variabilities south of Japan using TOPEX/Poseidon altimetric data. *J. Geophys. Res.*, **108**, doi:10.1029/2001JC001282.

Kondo, J., 1975: Air-sea bulk transfer coefficients in diabatic conditions. *Boundary-Layer Meteorol.*, **9**, 91–112.

Kuragano, T. and A. Shibata, 1997: Sea surface dynamic height of the Pacific Ocean derived from TOPEX/POSEIDON altimeter data: Calculation method and accuracy. *J. Oceanogr.*, **53**, 585–599.

Large, W. G. and S. Pond, 1981: Open ocean momentum flux measurements in moderate and strong winds. *J. Phys. Oceanogr.*, **11**, 324–336.

Levitus, S. and T. P. Boyer, 1994: World Ocean Atlas 1994, Volume 4: Temperature. Noaa atlas nesdis 4, U.S. Gov. Printing Office, Washington, D.C.

Levitus, S., R. Burgett, and T. P. Boyer, 1994: World Ocean Atlas 1994, Volume 3: Salinity. Noaa atlas nesdis 3, U.S. Gov. Printing Office, Washington, D.C.

Liu, W. T., W. Tang, and P. S. Polito, 1998: NASA scatterometer global ocean-surface wind fields with more structures than numerical weather prediction. *Geophys. Res. Lett.*, **25**, 761–764.

Lorenc, A. C., 1981: A global three-dimensional multivariate statistical interpolation scheme. *Mon. Wea. Rev.*, **109**, 701–721.

Lupton, R., 1993: Statistics in Theory and Practice. Princeton University Press, 188 pp.

Matsumoto, K., T. Takanezawa, and M. Ooe, 2000: Ocean tide models developed by assimilating TOPEX/POSEIDON altimeter data into hydrodynamical model: A global model and a regional model around Japan. *J. Oceanogr.*, **56**, 567–581.

Matsuyama, M. and S. Iwata, 1977: The Kyucho in Sagami Bay (I). *Bull. Fisheries Oceanogr. Jpn.*, **30**, 1–7, (in Japanese with English abstract).

Mellor, G. L., 2001: One-dimensional, ocean surface layer modeling, a problem and a solution. *J. Phys. Oceanogr.*, **31**, 790–809.

Mellor, G. L. and T. Ezer, 1991: A Gulf Stream model and an altimetry assimilation scheme. *J. Geophys. Res.*, **96**, 8779–8795.

Mellor, G. L., T. Ezer, and L.-Y. Oey, 1994: The pressure gradient conundrum of sigma coordinate ocean models. *J. Atmos. Oceanic Technol.*, **11**, 1126–1134.

Mellor, G. L., S. Häkkinen, T. Ezer, and R. Patchen: 2002, A generalization of a sigma coordinate ocean model and an intercomparison of model vertical grids. *Ocean Forecasting: Conceptual Basis and Applications*, N. Pinardi and J. D. Woods, eds., Springer, New York, 55–72.

Mellor, G. L. and T. Yamada, 1982: Development of a turbulence closure model for geophysical fluid problem. *Rev. Geophys. Space Phys.*, **20**, 851–875.

Menard, R. and L.-P. Chang, 2000: Assimilation of stratospheric chemical tracer observations using Kalman filter Part II: χ^2-validated results and analysis of variance and correlation dynamics. *Mon. Wea. Rev.*, **128**, 2672–2686.

Mitsudera, H., T. Waseda, Y. Yoshikawa, and B. Taguchi, 2001: Anticyclonic eddies and Kuroshio meander formation. *Geophys. Res. Lett.*, **28**, 2025–2028.

References

Miyazawa, Y., X. Guo, and T. Yamagata, 2004: Roles of meso-scale eddies in the Kuroshio paths. *J. Phys. Oceanogr.*, **34**, 2203–2222.
Miyazawa, Y., S. Yamane, X. Guo, and T. Yamagata, 2005: Ensemble forecast of the Kuroshio meandering. *J. Geophys. Res.*, **110**, doi:doi:10.1029/2004JC002426, doi:10.1029/2004JC002426.
Nakata, H., S. Funakoshi, and M. Nakamura, 2000: Alternating dominance of postlarval sardine and anchovy caught by coastal fishery in relation to the Kuroshio meander in the Enshu-nada Sea. *Fish. Oceanogr.*, **9**, 248–258.
Nonaka, M. and S.-P. Xie, 2003: Covariations of sea surface temperature and wind over the Kuroshio and its Extension: Evidence for ocean-to-atmosphere feedback. *J. Clim.*, **16**, 1404–1413.
Ohwaki, A., M. Matsuyama, and S. Iwata, 1991: Evidence for predominance of internal tidal currents in Sagami and Suruga Bays. *J. Oceanogr. Soc. Jpn.*, **47**, 194–206.
Paulson, E. A. and J. J. Simpson, 1977: Irradiance measurements in the upper ocean. *J. Phys. Oceanogr.*, **7**, 952–956.
Qiu, B. and K. A. Kelly, 1993: Upper-ocean heat balance in the Kuroshio Extension region. *J. Phys. Oceanogr.*, **23**, 2027–2041.
Qiu, B. and W. Miao, 2000: Kuroshio path variations south of Japan: Bimodality as a self-sustained internal oscillation. *J. Phys. Oceanogr.*, **30**, 2124–2137.
Reed, R. K., 1977: On estimating insolation over the ocean. *J. Phys. Oceanogr.*, **7**, 482–485.
Smagorinsky, J., 1963: General circulation experiments with the primitive equations. I. The basic experiment. *Mon. Wea. Rev.*, **91**, 99–164.
Suzuki, Y., K. Nadaoka, Y. Miyazawa, S. Harii, and N. Yasuda, 2004: Analysis of long-distance larval dispersal of corals and crow-of-thorns starfish using JCOPE and a coastal current model. *Ann. J. Coastal Eng., Jpn. Soc. Civil Eng.*, **51**, 1146–1150, (in Japanese).
Takeoka, H., H. Akiyama, and T. Kikuchi, 1993: The Kyucho in the Bungo Channel, Japan-periodic intrusion of oceanic warm water. *J. Oceanogr.*, **49**, 57–70.
Takeuchi, K. and T. Hibiya, 1997: Numerical simulation of baroclinic tidal currents in Suruga Bay and Uchiura Bay using a high resolution level model. *J. Oceanogr.*, **53**, 539–552.
Tetens, O., 1930: Über einige meteorologische Begriffe. *Z. Geophys.*, **6**, 293–309.
Uchimoto, K., H. Mitsudera, N. Ebuchi, and Y. Miyazawa, 2007: Anticyclonic eddy caused by the Soya Warm Current in an Okhotsk OGCM. *J. Oceanogr.*, **63**, 379–391.
Uda, M., 1953: On the stormy current (Kyucho) and its prediction in the Sagami Bay. *J. Oceanogr. Soc. Jpn.*, **9**, 15–22, (in Japanese with English abstract).
Yamaguchi, K., T. Yamashita, and K.-O. Kim, 2005. Effects of Kuroshio Warm Current SST on coastal wind and precipitation fields simulated by meso-scale meteorological model MM5. *Ann. J. Coastal Eng., Jpn. Soc. Civil Eng.*, **52**, 366–370 (in Japanese).

Chapter 14
High-Resolution Simulation of the Global Coupled Atmosphere–Ocean System: Description and Preliminary Outcomes of CFES (CGCM for the Earth Simulator)

Nobumasa Komori, Akira Kuwano-Yoshida, Takeshi Enomoto, Hideharu Sasaki, and Wataru Ohfuchi

Summary We have been developing a global, high-resolution, coupled atmosphere–ocean general circulation model, named CFES, which was designed to achieve efficient computational performance on the Earth Simulator. A brief description of CFES and some preliminary results obtained from 66-month integration are presented. Although some deficiencies are apparent in the results, realistically simulated small-scale structures such as extratropical cyclones and sea surface temperature fronts in the mid-latitudes, and seasonal variation of tropical sea surface temperature and polar sea-ice extent encourage us to study mechanism and predictability of high-impact phenomena and their relation to the global-scale circulations using CFES.

14.1 Introduction

Mid-latitude high-impact phenomena of the atmosphere and ocean on the seasonal to interannual time scale occur under the influence of the prominent climate variations in the tropics and high-latitudes, such as El Niño and Southern Oscillation, Indian Ocean Dipole, Arctic Oscillation, and so on. With the recent development of high-resolution satellite observations, several papers (e.g., Xie 2004) have pointed out the existence of local air–sea interactions over the sea surface temperature (SST) fronts associated with the western boundary currents in the mid-latitudes besides basin-scale interactions. Mid-latitude high-impact phenomena are possibly affected by these interactions but their mechanisms are not fully understood. Moreover, Nakamura et al. (2004) discussed the observed linkage between the mid-latitude SST fronts and polar-front jets via storm tracks. Thus, prediction of such phenomena is a scientifically challenging problem, and societal needs for reliable prediction are enormous due to large population and economic activity in the North America, western Europe, and eastern Asia.

In order to promote the studies on mechanism and predictability of high-impact phenomena especially in the mid-latitudes and their relation to the global-scale circulations, we have been developing a global, high-resolution, coupled atmosphere–ocean general circulation model (GCM), which can utilize the vector parallel architecture of the Earth Simulator efficiently, and was named CFES (Coupled GCM for the Earth Simulator) (Takahashi et al. 2003a). The current version of CFES consists of AFES 2 (Atmospheric GCM for the Earth Simulator version 2) (Enomoto et al. 2007) as an atmospheric component including land-surface process and OIFES (Coupled Ocean–Sea-Ice Model for the Earth Simulator) (Komori et al. 2005) as an oceanic component including sea-ice process. CFES uses a novel approach to coupling the atmospheric and oceanic components, and this approach minimizes communication overhead in massively parallel computations on the Earth Simulator.

In this article, we briefly describe the atmospheric and oceanic components and the coupling method of CFES in Sect. 14.2, and show some preliminary results obtained from our first trial of high-resolution simulation of the global coupled atmosphere–ocean system using CFES in Sect. 14.3.

14.2 Coupled Atmosphere–Ocean GCM: CFES

14.2.1 Atmospheric Component: AFES 2

AFES 2 is the improved version of AFES (Ohfuchi et al. 2004). AFES is adapted from CCSR/NIES AGCM version 5.4.02 (Numaguti et al. 1997) developed at Center for Climate System Research (CCSR) of the University of Tokyo and National Institute for Environmental Studies (NIES), which uses a spectral transform method for horizontal discretization and the Lorenz differencing method in the vertical σ-coordinate to solve primitive equations. The source code of AFES was totally rewritten in Fortran 90 with the Message Passing Interface (MPI) to achieve high computational efficiency on the Earth Simulator (Shingu et al. 2002; Ohfuchi et al. 2004).

For the coupled simulations with OIFES, a simple river model is introduced to AFES. It transports the surface runoff with no time lag from each land-surface point to the corresponding mouth of a river determined from the Total Runoff Integrating Pathways (TRIP) data (Oki and Sud 1998). It also carries excess snow to the ocean in order to mimic the effect of glacier.

Recently, the land-surface model used in AFES was improved by replacing the simple bucket-type model with MATSIRO (Minimal Advanced Treatments of Surface Interaction and RunOff) (Takata et al. 2003), which has a single layer of vegetation canopy with the effects of photosynthesis and five layers of the soil with complicated snow process. In addition, the original radiation transfer scheme (Nakajima and Tanaka 1986) has been updated to mstrnX (Sekiguchi 2004), which is the improved version of mstrn8 (Nakajima et al. 2000), and computationally optimized for the Earth Simulator. In detail about improvements of AFES 2 including a cumulus convection scheme and a new Legendre transform method, see Enomoto et al. (2007).

14.2.2 Oceanic Component: OIFES

OIFES is the extended version of OFES (Masumoto et al. 2004). OFES is based on Modular Ocean Model version 3 (MOM 3) (Pacanowski and Griffies 1999) developed at Geophysical Fluid Dynamics Laboratory (GFDL) of National Oceanic and Atmospheric Administration, which uses a finite-difference method to solve primitive equations and adopts z-coordinate in vertical with free surface. In OFES, one-dimensional domain decomposition in the meridional direction is employed for parallelization and the computational optimization for the Earth Simulator is significantly improved (Takahashi et al. 2003b; Masumoto et al. 2004). Note that OFES is developed mainly to study the ocean surface circulations in the tropical and subtropical regions, and therefore the computational domain used by Masumoto et al. (2004) excludes the Arctic Ocean.

OIFES contains sea-ice model to be used for global ocean simulations including the Arctic and Antarctic regions and for global coupled atmosphere–ocean simulations. The sea-ice model is based on that developed at International Arctic Research Center (IARC) of University of Alaska (Zhang and Zhang 2001). Their model employs two-category, zero-layer thermodynamics and viscous–plastic (VP) rheology in Cartesian coordinates (Hibler 1979; Parkinson and Washington 1979) with snow effects (Oberhuber et al. 1993) and is coupled with GFDL MOM 2 (Pacanowski 1995) for the use of the Arctic regional studies. However, VP rheology is not suitable for massively-parallel computation, so the sea-ice dynamics is extended from VP rheology in Cartesian coordinates to elastic–viscous–plastic (EVP) rheology in spherical coordinates, adopted from Hunke and Dukowicz (2002), for global simulations. Additionally, sublimation process is introduced to thermodynamics of the sea-ice model for coupled simulations with AFES. In detail about the sea-ice component of OIFES, see Komori et al. (2005).

14.2.3 Coupling Method

In CFES, component models (AFES 2 and OIFES) are mutually independent load modules and executed concurrently on different groups of processor nodes of the Earth Simulator, that is, so-called "multiple program multiple data" (MPMD) technique is utilized (Takahashi et al. 2003a). Therefore, parallel performance of CFES is optimized by changing the number of processors assigned to each group for component model to minimize the difference in execution time.

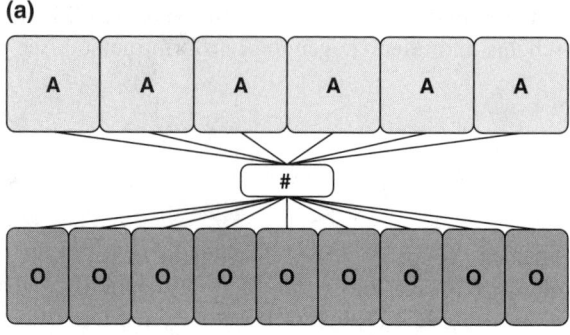

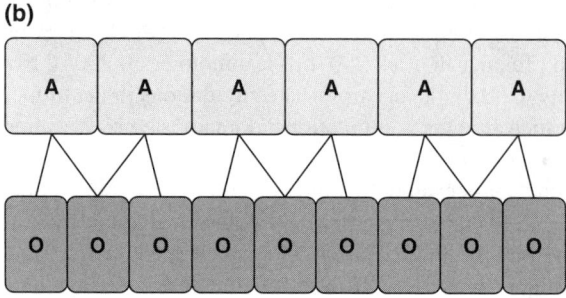

Fig. 14.1 Schematics of data communication between the atmospheric and oceanic processors. (**a**) Conventional CGCM using a flux coupler. (**b**) CFES

In conventional CGCMs using a flux "coupler," model variables communicated between AGCM and OGCM are firstly gathered from each processor to the root processor, then converted from the coordinate system used in AGCM (OGCM) to the other, and finally surface fluxes calculated from those variables are broadcast from the root processor to each processor (Fig. 14.1a). This communication cost becomes higher when the number of processors used is larger. In CFES, model variables necessary to calculate air–sea fluxes within each decomposed computational domain are directly transferred in parallel between corresponding processors for AFES 2 and OIFES (Fig. 14.1b) in order to reduce communication cost arising from gathering/broadcasting data (Takahashi et al. 2003a).

In the current version of CFES, momentum, heat, and freshwater fluxes over the ocean (including sea-ice region) are calculated in the atmospheric component. Therefore, SST, sea-ice concentration, sea-ice thickness, and snow depth over sea-ice (and sea surface velocity depending on the options selected) are transferred from OIFES to AFES 2 with a particular coupling interval, and temporally averaged air–sea fluxes, in addition to surface radiation fluxes and sea level pressure, are passed from AFES 2 to OIFES.

14.3 Preliminary Results

14.3.1 Simulation Setting

Model resolution of the atmospheric component is T239 (the triangle truncation at wave number 239, ~50 km) in horizontal and 48 layers in vertical (12 layers are assigned beneath $\sigma = 0.8$) with the top level placed at $\sigma = 0.003$ (about 3 hPa). The surface topography is based on the Global 30 Arc-Second Elevation (GTOPO30) data set. The monthly mean atmospheric ozone climatology is constructed from the data for Atmospheric Model Intercomparison Project 2 (AMIP2). The horizontal grid spacing of the land-surface component (MATSIRO) is the same as that of the atmospheric component, and the soil layers have thicknesses of 5, 20, 75, 100, and 200 cm from the surface. Climatological monthly leaf area index (LAI) data of Myneni et al. (1997) are prescribed in the land-surface component.

Horizontal resolution of the ocean/sea-ice component is 1/4° (~25 km at the equator) in both longitude and latitude. The ocean component contains 54 levels in vertical, with varying distance between the levels from 5 m at the surface to 330 m at the maximum depth of 6,065 m. The model topography is created from the 1/30° bathymetry data made by the Ocean Circulation and Climate Advanced Modelling (OCCAM) project at the Southampton Oceanography Centre (obtained through GFDL) and the partial-cell method is used. The Smagorinsky (1963) scheme and the Laplacian mixing with constant diffusivity are used for horizontal mixing of the momentum and tracers, respectively. The KPP boundary layer mixing scheme of Large et al. (1994) and the thickness diffusion scheme of Gent and McWilliams (1990) are also adopted. In the region poleward of 85°N, a finite-impulse-response filter is applied in order to relax the time step constraint imposed by convergence of meridians.

The atmosphere/land-surface component is integrated for 2 months before coupling with the ocean/sea-ice component. The atmospheric initial conditions are made from the European Centre for Medium-Range Weather Forecasts (ECMWF) Re-Analysis (ERA-40) (Uppala et al. 2005) of November 1, 1982. In the land-surface component, the initial soil temperature is set equal to the surface air temperature, the initial soil moisture is set to a spatially uniform value (0.3), and the initial snow amount is set to zero. Initial conditions of the ocean component are climatological temperature and salinity fields in January made from the World Ocean Atlas 1998 (WOA98) (Antonov et al. 1998a,b,c; Boyer et al. 1998a,b,c) with no motion. Sea-ice is initially set over the region where the initial SST is below the freezing temperature with its concentration of unity. The ocean/sea-ice component is coupled to the atmosphere/land-surface component without spin-up integration.

The first trial of our coupled simulation is started at January 1, 0001, with the coupling interval of 1 h without flux adjustment nor any restoring of sea surface variables, and the integration is stopped at July 17, 0006, due to a numerical instability (negative salinity) in the Ross Sea, Antarctica. This instability might be caused by a combination of factors such as a strong extratropical cyclone, inappropriate treatment of wind stress in coastal grids, sea-ice formation, and local bathymetry. In this case, we used 60 nodes (480 processors) of the Earth Simulator (30 nodes for each component) and it took about 20 wall-clock hours for 1-year integration.

14.3.2 Global View of Snapshots

Figure 14.2 shows a 10-min averaged field of 850-hPa specific humidity at 06 UTC January 1, 0006. Synoptic-scale phenomena such as extratropical cyclones with mesoscale structure are embedded in the global-scale circulation and realistically simulated. Additionally, tropical cyclones with modest intensity are formed automatically in this coupled system and one of them over the Indian Ocean (around 60°E, 10°S) is captured in this figure.

Figure 14.3 is a daily-mean field of 100-m depth current speed at the same day. Although the ocean simulation with the horizontal resolution of 1/4° is "eddy-permitting" rather than "eddy-resolving," various scales of eddies fill the simulated world ocean, and the major western boundary currents such as the Kuroshio and Oyashio in the North Pacific, the East Australian Current in the South Pacific, the Gulf Stream in the North Atlantic, the Brazil and Malvinas Currents in the South Atlantic, and the Agulhas Current in the Southern Indian are represented fairly well.

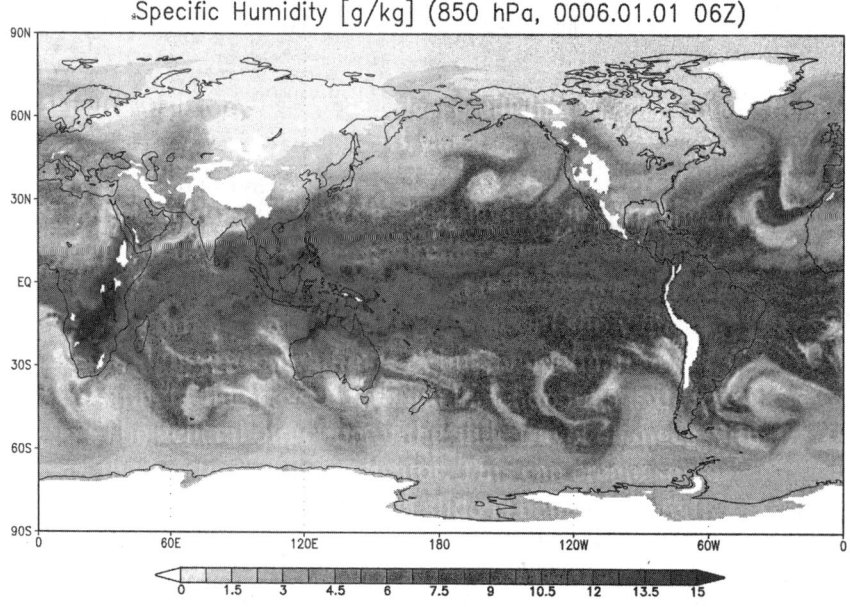

Fig. 14.2 10-min averaged field of 850-hPa specific humidity [g kg^{-1}]

14.3 Preliminary Results

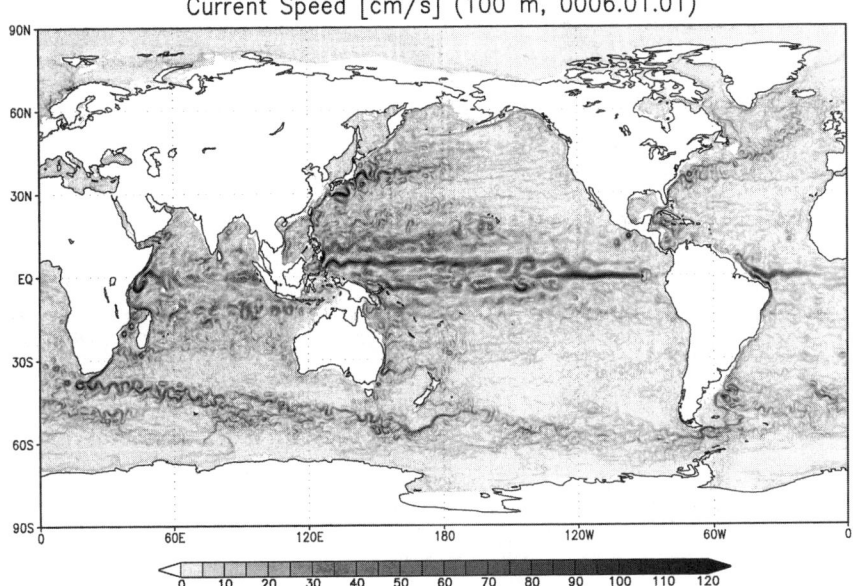

Fig. 14.3 Daily-mean field of 100-m depth current speed [cm s^{-1}]

Of course, integration time of several years is too short for these currents to spin up and integration of several decades would be necessary to evaluate them in detail.

Figure 14.4 shows latent heat flux averaged from 00 to 06 UTC. Not only the footprints of extratropical cyclones are recognized as local minima, the patterns of meandering western boundary currents (the Kuroshio, Gulf Stream, and Agulhas Current) are clearly seen in the figure. Furthermore, local maxima associated with small-scale phenomena such as a tropical cyclone over the Indian Ocean described above and gap winds over the eastern Pacific warm pool (off the Central American isthmus) are visible. The effects of gap winds are also manifested in Fig. 14.3 as small eddies there. To what extent these small-scale phenomena affect the global-scale general circulations of the atmosphere and ocean may be an issue in the future.

14.3.3 Local View Around Japan

As an example of high-impact weather events in the mid-latitudes, Fig. 14.5 shows a well-developed extratropical cyclone near Japan at February 7, 0006 and corresponding sea surface conditions. Sea level pressure (10-minute average) and outgoing longwave radiation (6-h average) are displayed in Fig. 14.5a, and daily-mean SST and sea-ice concentration are in Fig. 14.5b. Intensity and structure of this extratropical cyclone seem quite realistic. At the same time, SST fronts in the Kuroshio–Oyashio

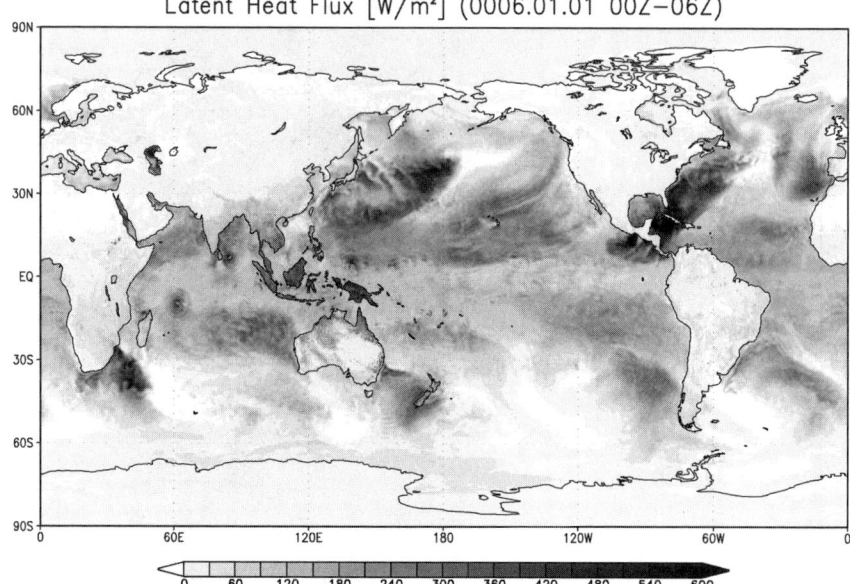

Fig. 14.4 6-h averaged field of latent heat flux [W m^{-2}]

Extension region and sea-ice extent in the Okhotsk Sea are well simulated. These results encourage us to use this coupled GCM for studies of high-impact phenomena in the mid-latitudes.

14.3.4 Annual-Mean Surface Climatologies

We now compare simulated surface "climatology" calculated from 4-year data (January 0002 through December 0005) with observation. Annual-mean fields of SST and sea surface salinity (SSS) are compared with WOA98 climatology in Figs. 14.6 and 14.7, respectively, and annual-mean precipitation rate is compared with Climate Prediction Center Merged Analysis of Precipitation (CMAP) (Xie and Arkin 1996) climatology in Fig. 14.8. Annual-mean sea surface velocity is also shown in Fig. 14.9.

The most apparent deficiency in this simulation is the so-called "double ITCZ" bias, which consists of warm SST (Fig. 14.6), fresh SSS (Fig. 14.7), and heavy precipitation (Fig. 14.8) biases formed around 5°S, and is a common problem among coupled GCMs in the world (e.g., Covey et al. 2003). In our simulation, this double ITCZ bias is accompanied by a swift zonal jet at the latitude (Fig. 14.9a), which may correspond to (equatorward-shifted) South Equatorial Counter Current, advects warm and fresh water in the western equatorial Pacific to the east, and contributes to maintain this bias to some extent.

14.3 Preliminary Results

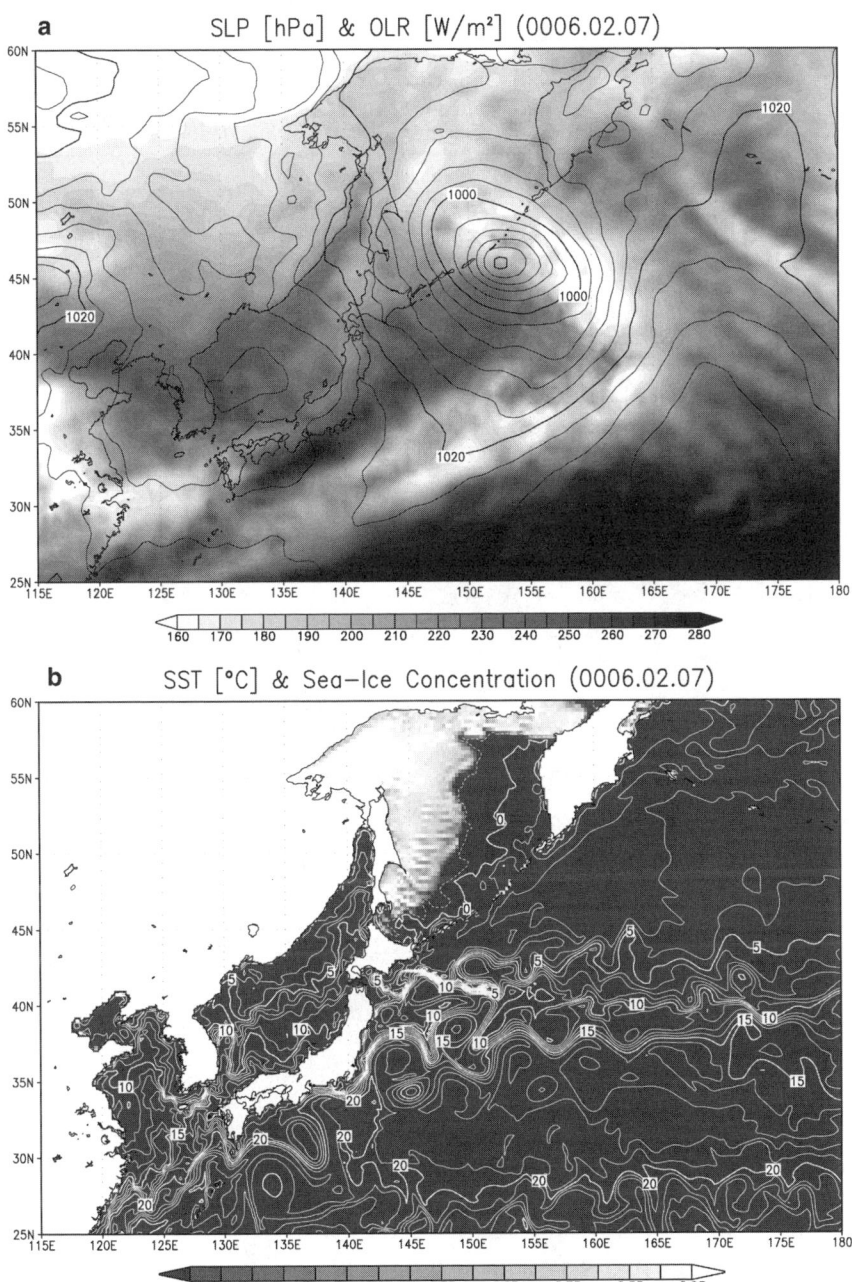

Fig. 14.5 (a) Sea level pressure (in contours drawn every 4 hPa) and outgoing longwave radiation (W m^{-2} in shade). (b) Sea surface temperature (in contours drawn every 1°C) and sea-ice concentration (in shade)

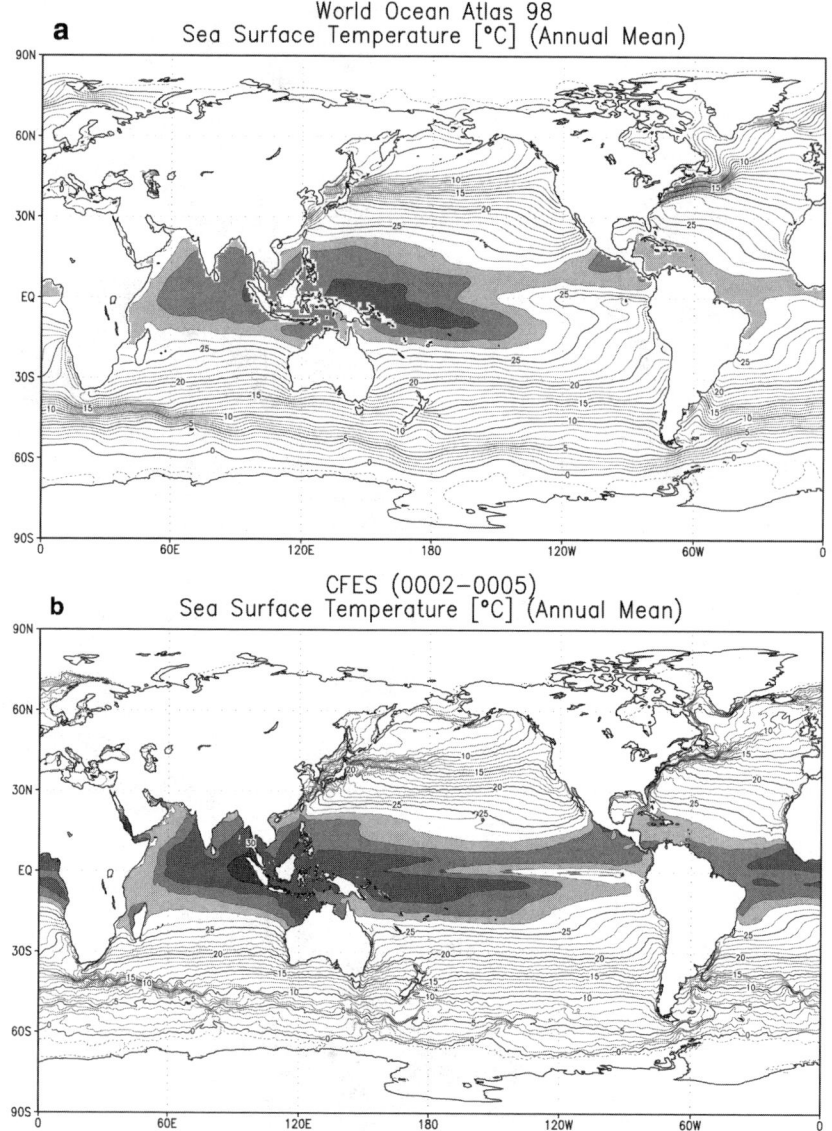

Fig. 14.6 (a) Observed (WOA98) and (b) simulated annual-mean sea surface temperature [°C]. Contour interval is 1°C and the region >27°C is shaded

14.3 Preliminary Results

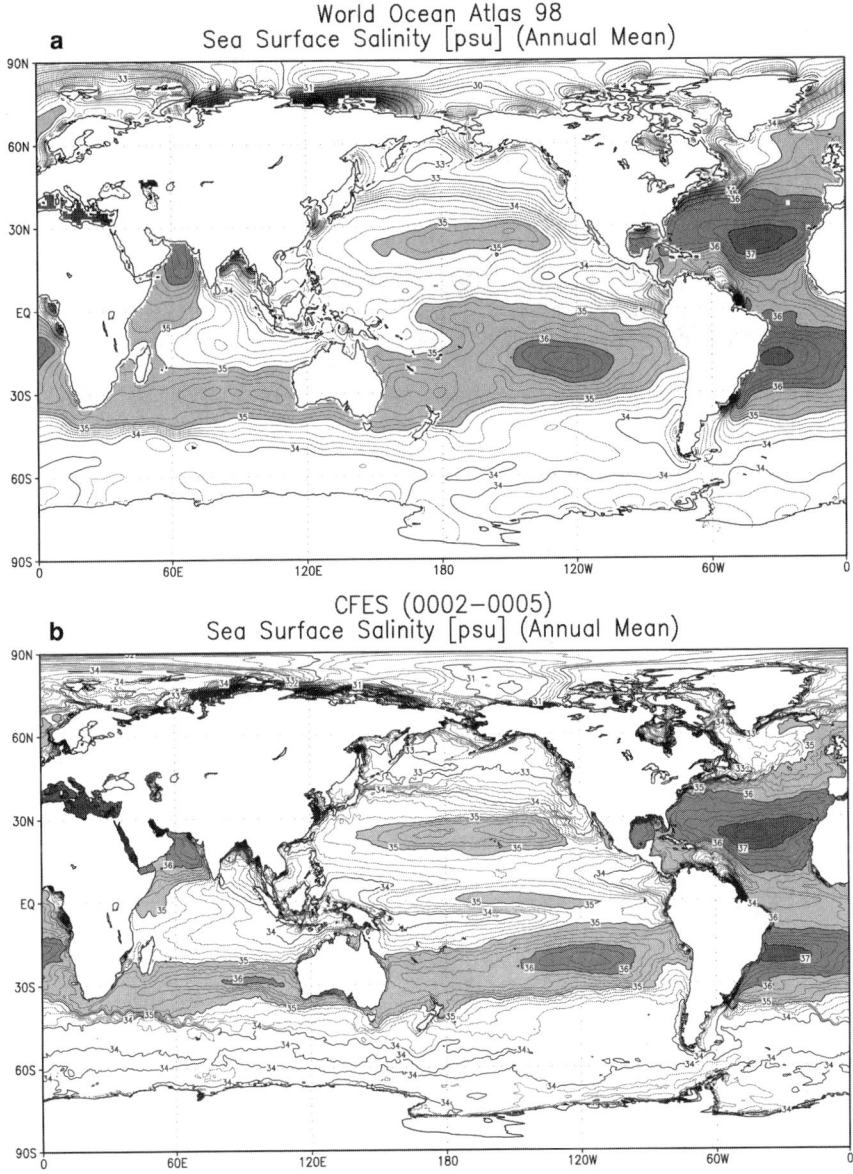

Fig. 14.7 (**a**) Observed (WOA98) and (**b**) simulated annual-mean sea surface salinity [psu]. Contour interval is 0.2 psu and the region >35 psu is shaded

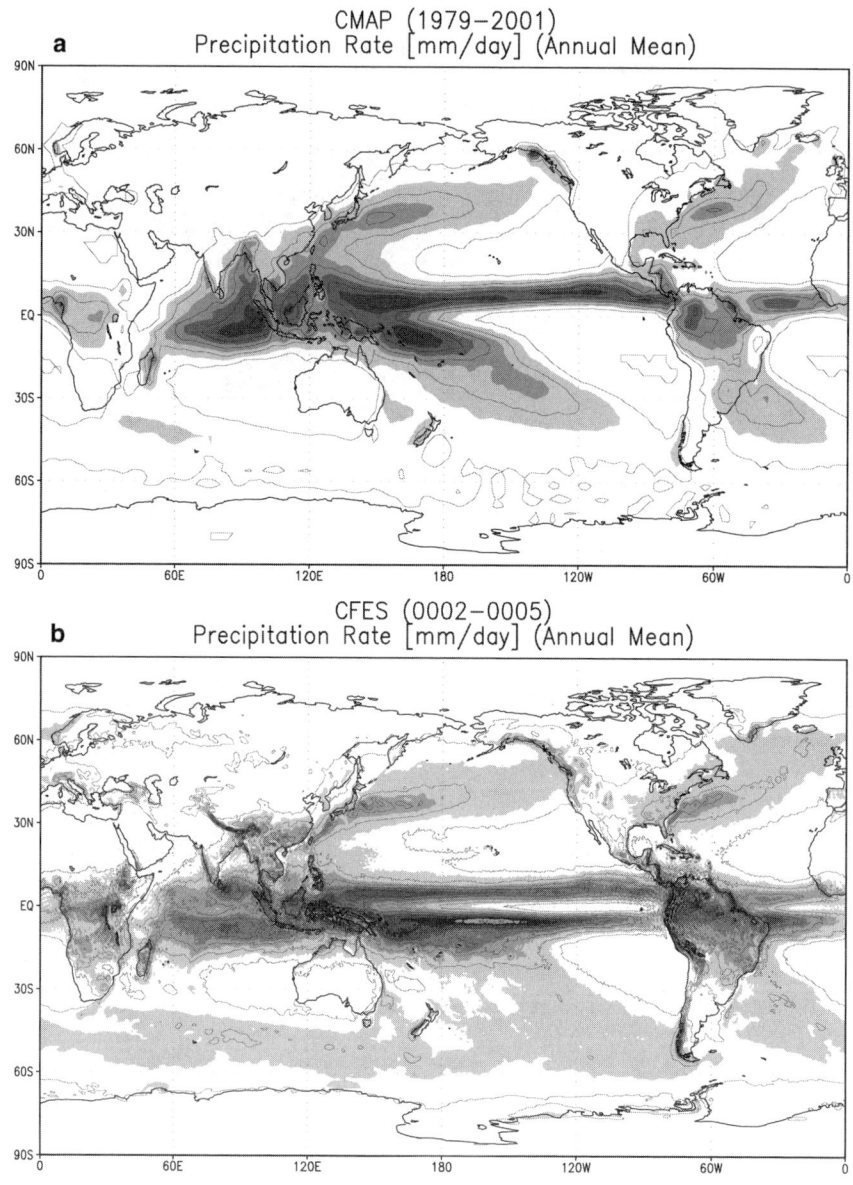

Fig. 14.8 (a) Observed (CMAP) and (b) simulated annual-mean precipitation rate [mm day^{-1}]. Contour interval is 2 mm day^{-1} and the region >3 mm day^{-1} is shaded

14.3 Preliminary Results

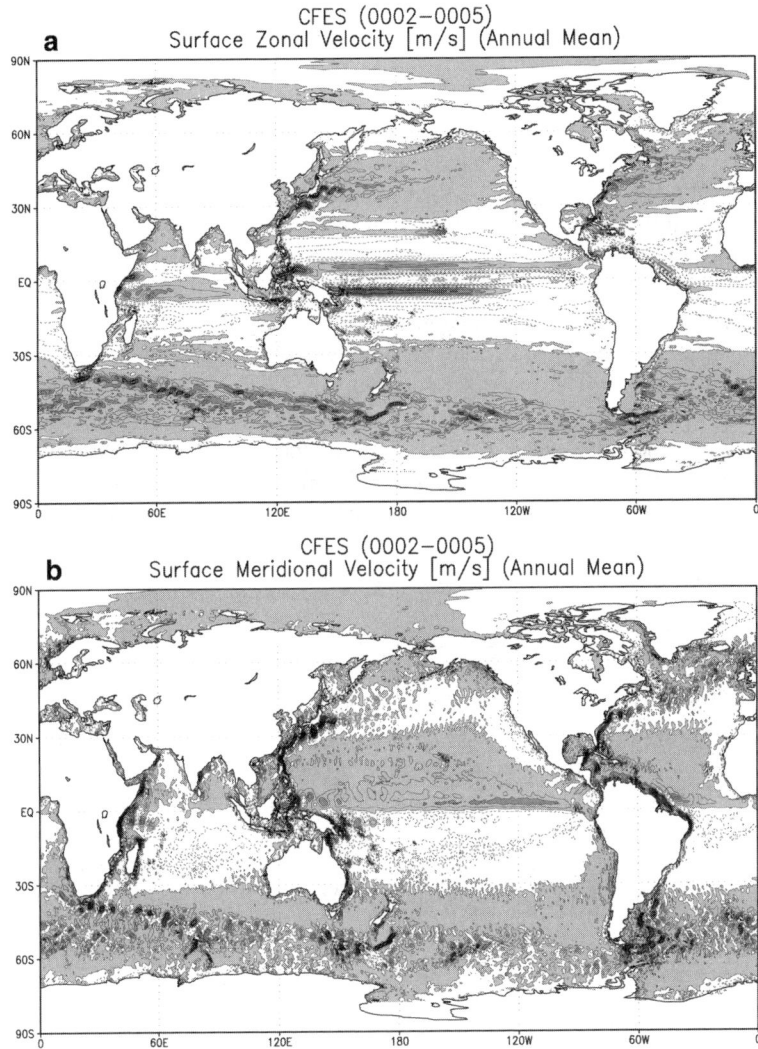

Fig. 14.9 Simulated (**a**) zonal and (**b**) meridional components of annual-mean sea surface velocity [m s^{-1}]. Contour intervals are 0.1 m s^{-1} for (**a**) and 0.05 m s^{-1} for (**b**), and the positive region is shaded

In the mid-latitudes, the amount of simulated precipitation (Fig. 14.8b) is somewhat larger than observation (Fig. 14.8a), but its spatial pattern is not so bad. In particular, localized precipitation bands over the SST fronts of the Kuroshio and the Gulf Stream are well reproduced in this coupled simulation. This may be due to relatively good representation of the SST fronts in our OGCM.

14.3.5 Seasonal Cycle of Tropical SST and Polar Sea-Ice Extent

The seasonal cycle in the tropical SST is manifested as the result of complex air–sea interactions, and suitable for evaluating a performance of coupled atmosphere–ocean models. On the other hand, sea-ice plays a very important role in the global coupled atmosphere–ocean system, and realistic reproduction of its variability is one of the key factors for successful coupled simulations.

The seasonal cycle of equatorial SST (averaged over the region 2°S–2°N) as deviation from the annual mean is shown in Fig. 14.10. Simulated seasonal cycle (Fig. 14.10b) well captures the overall features found in observed seasonal cycle (Fig. 14.10a). In particular, a predominance of the annual cycle and a westward propagation in the central and eastern Pacific are successfully reproduced in this simulation, although its amplitude is somewhat smaller than observation. Some portion of this achievement may be attributed to an improved representation of the steep orography near the coast (e.g., Andes), thanks to the high resolution of our AGCM.

Figure 14.11 shows comparison between observed and simulated sea-ice concentration in the Arctic Ocean in March (Fig. 14.11a, b) and in September (Fig. 14.11c, d). Climatological monthly observations are calculated using global sea-ice concentration data of GISST (Parker et al. 1995) from 1974 to 1999. Seasonal variation of simulated sea-ice extent well captures the observed feature such that a main portion of sea-ice in the Arctic Ocean remains throughout the year while sea-ice in the Okhotsk and Bering Seas vanishes in summer. However, too much sea-ice in the Barents Sea should be pointed out as a noticeable problem.

Simulated sea-ice concentration around the Antarctica is compared with observed climatology (GISST) in Fig. 14.12. Obviously, too much sea-ice remains in summer off the Weddell and Ross Seas in our simulation results (Fig. 14.12b) than in observations (Fig. 14.12a). On the other hand, simulated sea-ice extent in winter (Fig. 14.12d) looks very similar to the observed one (Fig. 14.12c). Some portions of this difference may be attributed to the difference in surface wind fields around the Antarctica: Simulated surface wind in September is close to observed one, while that in March has biases over the Weddell and Ross Seas (not shown).

14.3 Preliminary Results

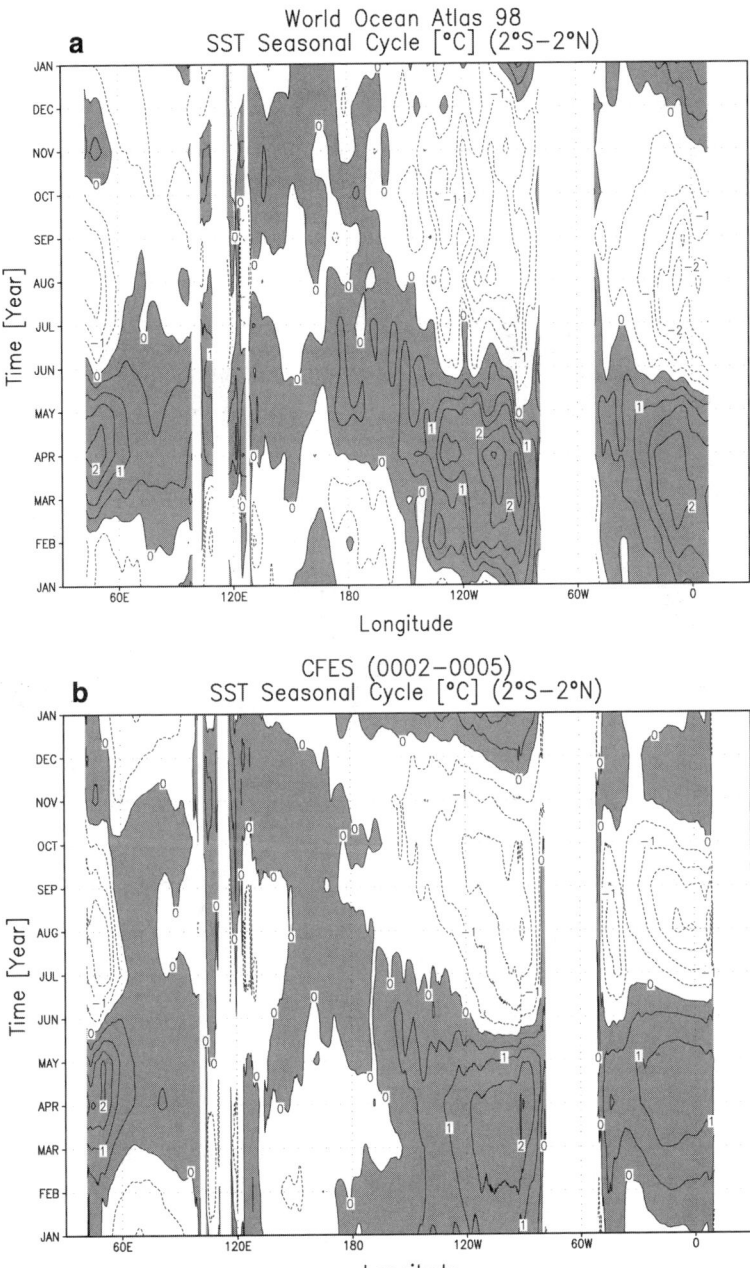

Fig. 14.10 (a) Observed (WOA98) and (b) simulated sea surface temperature annual cycle [°C] along the equator (2°S–2°N). Shown are the deviations from the annual mean. Contour interval is 0.5°C and positive values are shaded

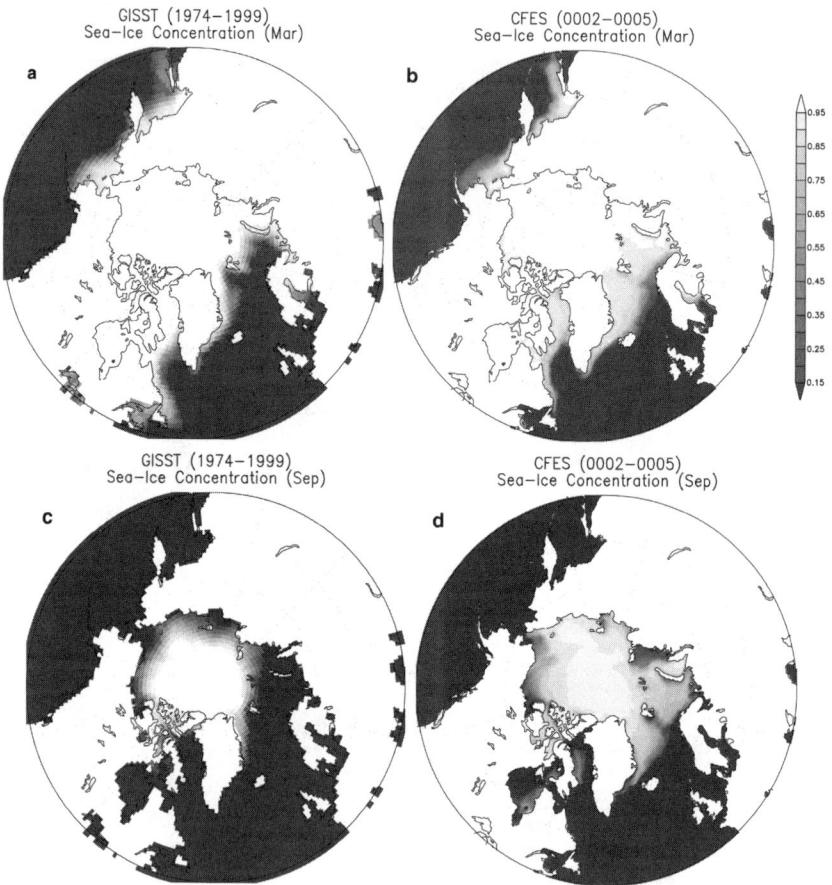

Fig. 14.11 (**a, c**) Observed (GISST) and (**b, d**) simulated sea-ice concentration in the Arctic region. (**a, b**) March. (**c, d**) September

14.4 Concluding Remarks

Now the computational power of the Earth Simulator enables us to treat the global-scale general circulations and various small-scale phenomena simultaneously even in a coupled atmosphere–ocean simulation, although high-resolution is not necessarily a panacea for all difficulties found in state-of-the-art CGCMs. The "virtual atmosphere and ocean" constructed in the Earth Simulator may give more information than observations in some respects.

In this article, preliminary results of a high-resolution coupled atmosphere–ocean simulation for only 66 months are presented. Some deficiencies are found there in comparison with observations, but small-scale structures such as extratropical

14.4 Concluding Remarks

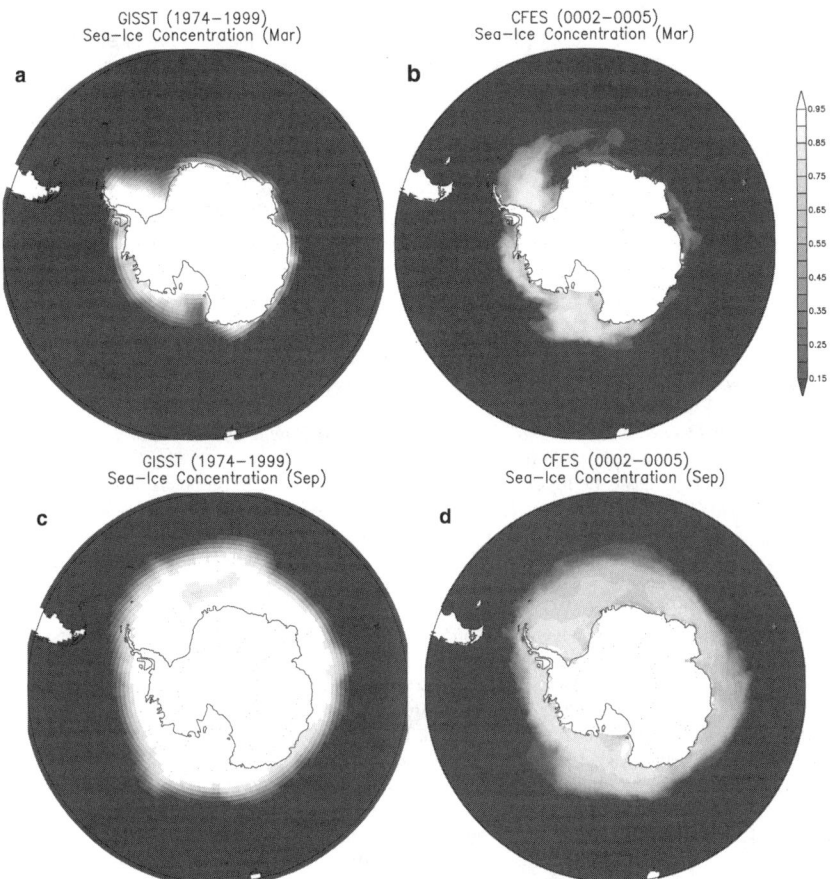

Fig. 14.12 Same as Fig. 14.11 but in the Antarctic region

cyclones and SST fronts in the mid-latitudes, and seasonal variation of tropical SST and polar sea-ice extent are simulated fairly well. After this first trial, we made much efforts at improving numerical schemes and tuning model parameters, and some portion of model biases are reduced in recent simulations.

In the near future, we will produce a data set from high-resolution coupled simulations for several decades in order to validate the simulation results and identify the weak points of the model in detail as well as to investigate the effects of mesoscale phenomena on the global-scale general circulations of the atmosphere and ocean. In addition, we will try seasonal prediction experiments of the coupled atmosphere–ocean system focusing on the mid-latitudes using our high-resolution coupled GCM, CFES.

Acknowledgments We thank Drs. Keiko Takahashi, Kenji Komine, Toru Miyama, Takashi Mochizuki, Hiromichi Igarashi, Genki Sagawa, and many other researchers relating to AFES, OFES, and CFES projects. Our thanks are extended to Messrs. Takashi Abe, Akira Azami, and Masayuki Yamada for their help in optimizing the source code of CFES for the Earth Simulator. Comments by Prof. Kevin P. Hamilton and the anonymous reviewer are greatly acknowledged. The radiation model mstrnX and the land-surface model MATSIRO are provided by courtesy of Prof. Teruyuki Nakajima and Dr. Seita Emori, respectively. Development of CFES is partly supported by MEXT (Ministry of Education, Culture, Sports, Science and Technology) RR2002 Project for Sustainable Co-existence of Human, Nature and the Earth, category 7 (PI: Prof. Toshiyuki Awaji). The numerical calculation was carried out on the Earth Simulator.

References

Antonov, J. I., S. Levitus, T. P. Boyer, M. E. Conkright, T. O'Brien, and C. Stephens, 1998a: *World Ocean Atlas 1998, Vol. 1: Temperature of the Atlantic Ocean*. NOAA Atlas NESDIS 27, U.S. Government Printing Office, Washington, D.C.

— 1998b: *World Ocean Atlas 1998, Vol. 2: Temperature of the Pacific Ocean*. NOAA Atlas NESDIS 28, U.S. Government Printing Office, Washington, D.C.

Antonov, J. I., S. Levitus, T. P. Boyer, M. E. Conkright, T. O'Brien, C. Stephens, and B. Trotsenko, 1998c: *World Ocean Atlas 1998, Vol. 3: Temperature of the Indian Ocean*. NOAA Atlas NESDIS 29, U.S. Government Printing Office, Washington, D.C.

Boyer, T. P., S. Levitus, J. I. Antonov, M. E. Conkright, T. O'Brien, and C. Stephens, 1998a: *World Ocean Atlas 1998, Vol. 4: Salinity of the Atlantic Ocean*. NOAA Atlas NESDIS 30, U.S. Government Printing Office, Washington, D.C.

— 1998b: *World Ocean Atlas 1998, Vol. 5: Salinity of the Pacific Ocean*. NOAA Atlas NESDIS 31, U.S. Government Printing Office, Washington, D.C.

Boyer, T. P., S. Levitus, J. I. Antonov, M. E. Conkright, T. O'Brien, C. Stephens, and B. Trotsenko, 1998c: *World Ocean Atlas 1998, Vol. 6: Salinity of the Indian Ocean*. NOAA Atlas NESDIS 32, U.S. Government Printing Office, Washington, D.C.

Covey, C., K. M. AchutaRao, U. Cubasch, P. Jones, S. J. Lambert, M. E. Mann, T. J. Phillips, and K. E. Taylor, 2003: An overview of results from the Coupled Model Intercomparison Project. *Global Planet. Change*, **37**, 103–133.

Enomoto, T., A. Kuwano-Yoshida, N. Komori, and W. Ohfuchi, 2007: Description of AFES 2: Improvements for high-resolution and coupled simulations. *High Resolution Numerical Modelling of the Atmosphere and Ocean*, W. Ohfuchi and K. Hamilton, eds., Springer, New York, this volume, Chapter 5.

Gent, P. R. and J. C. McWilliams, 1990: Isopycnal mixing in ocean circulation models. *J. Phys. Oceanogr.*, **20**, 150–155.

Hibler, W. D., 1979: A dynamic thermodynamic sea ice model. *J. Phys. Oceanogr.*, **9**, 815–846.

Hunke, E. C. and J. K. Dukowicz, 2002: The elastic–viscous–plastic sea ice dynamics model in general orthogonal curvilinear coordinates on a sphere—incorporation of metric terms. *Mon. Wea. Rev.*, **130**, 1848–1865.

Komori, N., K. Takahashi, K. Komine, T. Motoi, X. Zhang, and G. Sagawa, 2005: Description of sea-ice component of Coupled Ocean–Sea-Ice Model for the Earth Simulator (OIFES). *J. Earth Simulator*, **4**, 31–45.

Large, W. G., J. C. McWilliams, and S. C. Doney, 1994: Oceanic vertical mixing: A review and a model with a nonlocal boundary layer parameterization. *Rev. Geophys.*, **32**, 363–404.

Masumoto, Y., H. Sasaki, T. Kagimoto, N. Komori, A. Ishida, Y. Sasai, T. Miyama, T. Motoi, H. Mitsudera, K. Takahashi, H. Sakuma, and T. Yamagata, 2004: A fifty-year eddy-resolving simulation of the world ocean: Preliminary outcomes of OFES (OGCM for the Earth Simulator). *J. Earth Simulator*, **1**, 35–56.

Myneni, R. B., R. R. Nemani, and S. W. Running, 1997: Estimation of global leaf area index and absorbed PAR using radiative transfer models. *IEEE Trans. Geosci. Remote Sens.*, **35**, 1380–1393.

Nakajima, T. and M. Tanaka, 1986: Matrix formulations for the transfer of solar radiation in a plane-parallel scattering atmosphere. *J. Quant. Spectrosc. Radiat. Transfer*, **35**, 13–21.

References

Nakajima, T., M. Tsukamoto, Y. Tsushima, A. Numaguti, and T. Kimura, 2000: Modeling of the radiative process in an atmospheric general circulation model. *Appl. Opt.*, **39**, 4869–4878.

Nakamura, H., T. Sampe, Y. Tanimoto, and A. Shimpo, 2004: Observed associations among storm tracks, jet streams and midlatitude oceanic fronts. *Earth's Climate: The Ocean–Atmosphere Interaction*, C. Wang, S.-P. Xie, and J. A. Carton, eds., American Geophysical Union, Washington, D.C., U.S.A., Geophys. Monogr. 147, 329–346.

Numaguti, A., M. Takahashi, T. Nakajima, and A. Sumi, 1997: Description of CCSR/NIES atmospheric general circulation model. *Study on the Climate System and Mass Transport by a Climate Model*, A. Numaguti, S. Sugata, M. Takahashi, T. Nakajima, and A. Sumi, eds., Center for Global Environmental Research, National Institute for Environmental Studies, CGER's Supercomputer Monograph Report 3, 1–48.

Oberhuber, J. M., D. M. Holland, and L. A. Mysak, 1993: A thermodynamic-dynamic snow sea-ice model. *Ice in the Climate System*, W. R. Peltier, ed., Springer-Verlag, NATO ASI Series, Series I: Global Environmental Change, 653–673.

Ohfuchi, W., H. Nakamura, M. K. Yoshioka, T. Enomoto, K. Takaya, X. Peng, S. Yamane, T. Nishimura, Y. Kurihara, and K. Ninomiya, 2004: 10-km mesh meso-scale resolving simulations of the global atmosphere on the Earth Simulator: Preliminary outcomes of AFES (AGCM for the Earth Simulator). *J. Earth Simulator*, **1**, 8–34.

Oki, T. and Y. C. Sud, 1998: Design of Total Runoff Integrating Pathways (TRIP)—a global river channel network. *Earth Interactions*, **2**, paper No. 1.

Pacanowski, R. C., 1995: MOM2 documentation, user's guide and reference manual. GFDL Ocean Group Technical Report 3, NOAA/Geophysical Fluid Dynamics Laboratory, Princeton, NJ, U.S.A.

Pacanowski, R. C. and S. M. Griffies, 1999: The MOM 3.0 manual. GFDL Ocean Group Technical Report 4, NOAA/Geophysical Fluid Dynamics Laboratory, Princeton, NJ, U.S.A.

Parker, E., M. Jackson, and E. B. Horton, 1995: The 1961–1990 GISST2.2 sea surface temperature and sea ice climatology. Climate Research Technical Note 63, Hadley Centre, U.K. Met Office, Bracknell, U.K.

Parkinson, C. L. and W. M. Washington, 1979: A large scale numerical model of sea ice. *J. Geophys. Res.*, **84**, 311–337.

Sekiguchi, M., 2004: *A study on evaluation of the radiative flux and its computational optimization in the gaseous absorbing atmosphere*. Science doctoral dissertation, University of Tokyo, 121 pp., in Japanese.

Shingu, S., H. Takahara, H. Fuchigami, M. Yamada, Y. Tsuda, W. Ohfuchi, Y. Sasaki, K. Kobayashi, T. Hagiwara, S. Habata, M. Yokokawa, H. Itoh, and K. Otsuka, 2002: A 26.58 Tflops global atmospheric simulation with the spectral transform method on the Earth Simulator. *Proc. ACM/IEEE SC2002 Conference*, Baltimore, Maryland.

Smagorinsky, J., 1963: General circulation experiments with the primitive equations: I. The basic experiment. *Mon. Wea. Rev.*, **91**, 99–164.

Takahashi, K., S. Shingu, A. Azami, T. Abe, M. Yamada, H. Fuchigami, M. Yoshioka, Y. Sasaki, H. Sakuma, and T. Sato, 2003a: Coupling strategy of atmospheric–oceanic general circulation model with ultra high resolution and its performance on the Earth Simulator. *Parallel Computational Fluid Dynamics—New Frontiers and Multi-Disciplinary Applications*, K. Matsuno, A. Ecer, J. Periaux, N. Satofuka, and P. Fox, eds., Elsevier Science, 93–100.

Takahashi, K., Y. Tsuda, M. Kanazawa, S. Kitawaki, H. Sasaki, and T. Sato, 2003b: Parallel architecture and its performance of oceanic global circulation model based on MOM3 to be run on the Earth Simulator. *Parallel Computational Fluid Dynamics—New Frontiers and Multi-Disciplinary Applications*, K. Matsuno, A. Ecer, J. Periaux, N. Satofuka, and P. Fox, eds., Elsevier Science, 101–108.

Takata, K., S. Emori, and T. Watanabe, 2003: Development of the minimal advanced treatments of surface interaction and runoff. *Global Planet. Change*, **38**, 209–222.

Uppala, S. M., P. W. Kållberg, A. J. Simmons, U. Andrae, V. da Costa Bechtold, M. Fiorino, J. K. Gibson, J. Haseler, A. Hernandez, G. A. Kelly, X. Li, K. Onogi, S. Saarinen, N. Sokka, R. P. Allan, E. Andersson, K. Arpe, M. A. Balmaseda, A. C. M. Beljaars, L. van de Berg, J. Bidlot, N. Bormann, S. Caires, F. Chevallier, A. Dethof, M. Dragosavac, M. Fisher, M. Fuentes, S. Hagemann, E. Hólm, B. J. Hoskins, L. Isaksen, P. A. E. M. Janssen, R. Jenne, A. P. McNally,

J.-F. Mahfouf, J.-J. Morcrette, N. A. Rayner, R. W. Saunders, P. Simon, A. Sterl, K. E. Trenberth, A. Untch, D. Vasiljevic, P. Viterbo, and J. Woollen, 2005: The ERA-40 re-analysis. *Q. J. R. Meteorol. Soc.*, **131**, 2961–3012.

Xie, P. and P. A. Arkin, 1996: Analyses of global monthly precipitation using gauge observations, satellite estimates, and numerical model predictions. *J. Climate*, **9**, 840–858.

Xie, S.-P., 2004: Satellite observations of cool ocean–atmosphere interaction. *Bull. Amer. Meteor. Soc.*, **85**, 195–208.

Zhang, X. and J. Zhang, 2001: Heat and freshwater budgets and pathway in the Arctic Mediterranean in a coupled ocean/sea-ice model. *J. Oceanogr.*, **57**, 207–234.

Chapter 15
Impact of Coupled Nonhydrostatic Atmosphere–Ocean–Land Model with High Resolution

Keiko Takahashi, Xindong Peng, Ryo Onishi, Mitsuru Ohdaira, Koji Goto, Hiromitsu Fuchigami, and Takeshi Sugimura

Summary This chapter presents basic formulation of Multi-Scale Simulator for the Geoenvironment (MSSG) which is a coupled non-hydrostatic AGCM-OGCM developed in Earth Simulator Center. MSSG is characterized by Yin-Yang grid system for both of the components, computational schemes with high accuracy in the dynamical core and high computational performance on the Earth Simulator. In particular some preliminary results from 120-h forecast experiments with MSSG are presented.

15.1 Introduction

It is widely accepted that the most powerful tools available for assessing future weather/climate are fully coupled ocean–atmosphere models. Earth Simulator Center (ESC) has developed MSSG which is a coupled atmosphere–ocean–land general circulation model to be run on the Earth Simulator with high computational performance. MSSG is composed of non-hydrostatic atmosphere global circulation model, nonhydrostatic/hydrostatic ocean general circulation model, and a simple land surface model.

The purpose of this paper is to introduce the basic formulation of the MSSG and show preliminary simulation results. We have performed 48 h integration to confirm its physical performance of MSSG-A which is an AGCM component of MSSG with only microcloud physics. A 72-h forecasting experiments have been performed to validate physical performance of MSSG-A. For validation of physical performance of the ocean component; MSSG-O, 15-years integration of the north Pacific basin has done with 11 km horizontal resolution and 40 vertical levels. Furthermore, tracking and intensity forecasting experiments of typhoon ETAU in 2003 has been performed with the MSSG with high resolution. Although those results are preliminary ones, results of high resolution simulations suggest some important features by high resolution simulations.

Outline of the MSSG, coupled non-hydrostatic atmosphere–ocean–land GCM is described in Sect. 15.2. In Sect. 15.3, preliminary validation results of simulations with the MSSG are shown. Future work is presented in Sect. 15.4.

15.2 Model Configuration

15.2.1 MSSG-A Non-hydrostatic Atmosphere Global Circulation Model

MSSG-A which is a non-hydrostatic global atmosphere component is described of fully compressive flux form of dynamic Satomura and Akiba (2003) and Smagorinsky–Lilly type parameterizations in Lilly (1962) and Smagorinsky et al. (1965) for subgrid scale mixing, surface fluxes presented by Zhang and Anthes (1982) and Blackadar (1979), cloud microphysics with mixed phases by Reisner et al. (1998), cumulus convective processes presented by Kain and Fritsch (1993) and Fritsch and Chappell (1980) and cloud-radiation scheme longwave and shortwave interactions with explicit cloud and clear-air.

The set of the prognostic equations are presented as follows:

$$\frac{\partial \rho'}{\partial t} + \frac{1}{G^{1/2}a\cos\varphi}\frac{\partial(G^{1/2}G^{13}\rho u)}{\partial \lambda} + \frac{1}{G^{1/2}a\cos\varphi}\frac{\partial(G^{1/2}G^{23}\cos\varphi\rho v)}{\partial \varphi}$$
$$+ \frac{1}{G^{1/2}}\frac{\partial(\rho w^*)}{\partial z^*} = 0, \qquad (15.1)$$

$$\frac{\partial \rho u}{\partial t} + \frac{1}{G^{1/2}a\cos\varphi}\frac{\partial(G^{1/2}G^{13}P')}{\partial \lambda} = -\nabla\cdot(\rho u \mathbf{v}) + 2f_r\rho v - 2f_\varphi\rho w$$
$$+ \frac{\rho v u \tan\varphi}{a} - \frac{\rho w u}{a} + F_\lambda, \qquad (15.2)$$

$$\frac{\partial \rho v}{\partial t} + \frac{1}{G^{1/2}a}\frac{\partial(G^{1/2}G^{23}P')}{\partial \varphi} = -\nabla\cdot(\rho v \mathbf{v}) + 2f_\lambda\rho w - 2f_r\rho u$$
$$- \frac{\rho u u \tan\varphi}{a} - \frac{\rho w v}{a} + F_\varphi, \qquad (15.3)$$

$$\frac{\partial \rho w}{\partial t} + \frac{1}{G^{1/2}}\frac{\partial P'}{\partial z^*} + \rho'\mathbf{g} = -\nabla\cdot(\rho w \mathbf{v}) + 2f_\varphi\rho u - 2f_\lambda\rho v + \frac{\rho u u}{a}$$
$$+ \frac{\rho v v}{a} + F_r, \qquad (15.4)$$

$$\frac{\partial P'}{\partial t} + \nabla\cdot(P\mathbf{v}) + (\gamma-1)P\nabla\cdot\mathbf{v} = (\gamma-1)\nabla\cdot(\kappa\nabla T) + (\gamma-1)\phi, \quad (15.5)$$

$$P = \rho RT, \qquad (15.6)$$

$$\rho w^* = \frac{1}{G^{1/2}}(G^{1/2}G^{13}\rho u + G^{1/2}G^{23}\rho v + \rho w). \qquad (15.7)$$

In (15.1)–(15.7), prognostic valuables are momentum $\rho\mathbf{v} = (\rho v, \rho v, \rho\omega)$, ρ' which is calculated as $\rho' = \rho - \bar{\rho}$ and P' defined by $P' = P - \bar{P}$. ρ is the density; P

15.2 Model Configuration

is the pressure; $\bar{P}$ is a constant reference pressure. $\mathbf{f}, \mu, \kappa, \gamma$, and ϕ are the Coriolis force, the viscosity coefficient, the diffusion coefficient, and the ratio of specific heat, and diabatic heating, respectively. F is subscripted as the three components of the viscous drag force. G is the metric term for vertical coordinate; λ is the latitude; φ is the longitude.

The treatment of cloud and precipitation is controlled by selecting a parameterization scheme due to horizontal resolution. For grid spacing greater than 10 km, Kain and Fritsch scheme in Kain and Fritsch (1993) and Fritsch and Chappell (1980) is used and cloud microphysics based on mixed phase microcloud physics in Reisner et al. (1998) is used for below 5 km spacing.

Over land, the ground temperature and ground moisture are computed by using a bucket model as a simplified land model. As upper boundary condition, Rayleigh friction refers top three layers to damp the full velocity fields with weighted function of height.

Regional version of the atmosphere component is utilized with one-way nesting scheme by choosing the target region on the sphere, although two-way nesting is available as an option. Any large regions can be selected from the global, because both Coriolis and metric terms have been introduced in the regional formulation.

15.2.2 MSSG-O; Non-hydrostatic/Hydrostatic Ocean Global Circulation Model

In the ocean component; MSSG-O, the in-compressive and hydrostatic/non-hydrostatic equations with the Boussinesq approximation are introduced based on describing in Marshall et al. (1997a,b). Before starting experiments, non-hydrostatic or hydrostatic configuration has to be selected due to horizontal resolution. In this chapter, we describe a formulation of hydrostatic version. In addition, either explicit free surface solver or rigid lid solver is available as one of options.

The set of hydrostatic equations in the ocean component becomes as follows,

$$\frac{\partial c}{\partial t} = -\mathbf{v}\,\text{grad}\,c + F_c \tag{15.8}$$

$$\frac{\partial T}{\partial t} = -\mathbf{v}\,\text{grad}\,T + F_T \tag{15.9}$$

$$0 = \nabla \cdot \mathbf{v} = \left(\frac{1}{r\cos\varphi} \frac{\partial u}{\partial \lambda} + \frac{1}{r\cos\varphi} \frac{\partial(\cos\varphi v)}{\partial \varphi} + \frac{1}{r^2} \frac{\partial(r^2 w)}{\partial r} \right) \tag{15.10}$$

$$\frac{\partial u}{\partial t} = -\mathbf{v}\,\text{grad}\,u + 2f_r v - 2f_\varphi w + \frac{vu\tan\varphi}{r}$$
$$-\frac{wu}{r} - \frac{1}{\rho_0 r \cos\varphi} \frac{\partial P'}{\partial \lambda} + F_\lambda \tag{15.11}$$

$$\frac{\partial v}{\partial t} = -\mathbf{v}\operatorname{grad} v + 2f_\lambda w - 2f_r u - \frac{uu \tan \varphi}{r} - \frac{wv}{r}$$

$$-\frac{1}{\rho_0 r}\frac{\partial P'}{\partial \varphi} = +F_\varphi \tag{15.12}$$

$$\frac{\partial w}{\partial t} = -\mathbf{v}\operatorname{grad} w + 2f_\varphi u - 2f_\lambda v + \frac{uu}{r} + \frac{vv}{r} - \frac{1}{\rho_0}\frac{\partial P'}{\partial r}$$

$$-\frac{\rho'}{\rho_0}\mathbf{g} + F_r \tag{15.13}$$

$$\rho = \rho(T, c, P_0) \tag{15.14}$$

$$\frac{\mathrm{d}}{\mathrm{d}r} P_0 = -\rho_0 g(r) \tag{15.15}$$

where the Boussinesq approximation is adopted in (15.9) and all variables are defined as above for the atmospheric component. In (15.14), UNESCO scheme in Gill (1982) is used.

Smagorinsky type scheme presented by Lilly (1962) and Smagorinsky et al. (1965) is used as the subgrid scale mixing in an ocean component in this study. In the ocean component, the level-2 turbulence closure of Mellor Yamada in Mellor and Yamada (1974) has been also introduced as one of optional schemes.

In the ocean component, sponge layers are used for lateral boundary in the open ocean. The lateral boundary condition between ocean and land is defined as $\partial T/\partial t = \partial S/\partial t = 0$ and $\mathbf{v} = 0$. Bottom condition is defined by Neumann condition without vertical velocity, and the ocean component uses realistic high resolution bottom topography. The upper boundary conditions are given as momentum fluxes by wind and heat fluxes from observational data of atmosphere.

15.2.3 Grid System Configuration

Yin-Yang grid system presented in Kageyama and Sato (2004) is used both for MSSG-A and MSSG-O components. Yin-Yang gird system as shown Fig. 15.1 is characterized by overlapped two panels to cover the sphere. Basically, one component grid is defined as a part of low-latitude region covered between 45° N and 45° S, and 270° in longitude of the usual latitude–longitude grid system and the other component of the grid system is defined in the same way but in different spherical coordinates. The region covered by a panel is able to change by rotating axes of the panels.

By using Yin-Yang grid system, we can find a solution on an issue of how to avoid singular points such as the south and north poles on a latitude/longitude grid system. In addition, the advantage to enlarge a time step is compared to conventionally utilized latitude/longitude grid system.

Conservation scheme was discussed in Peng et al. (2006) and no side effects of over lapped grid system such as Yin-Yang grid were presented due to validations

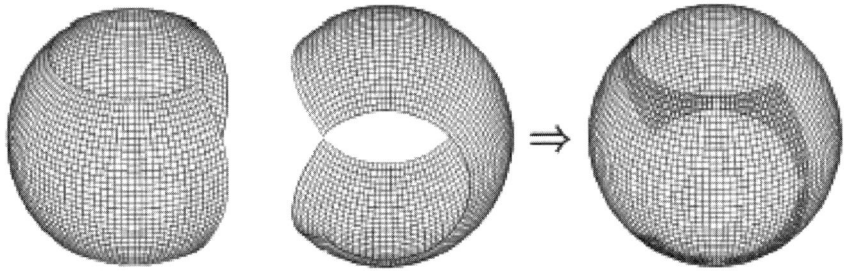

Fig. 15.1 Yin-Yang grid system which is composed of two panels on the sphere

results of various benchmark experiments in Komine et al. (2004), Ohdaira et al. (2004), and Takahashi et al. (2004a,b, 2005).

15.2.4 Differencing Schemes

In both the atmospheric and ocean components, the Arakawa-C grid is used. The atmospheric component utilizes the terrain following vertical coordinate with Lorenz type variables distribution in Gal-Chen and Somerville (1975). The ocean component uses the z-coordinate system for the vertical direction. In discritization of time, the second, third, and fourth Runge–Kutta schemes and leap-flog schemes with Robert–Asselin time filter are available. The third Runge–Kutta scheme presented in Wicker and Skamarock (2002) is adopted for the atmosphere component. In the ocean component, leap-flog schemes with Robert–Asselin time filter is used for the ocean component.

For momentum and tracer advection computations, several discritization schemes are available described in Peng et al. (2004). In this study, the fifth order upwind scheme is used for the atmosphere and central difference is utilized in the ocean component.

The vertical speed of sound in the atmosphere is dominant comparing horizontal speed, because vertical discritization is tend to be finer than horizontal discritization. From those reasons, horizontally explicit vertical implicit (HEVI) scheme presented in Durran (1991) is adopted in the atmosphere component. The speed of sound in the ocean is three times faster than it in the atmosphere, implicit method is introduced and Poisson equation (15.16) is solved in the method. Poisson equation is described as

$$\nabla \bullet \text{grad } P' = B,$$
$$B = \rho_0 \nabla \cdot \mathbf{G_v} = \frac{\rho_0}{r \cos \varphi} \frac{\partial}{\partial \lambda} G_u + \frac{\rho_0}{r \cos \varphi} \frac{\partial}{\partial \varphi} (\cos \varphi G_v) + \frac{\rho_0}{r^2} \frac{\partial}{\partial r} (r^2 G_w),$$

which are solved under Neumann boundary condition of

$$n \bullet grad P = n \bullet \mathbf{G_v}$$

Algebraic multigrid (AMG) method in Stuben (1999) is used in order to solve a Poisson equation. AMG is well known as an optimal solution method. We used the AMG library which has been developed by Fuji Research Institute Corporation. The AMG library is characterized in terms of following points,

- AGM in the library has been developed based on aggregation-type AMG in Davies (1976).
- In the library, AMG is used as a preconditioner in Krylov subspace algorithms.
- Incomplete LU decomposition (ILU) is adopted as a smoother in the library, which shows good computational performance even for ill-structured matrixes
- Local ILU is used for parallelization; in addition, fast convergence speed has been kept.
- Aggregation without smoothing is adopted with recalling procedure, because remarkably fast convergence has been performed by using the aggregation
- This AMG library is optimized for parallel computation.

15.2.5 Regional Coupled Model, Coupling and Nesting Schemes

Regional models for each of MSSG-A and MSSG-O components are available. For a lateral boundary condition of the regional model of each components, sponge type boundary condition in Davies (1976) is used. Regional version is utilized with one way nesting scheme. Two-way nesting scheme is now being implemented. Any regions can be selected from the global, because both Coriolis and metric terms have been introduced in the regional formulation defined as a part of global formulation. When wider regional/global model integration is performed, lateral boundary condition in regional component models is given every time step from results of those integrations. Here time step refers to the large time step in the Runge–Kutta scheme. This implementation enables online one way nesting without file IO.

Not only regional model for each of components but also regional coupled model, MSSG has been introduced. In the regional MSSG, regional MSSG-A component is coupled to regional ocean component, MSSG-O, with the same horizontal resolution. Lateral boundary condition for the regional components is given from a wider regional or global model with one way nesting (Fig. 15.2). In the region, interface between atmosphere and ocean components is coupled through transferring momentum, heat fluxes and sea surface temperature (SST). Generally, time step in the ocean component is set longer than it of atmosphere component. The heat fluxes, fresh water and momentum fluxes are averaged over defined time steps. The averaged fluxes are transferred during the defined time steps as upper boundary condition of the ocean component. SST is defined in the top layer of the MSSG-O and is given to the MSSG-A as constant heat source during the time steps.

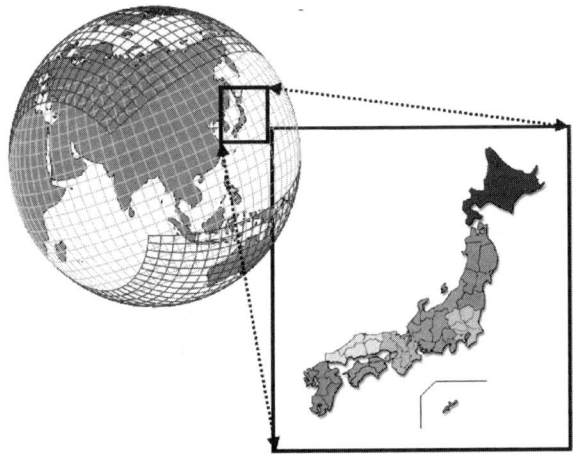

Fig. 15.2 Schematic figure of global and regional model with one way nesting

15.3 Results of High Resolution Simulations

15.3.1 Validation of the MSSG-A

Global atmosphere simulations have been performed to validate physical performance under the condition of 5.5 km horizontal resolution and 32 vertical layers. 48 h integration was executed with the non-hydrostatic atmospheric component with only cloud microphysics. Initialized data were interpolated to be fit to global 5.5 km horizontal resolution 00UTC08Aug2003 from grid point value (GPV) data provided by Japan Meteorological Business Support Center. Global spectral model data in GPV data set, which is provided with 1.25° of horizontal grid points and 16 vertical layers, was used over the entire region of the nested grid.

Sea surface data were also made by GPV data at 00UTC08Aug2003 and fixed during experiments. Figure 15.3 shows global precipitation distribution every 6 h. Precipitation distribution is corresponding to cloud distribution of GOES-9 satellite data (data not shown).

Typhoon tracking forecasting has been performed with both 11 and 5.5 km global simulations. The model was run with both 64 and 32 vertical layers in 11 km horizontal resolution version of global, and 32 vertical layers in the 5.5 km horizontal resolution version of global. Initial data in all of cases were interpolated from GPV data, which are provided by Japan Meteorological Business Support Center, to each resolution at 00UTC08Aug2003. SST was given as 1 day cyclic data from 00UTC07Aug2003 to 00UTC08Aug2003 during the simulation. A 72-h forecasting experiments have been performed. Figure 15.4 shows results of tracking forecast. Tracking in above three cases showed similar course even though experiments were set with different horizontal/vertical resolution.

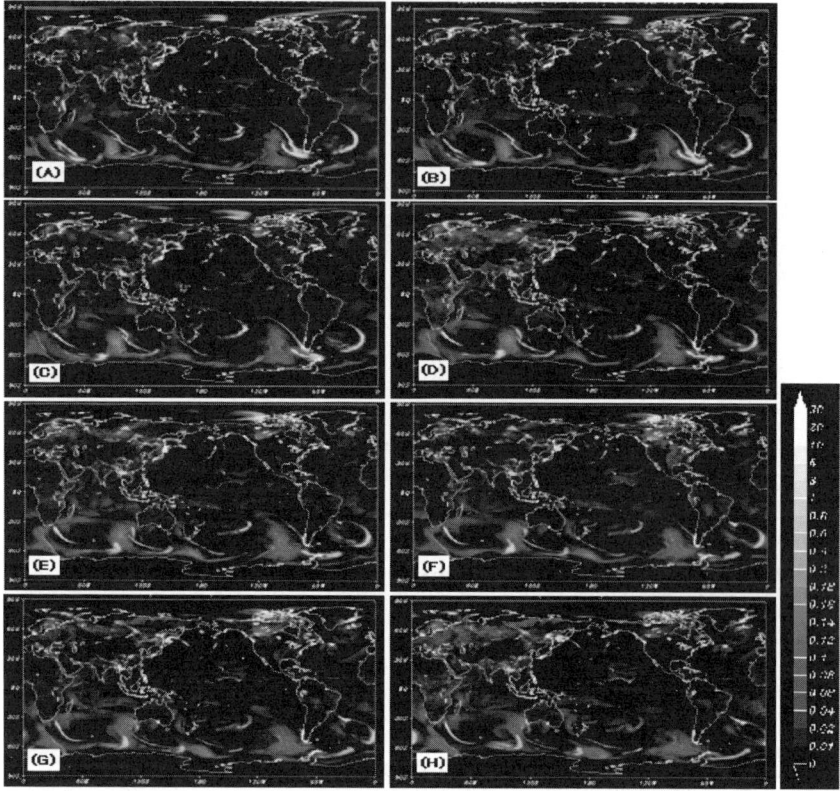

Fig. 15.3 Instantaneous global precipitation distribution (mm/h) plotted every 6 h obtained by validation experiments: (**a**)–(**h**)

Figure 15.5 shows precipitation distribution which sum up during the term from 09UTC07Aug2003 to 15UTC09Aug2003 after 72 h forecasting initialized at 00UTC07Aug2003. Tracking forecasting was similar to previous cases (data not shown); however, precipitation distribution with higher resolution simulation was distinct from it with coarser resolution. Figure 15.5a shows results of global simulation with 5.5 km resolution and 32 vertical layers. Figure 15.5b presents results of regional simulations with one way nesting from global simulation with 5.5 km horizontal resolution. The region was defined as the region of Japan region with 1.13 km horizontal resolution. Boundary condition was given every time step by interpolated from results of the global model with 5.5 km horizontal resolution. Figure 15.5c is observed data which are announced by Tokyo District Meteorological Observatory. Precipitation distribution is very sensitive to orography as apparent from the forecasting results obtained with high resolution. Further analysis is required to be clear mechanisms such as turbulent effect on sensitivity of precipitation, orography, and simulation resolution.

15.3 Results of High Resolution Simulations

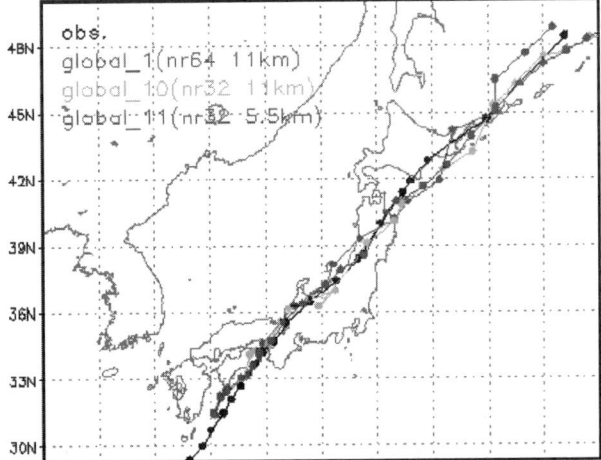

Fig. 15.4 A 72-h tracking forecasting with different horizontal/vertical resolution. *Red line* and *green line* show tracking forecasting results with 11 km horizontal resolution for global and 64/32 vertical layers, respectively. *Blue line* shows tracking forecast with 5.5 km horizontal resolution for global and 32 vertical layers

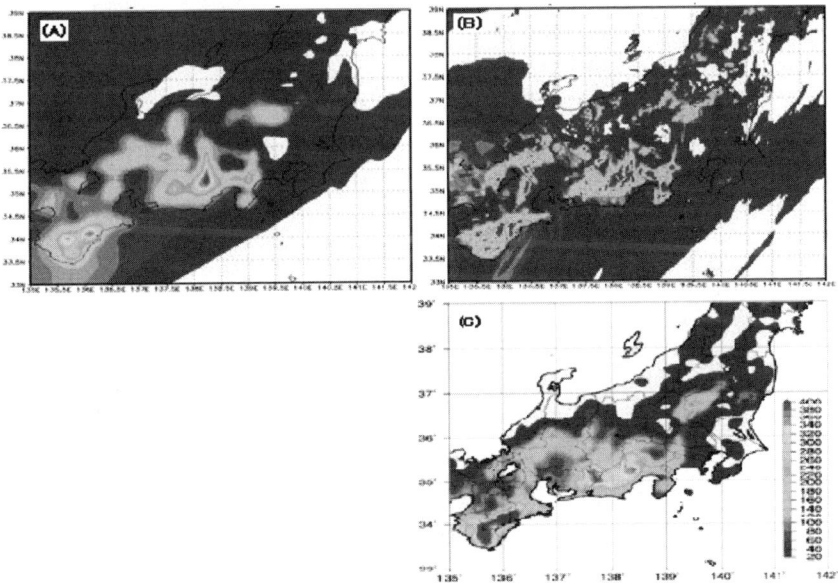

Fig. 15.5 Precipitation distribution with different horizontal resolution. (**a**) 5.5 km horizontal resolution for global forecasting simulation, (**b**) results of regional simulation with 1.13 km horizontal resolution, (**c**) observational data

15.3.2 Preliminary Validation Results of the MSSG-O

In MSSG-O as validation simulation, 15-years integration with 11 km horizontal resolution and 40 vertical layers has been done for the North Pacific basin. Surface heat fluxes and lateral boundary data are computed from climatological data provided by World Ocean Atlas (WOA). Momentum flux is obtained by interpolating from climatological data by NCAR. Figure 15.6 shows a snap shot in April after 15 years integration. Figure 15.6a shows temperature distribution at 15 m depth from the surface, which is corresponding to the second layer from a surface. Surface fields of SST such

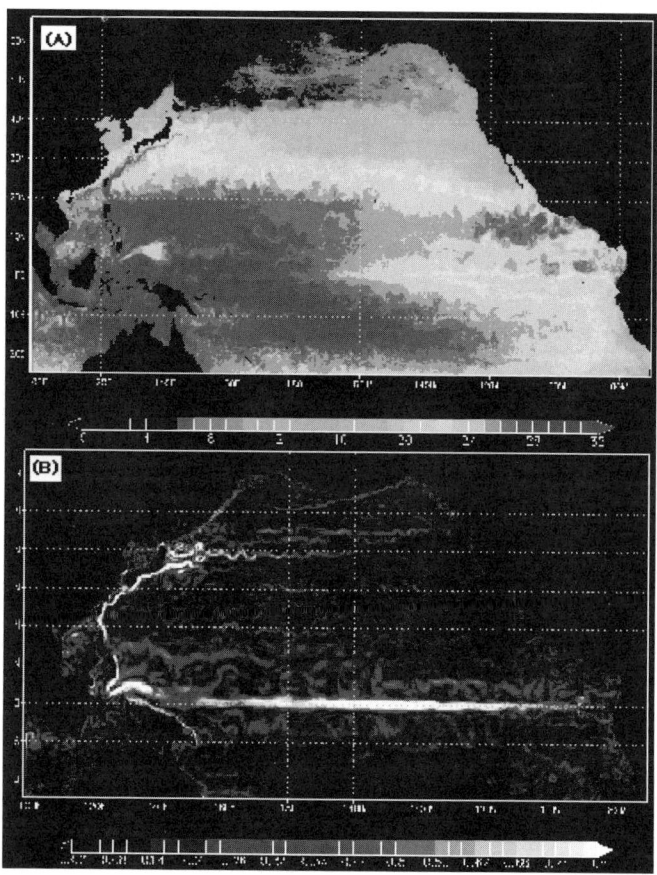

Fig. 15.6 (a) Temperature distribution at 15 m depth, (b) distribution of absolute value of horizontal velocity at 105 m depth

as warm pool and eddies in the east Pacific is resolved in Figure 15.6a. Large-scale SST fields and eddy-resolved structure are comparable to satellite data. Figure 15.6b shows distribution of absolute value of horizontal velocity at 105 m depth. Kuroshio current and the meander of Kuroshio with small scale eddies are represented as fine structure. Distribution of velocity is also comparable to previous studies and observations.

15.3.3 Physical Performance of the MSSG

The MSSG; coupled non-hydrostatic atmosphere–ocean–land model was tested by 120 h forecasting experiments for tracking of typhoon ETAU during the term from 15UTC06Aug2003 to 15UTC11Aug2003. The 120 h forecasting has performed with the coupled nonhydrostatic atmosphere–ocean–land model in Japan region (19.582°N–49.443°N, 123.082°E–152.943°E). In Japan region, both the nonhydrostatic atmosphere with only cloud microphysics and ocean components were resolved with 2.78 km horizontal grid-spacing. For vertical layers, 32 layers and 44 layers were used for the atmosphere and ocean components, respectively.

The initial atmospheric fields were interpolated from the GPV data at 15UTC06Aug2003 for the atmosphere. For the ocean component, further 24 h integration has been performed. After these global atmospheric simulations with 5.5 km horizontal resolution and 32 vertical layers, regional boundary data for Japanese region has been obtained from GPV data. In the oceanic component, initial data at forecast beginning date of 15UTC06Aug2003 was obtained by doing 10 days spin-up integration from 27th July in 2003 based on July climatological data of previous 15 years integration. During the 10 days spin-up integration, surface boundary data are given by six-hourly data by NCAR. Outside of the focused Japan region, global atmosphere simulation with 5.5 km horizontal resolution performed and its results was used as lateral boundary condition of the atmosphere component in Japan region. Lateral condition of the ocean component in Japan region was given by results of climatological data from previous 15 years integration. Coupling was done without any flux correction.

Figure 15.7a–f shows time series of 1-h average forecast fields obtained with the coupled model in the region of Japan. Distribution of blue gradation shows precipitation distribution. Fine structure like a rain band has been represented in Fig. 15.7a. Those distribution structure showed drastic change as a typhoon attacked Japan and went through Japan. In the ocean, SST response to a typhoon was simulated in Fig. 15.7b–f. Oscillation due to disturbance of a typhoon was recognized in not only SST but also vertical velocity and Kuroshio. Detailed analysis on eye core structure or ocean responses is still going on.

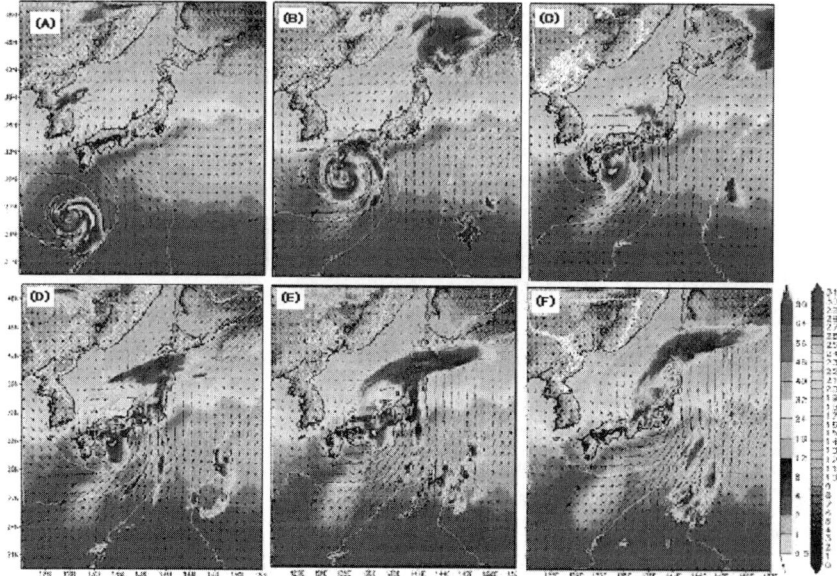

Fig. 15.7 Precipitation distribution (mm/h), wind velocity with black allow and SST distribution during typhoon ETAU attacked Japan region. *Left-hand side color bar* shows volume of precipitation and *right-hand side color bar* presents SST temperature

15.4 Conclusion and Future Work

In this paper, preliminary results of high resolution simulations have shown with some impact. Especially, it was clear that process of precipitation was sensitive to high resolution, as previous studies pointed out. In the coupled model simulations, fine structure and detail processes were represented in both atmosphere and ocean. Further detail analysis is required. In addition of various forecast simulation of forecasting such as heavy rain and other cases of typhoon, longer integration with the coupled model will be performed in near future. Those challenges might be one of the ways to understand mechanisms of weather or longer scale meteorological phenomena.

Acknowledgments This study has been performed as a part of Collaboration Project 2005 in ESC of JAMSTEC. This work partially supported by CREST 2005, JST.

References

Blackadar, A. K., 1979: High resolution models of the planetary boundary layer. *Advances in Environmental Science and Engineering, 1*, Pfafflin and Ziegler, Eds., Gordon and Breach Publisher Group, Newark, pp. 50–85.

References

Davies, H. C., 1976: A lateral boundary formulation for multi-level prediction models, *Q. J. R. Met. Soc., 102*, 405–418.

Durran, D., 1991: Numerical methods for wave equations in Geophysical Fluid Dynamics, Springer, Berlin, Heidelberg, New York.

Fritsch, J. M., and Chappell, C. F., 1980: Numerical prediction of convectively driven mesoscale pressure systems. Part I. Convective parameterization, *J. Atmos. Sci., 37*, 1722–1733.

Gal-Chen, T., and Somerville, R. C. J., 1975: On the use of a coordinate transformation for the solution of the Navier–Stokes equations. *J. Compu. Phys., 17*, 209–228.

Gill, A., 1982: Atmosphere–Ocean dynamics, Academic Press, Inc., New York.

Kageyama, A., and Sato, T., 2004: The Yin-Yang Grid: An overset grid in spherical geometry. *Geochem. Geophys. Geosyst., 5*, Q09005, doi:10.1029/2004GC000734.

Kain, J. S., and Fritsch, J. M., 1993: Convective parameterization for mesoscale models: The Kain–Fritsch scheme. *The Representation of Cumulus Convection in Numerical Models of the Atmosphre, Meteor. Monogr., 46*, Am. Meteorol. Soc., 165–170.

Komine, K. et al., 2004: Development of a global non-hydrostatic simulation code using Yin-Yang grid system, *Proc. The 2004 Workshop on the Solution of Partial Differential Equations on the Sphere, 67–69,* http://www.jamstec.go.jp/frcgc/eng/workshop/pde2004/pde2004_2/Poster/54poster_ohdaira.zip.

Lilly, D. K., 1962: On the numerical simulation of buoyant convection. *Tellus, 14*, 148–172.

Marshall, J., Hill, C., Perelman, L. and Adcroft, A., 1997a: Hydrostatic, quasi-hydrostatic, and non-hydrostatic ocean modeling. *J. Geophys. Res., 102*, 5733–5752.

Marshall, J., Adcroft, A., Hill, C., Perelman, L. and Heisey, C., 1997b: A finite-volume, incompressible Navier-Stokes model for studies of the ocean on parallel computers. *J. Geophys. Res., 102*, 5753–5766.

Mellor, G. L. and Yamada, T., 1974, A hierarchy of turbulence closure models for planetary boundary layers. *J. Atmos. Sci., 31*, 1791–1806.

Ohdaira, M. et al, 2004: Validation for the Solution of Shallow Water Equations in Spherical Geometry with Overset Grid System in Spherical Geometry. Proc. *The 2004 Workshop on the Solution of Partial Differential Equations on the Sphere,* http://www.jamstec.go.jp/frcgc/eng/workshop/pde2004/pde2004_2/Poster/56poster_ohdaira.zip.

Peng, X., Xiao, F., Takahashi, K. and Yabe T., 2004: CIP transport in meteorological models. *JSME Int. J. (Series B), 47(4)*, 725–734.

Peng, X., Xiao, F., Takahashi, K. and Xiao, F., 2006: Conservative constraint for a quasi-uniform overset grid on sphere. *Q. J. R. Meteorl. Soc. 132*, 979–996.

Reisner, J., Ramussen R. J., and Bruintjes, R. T., 1998: Explicit forecasting of supercoolez liquid water in winter storms using the MM5 mesoscale model. Q. J. R. Meteorol. Soc. 124, 1071–1107.

Satomura, T. and Akiba, S., 2003: Development of high-precision nonhydrostatic atmospheric model (1): Governing equations, *Annu. Disas. Prev. Res. Inst.,* Kyoto Univ., 46B, 331–336.

Smagorinsky, J., Manabe, S. and Holloway, J. L. Jr., 1965: Numerical results from a nine level general circulation model of the atmosphere., *Mon. Weather Rev., 93*, 727–768.

Stuben, K., 1999: A Review of Algebraic Multigrid, *GMD Report 96.*

Takahashi, K. et al., 2004a: Proc. 7th International Conference on High Performance Computing and Grid in Asia Pacific Region, 487–495.

Takahashi, K. et al., 2004b: Non-hydrostatic Atmospheric GCM Development and its computational performance, http://www.ecmwf.int/newsevents/meetings/workshops/2004/high_performance_computing-11th/presentations.html

Takahashi, K., et al. 2005: Non-hydrostatic atmospheric GCM development and its computational performance, *Use of High Performance computing in meteorology, Walter Zwieflhofer and George Mozdzynski Eds., World Scientific,* 50–62.

Wicker, L. J. and Skamarock, W.C., 2002: Time-splitting methods for elastic models using forward time schemes, *Mon. Weather Rev., 130*, 2088–2097.

Zhang, D. and Anthes, R. A., 1982: A High-Resolution Model of the Planetary Boundary Layer – Sensitivity Tests and Comparisons with SESAME-79 Data. *J. Appl. Meteorol., 21*, 1594–1609.

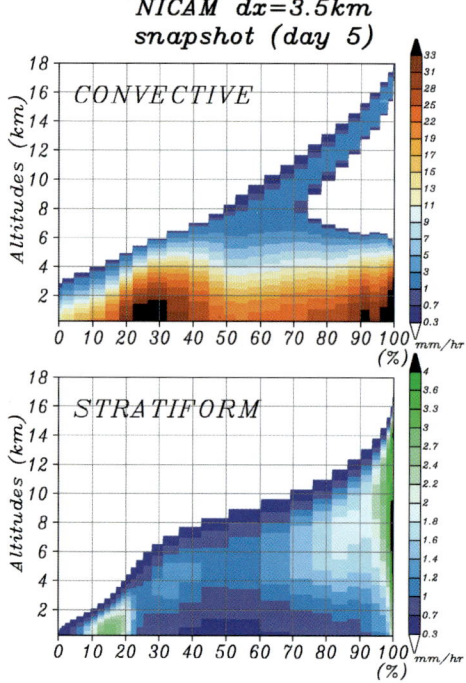

Fig. 6.5 Spectral representations of rainfall rate of NICAM. Rainfall rate is defined by the sum of rain and snow fall rate relative to the ground

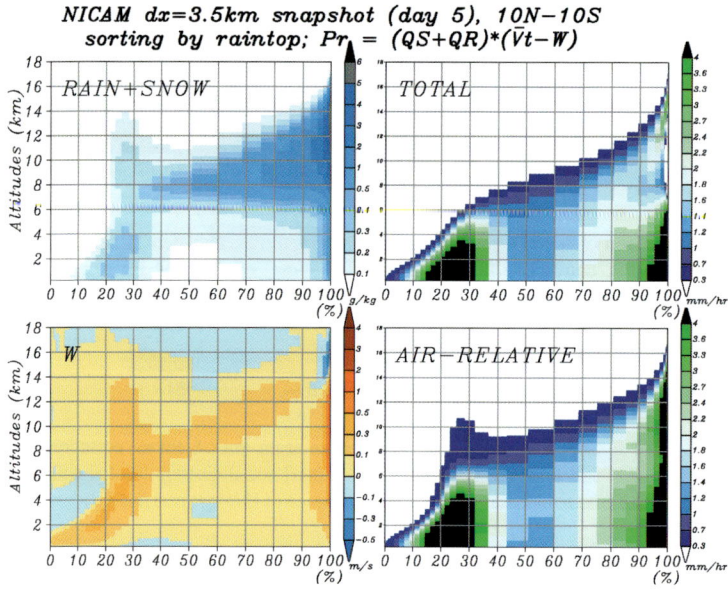

Fig. 6.6 Spectral representations of the sum of rain and snow concentrations (*top left*), vertical velocity (*bottom left*), total precipitation relative to the ground (*top right*), and total precipitation relative to the air (*bottom right*)

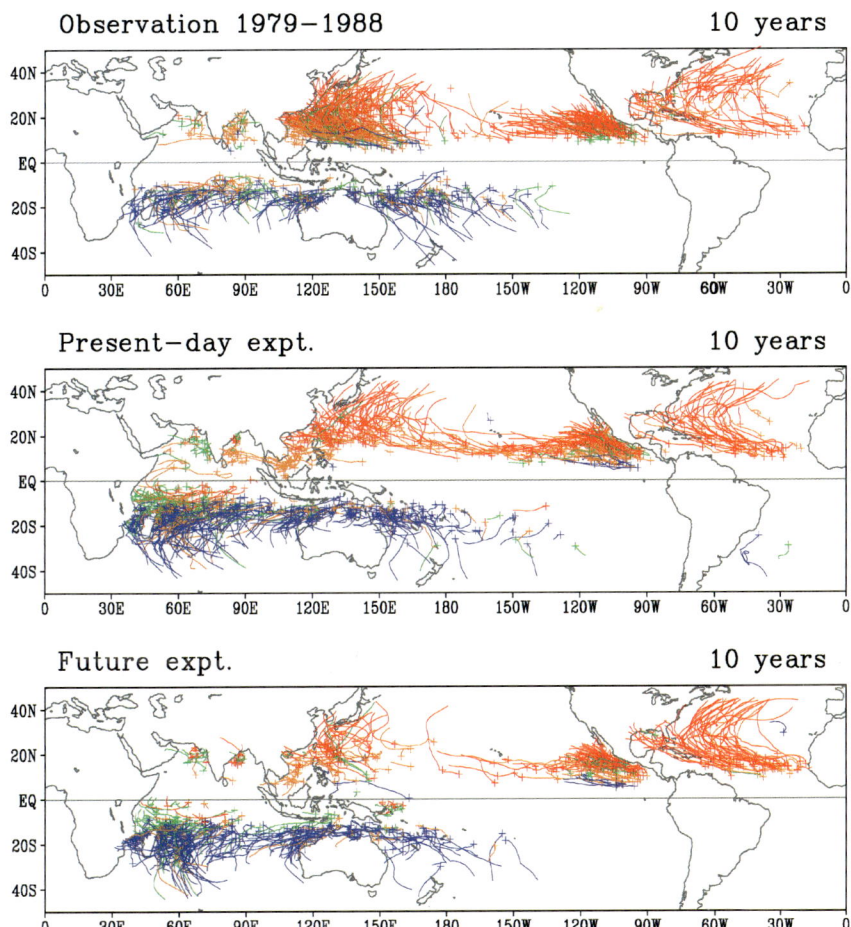

Fig. 7.2 Tropical cyclone tracks of the observational data (*top*), the present-day (*middle*), and the future climate experiments (*bottom*). The initial positions of tropical cyclones are marked with "*plus*" signs. The tracks detected at different seasons of each year is in *different colors* (*blue* for January, February, and March; *green* for April, May, and June; *red* for July, August, and September; *orange* for October, November, and December) (Oouchi et al. 2006)

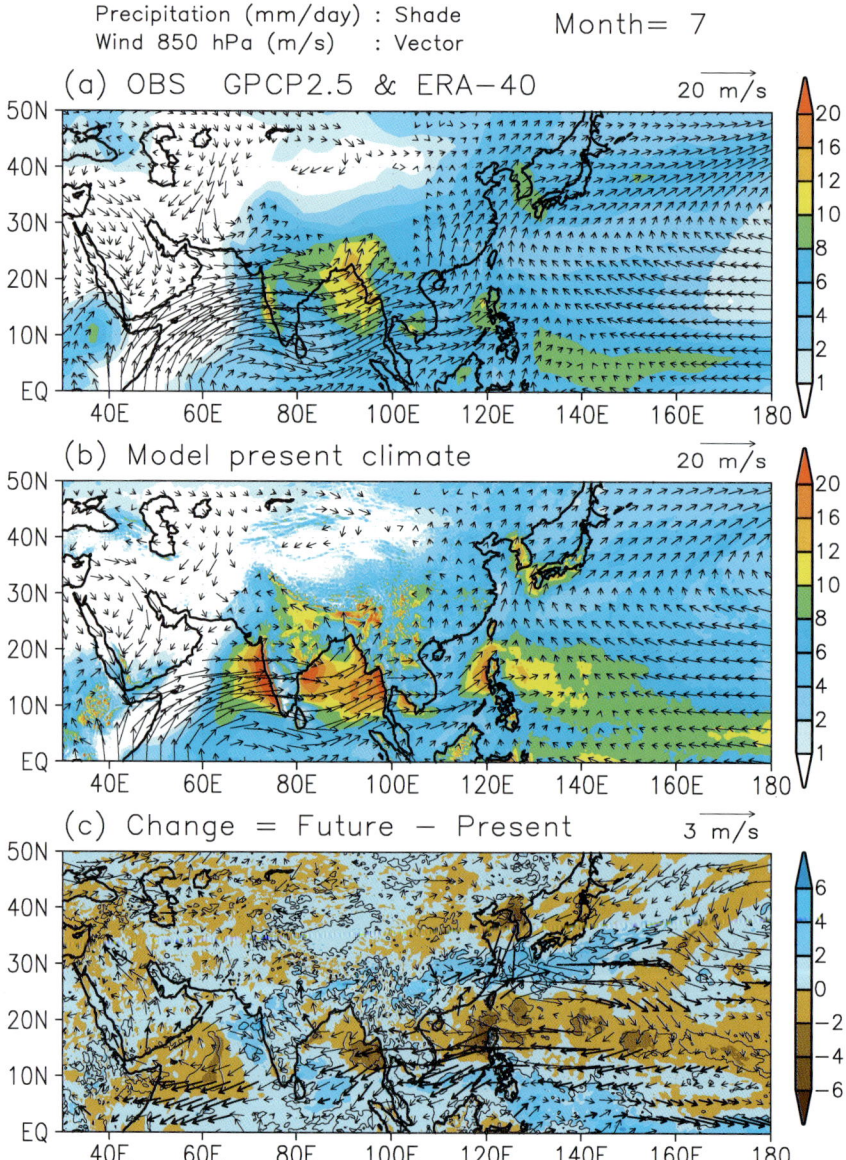

Fig. 7.5 Distribution of climatological precipitation (*color*, mm day^{-1}) and 850-hPa wind vector (*arrow*, m s^{-1}) for July. (**a**) Observed precipitation by GPCP 2.5° data (Adler et al. 2001) for 12 years from 1982 to 1993. Observed wind by ERA-40 data (Simmons and Gibson 2000) for 30 years from 1971 to 2000. (**b**) Model's present-day climate simulation. (**c**) Change as future minus present-day simulation. The *contour* and *thick arrow* show a 90% significance level of the precipitation and wind, respectively

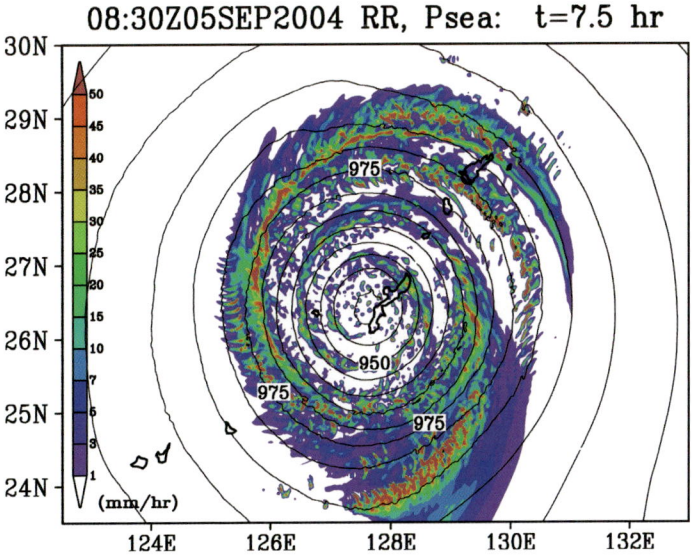

Fig. 9.5 Surface pressure (*contour lines*; hPa) and rainfall intensity (*color levels*; mm hr^{-1}) of the simulated Typhoon T0418 at 0830 UTC, September 5, 2004

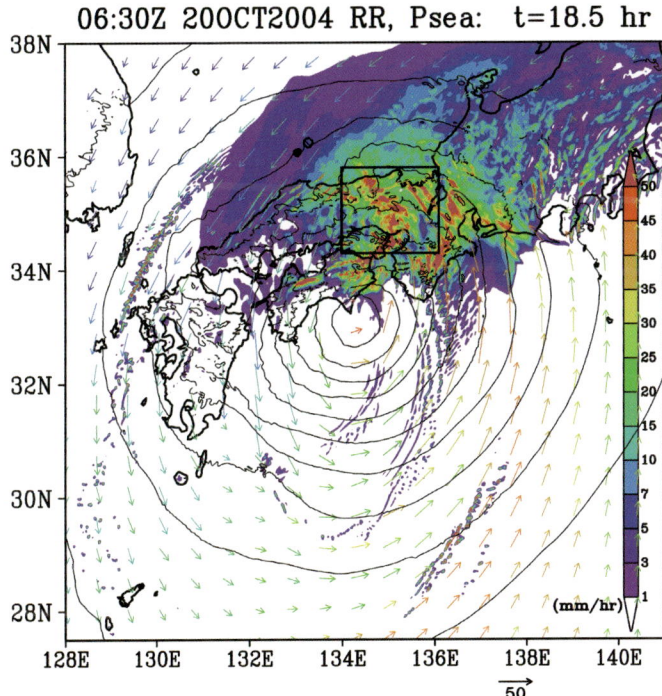

Fig. 9.6 Same as Fig. 9.5 but for the Typhoon T0423 at 0630 UTC, September 20, 2004. *Arrows* are horizontal wind velocity at a height of 974 m and *warmer-colored arrows* mean moister air. The *rectangle* indicates the region of Fig. 9.7

284   Color Plates

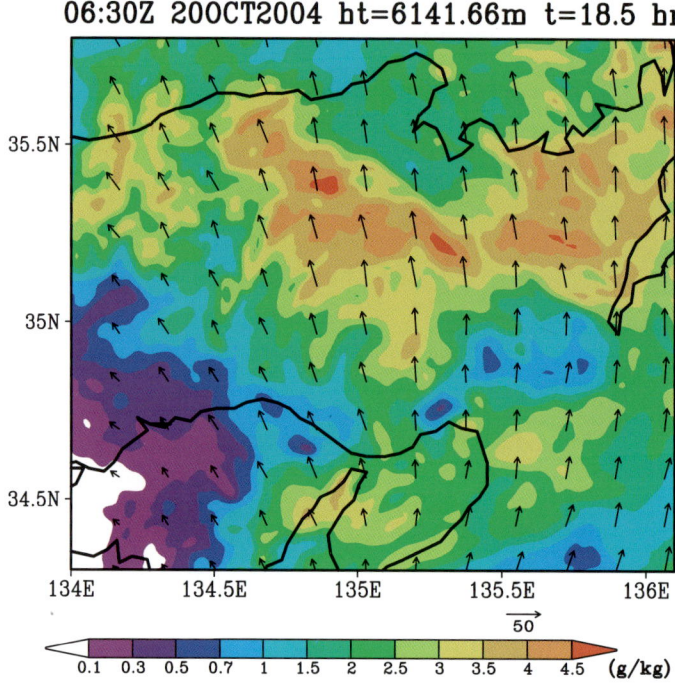

Fig. 9.7 Mixing ratio of precipitation (*color levels*; g kg^{-1}) and horizontal velocity (*arrows*) at a height of 6 142 m at 0630 UTC, September 20, 2004

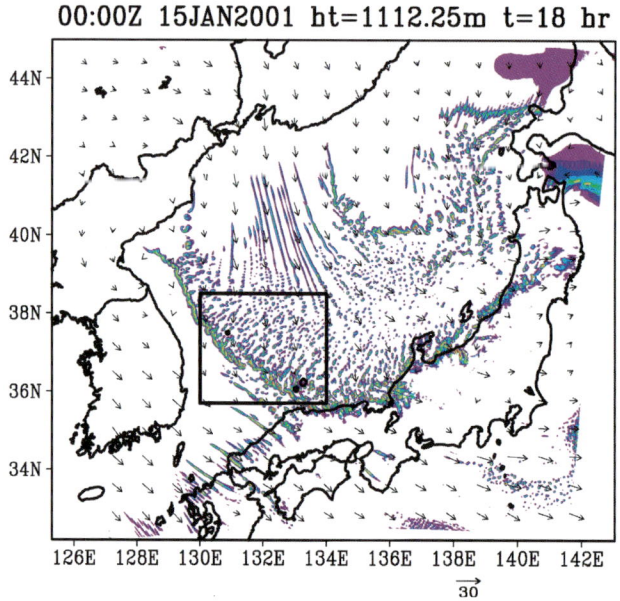

Fig. 9.12 Mixing ratio of precipitation (color levels; g kg^{-1}) and horizontal velocity (*arrows*) at a height of 1 112 m at 0000 UTC, January 15, 2001

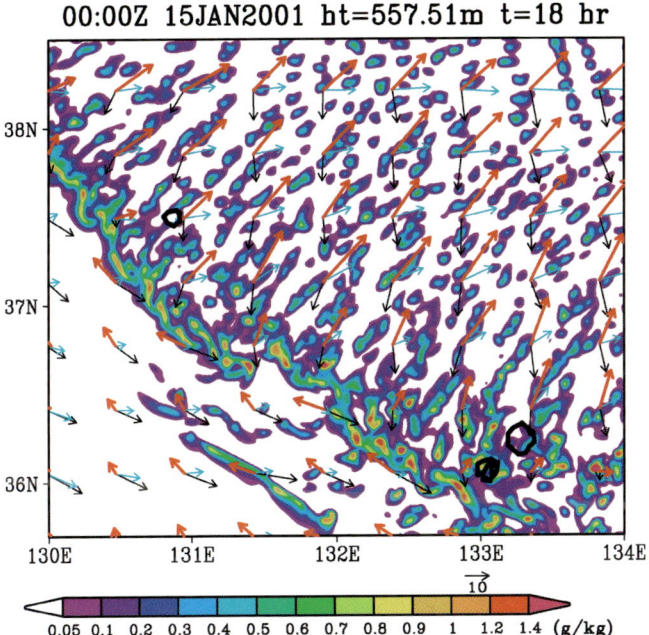

Fig. 9.13 Mixing ratio of precipitation (color levels; g kg^{-1}). *Black* and *blue arrows* are horizontal wind velocity at a height of 315 and 2 766 m, respectively. *Red arrows* are wind shear between these levels

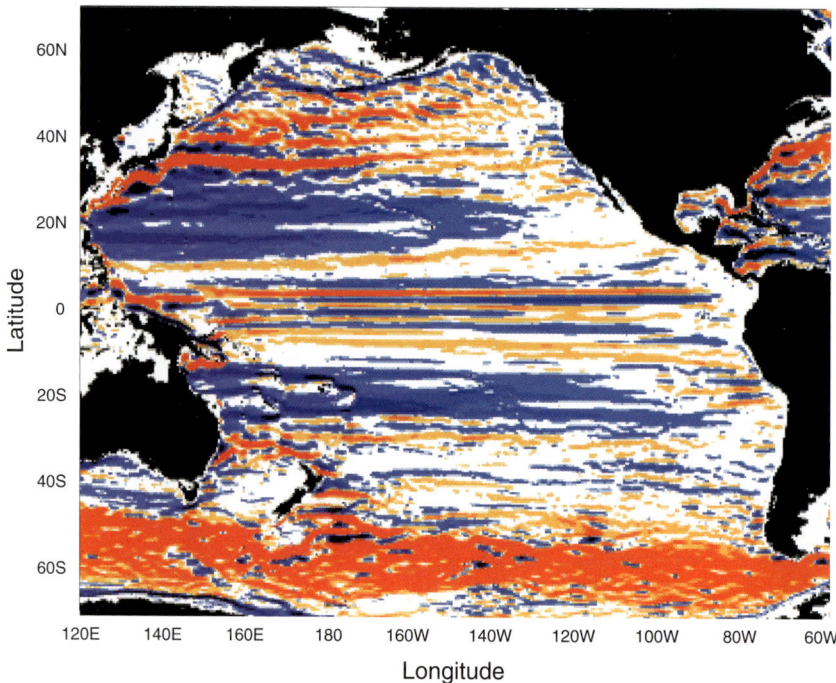

Fig. 11.1 Zonal component of velocity at 38 m depth averaged over 3 years from the climatological run of the POP model. Color saturates at $-0.06\,\mathrm{m\,s^{-1}}$ (*blue*) and $0.06\,\mathrm{m\,s^{-1}}$ (*red*)

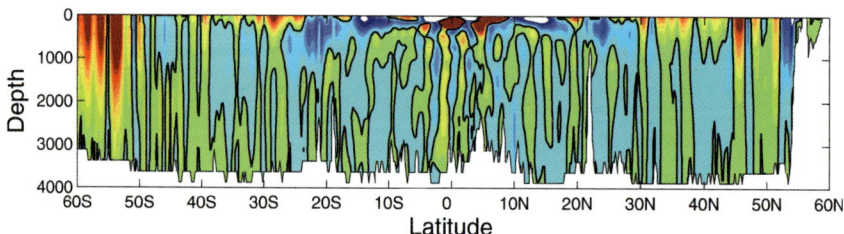

Fig. 11.2 Zonal component of velocity along 180°E averaged over 3 years from the climatological run of the POP model as a function of latitude and depth. Model run the same as in Fig. 11.1. Color saturates at $-0.1\,\mathrm{m\,s^{-1}}$ (*blue*) and $0.1\,\mathrm{m\,s^{-1}}$ (*red*). Zero contour is given by *black line*

Color Plates 287

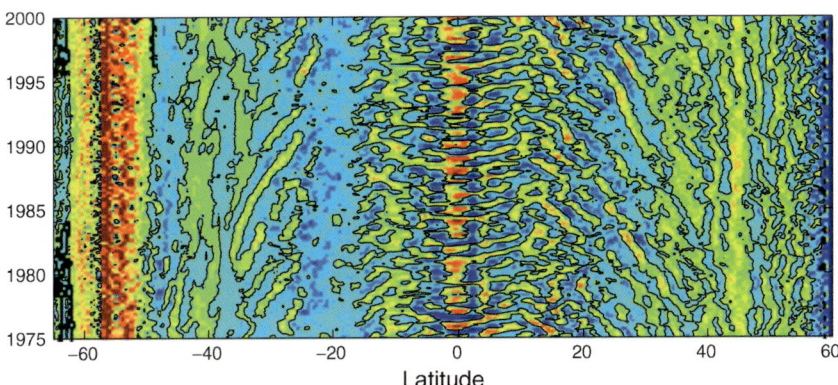

Fig. 11.5 Monthly averaged zonal component of velocity at 380 m depth averaged between 140°W and 150°W, as a function of latitude and time, from OFES. Color saturates at -0.2 m s^{-1} (*blue*) and 0.2 m s^{-1} (*red*). Zero contour is given by *black line*

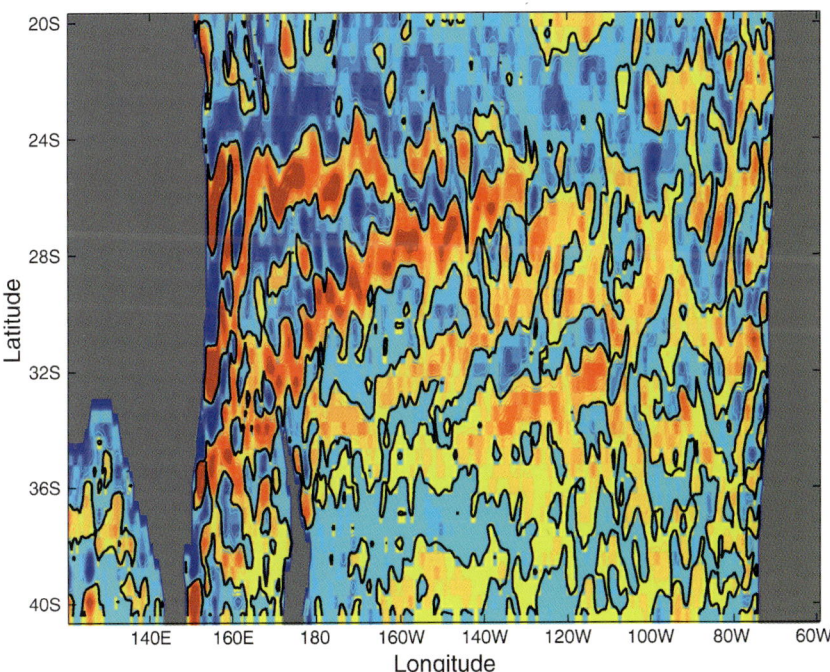

Fig. 11.6 Monthly averaged zonal component of velocity in the South Pacific for January 1980 at 400 m depth, from OFES. Color saturates at -0.2 m s^{-1} (*blue*) and 0.2 m s^{-1} (*red*). Zero contour is given by black line

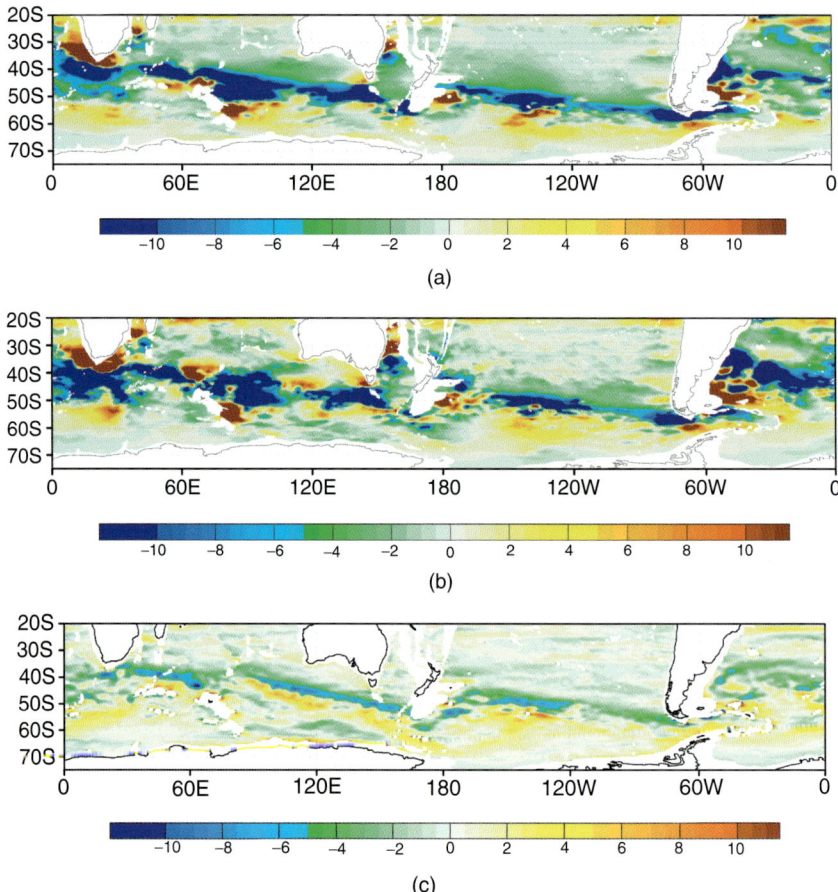

Fig. 12.1 The EBFC of (**a**) the SO6, (**b**) SO12, and (**c**) the Gent–McWilliams flux convergence (unit : 4×10^{-13} m s^{-3}) at 2,000 m depth

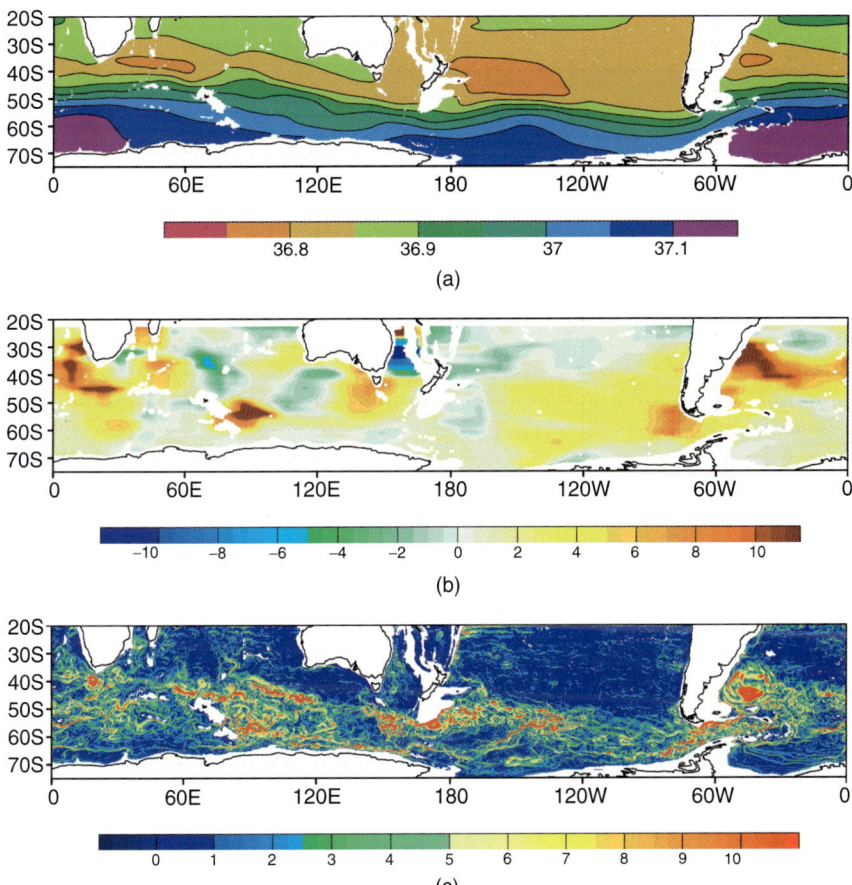

Fig. 12.2 (a) The potential density distribution (σ_2), (b) the coefficient $\bar{\kappa}$ calculated with the summation area size of $10° \times 5°$ (unit : 10^2 m^2 s^{-1}), and (c) the current strength at 2,000 m depth (unit : 10^{-2} m s^{-1})

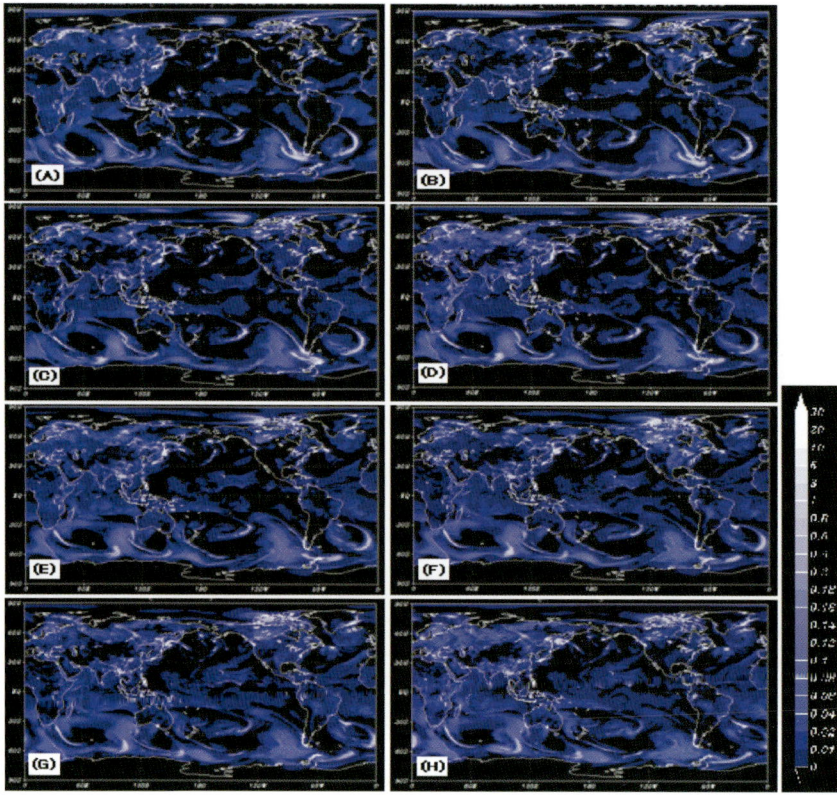

Fig. 15.3 Instantaneous global precipitation distribution (mm/h) plotted every 6 h obtained by validation experiments: (**a**)–(**h**)

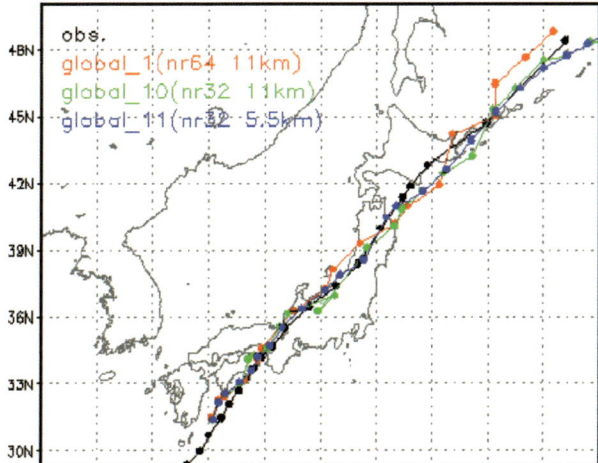

Fig. 15.4 A 72-h tracking forecasting with different horizontal/vertical resolution. *Red line* and *green line* show tracking forecasting results with 11 km horizontal resolution for global and 64/32 vertical layers, respectively. *Blue line* shows tracking forecast with 5.5 km horizontal resolution for global and 32 vertical layers

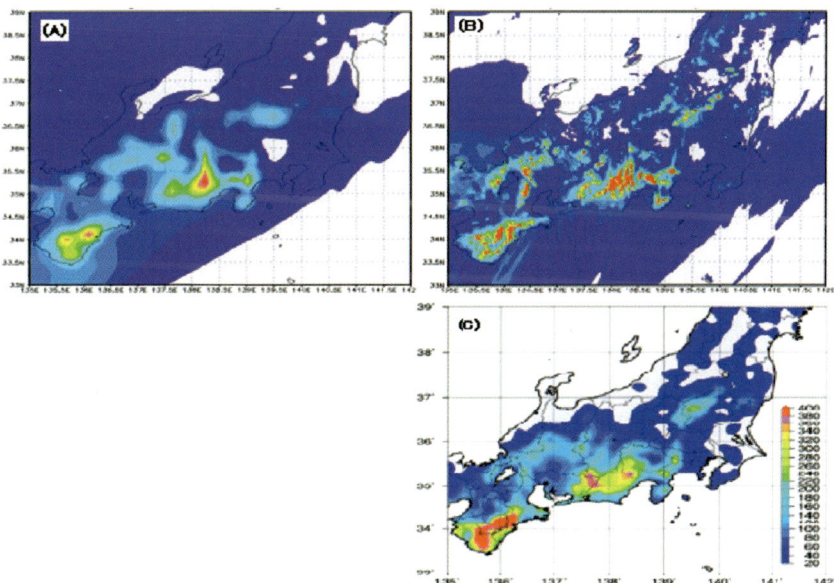

Fig. 15.5 Precipitation distribution with different horizontal resolution. (**a**) 5.5 km horizontal resolution for global forecasting simulation, (**b**) results of regional simulation with 1.13 km horizontal resolution, (**c**) observational data

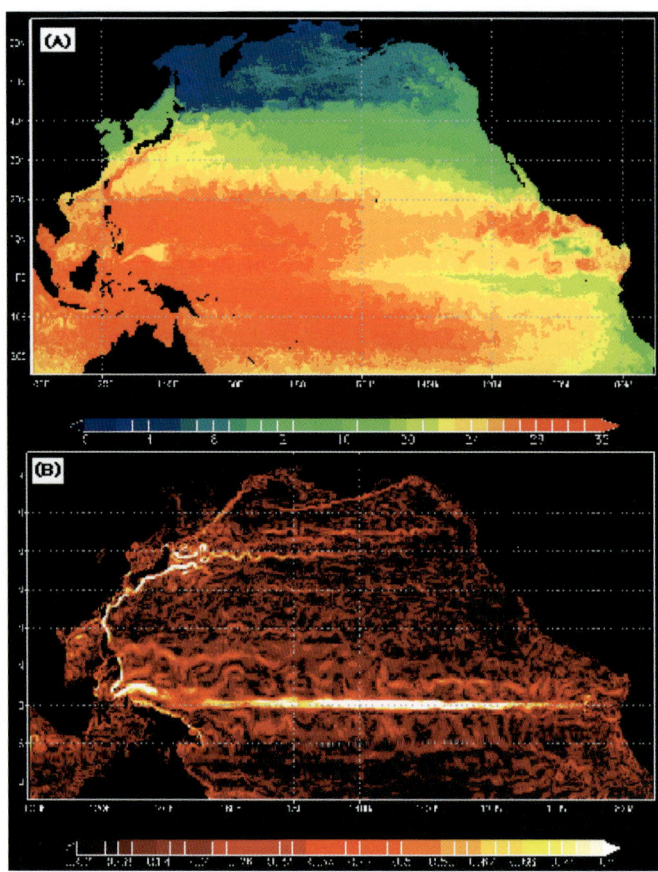

Fig. 15.6 (**a**) Temperature distribution at 15 m depth, (**b**) distribution of absolute value of horizontal velocity at 105 m depth

Color Plates 293

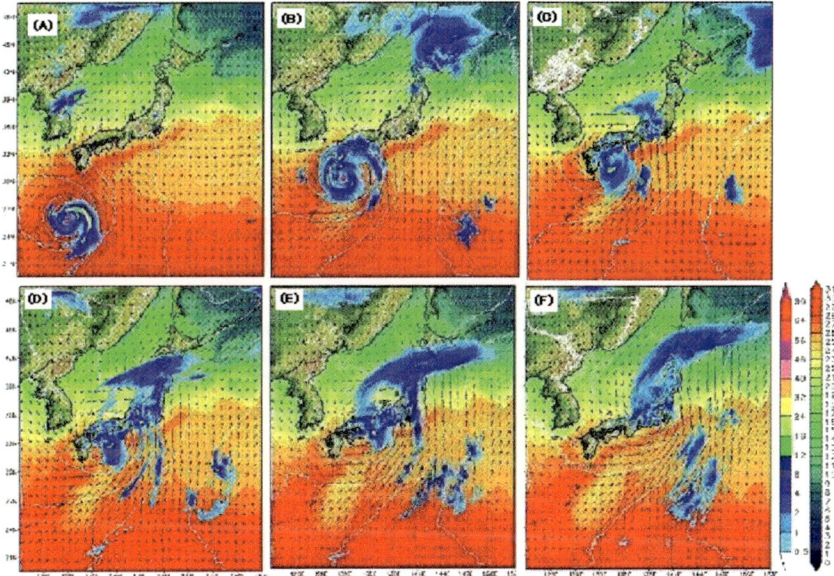

Fig. 15.7 Precipitation distribution (mm/h), wind velocity with black allow and SST distribution during typhoon ETAU attacked Japan region. *Left-hand side color bar* shows volume of precipitation and *right-hand side color bar* presents SST temperature